ꙅACRED BULL, HOLY COW

"*Sacred Bull, Holy Cow* is a captivating account of the dramatic impact cattle have had on religion, art, history, cuisine, entertainment, and the very survival of humankind throughout the world. One can't help being bullish about this book and the extraordinary scholarship it took to produce it. It is a fun and fascinating read."

Judson H. Taylor, President Emeritus,
State University of New York, Cortland

"This book is an engaging record of the importance of cattle in the world and throughout history, a fascinating story of civilization's reliance on this one animal. The scholarship of early myths and fables is impressive and the writing stimulating. No other book synthesizes so comprehensively how humanity has survived and how the importance of cattle continues to capture the news."

L. R. Schou, Professor,
Danish University of Education, Copenhagen

"To read this book is to go on a surprising and completely engrossing journey through mankind's time on earth, paying attention as never before to the extraordinary importance of cattle. The scope and depth of the research draws us into the adventure and refreshes our interest in man's ingenuity and capacity for reverence.

Donald K. Sharpes's story is as interesting as it is important. He is the latest in a line of teacher scholars who are also storytellers in the richest sense of the tradition. *Sacred Bull, Holy Cow* is an important book —brilliantly conceived and wonderfully written."

Steven G. Pappas, Associate Editor,
New Columbia Encyclopedia, *4th Edition*

SACRED BULL,
HOLY COW

PETER LANG
New York • Washington, D.C./Baltimore • Bern
Frankfurt am Main • Berlin • Brussels • Vienna • Oxford

DONALD K. SHARPES

SACRED BULL,
HOLY COW

A Cultural Study of Civilization's
Most Important Animal

PETER LANG
New York • Washington, D.C./Baltimore • Bern
Frankfurt am Main • Berlin • Brussels • Vienna • Oxford

Library of Congress Cataloging-in-Publication Data

Sharpes, Donald K.
Sacred bull, holy cow: a cultural study of civilization's
most important animal / Donald K. Sharpes.
p. cm.
Includes bibliographical references.
1. Cattle—History. 2. Cattle—Religious aspects.
3. Cattle—Social aspects. I. Title.
SF195.S53   636.2—dc22   2005029929
ISBN 0-8204-7902-0

Bibliographic information published by **Die Deutsche Bibliothek**.
Die Deutsche Bibliothek lists this publication in the "Deutsche
Nationalbibliografie"; detailed bibliographic data is available
on the Internet at http://dnb.ddb.de/.

Cover design by Joni Holst

The paper in this book meets the guidelines for permanence and durability
of the Committee on Production Guidelines for Book Longevity
of the Council of Library Resources.

© 2006 Peter Lang Publishing, Inc., New York
29 Broadway, New York, NY 10006
www.peterlangusa.com

Printed in Germany

TABLE OF CONTENTS

FIGURES

ABBREVIATIONS

ARCH	Archaeology Magazine
BAR	Biblical Archaeology Review
CBAA	Catholic Biblical Association of America
IAEA	International Atomic Energy Agency
FAO	Food and Agricultural Organization of the United Nations
NAS	National Academy of Sciences
CDC	National Centers for Disease Control
NG	National Geographic
NIH	National Institutes of Health
NYT	The New York Times
PNAS	Proceedings of the National Academy of Sciences
SA	Scientific American
WSJ	The Wall Street Journal

ACKNOWLEDGEMENTS

A special thanks to those gentlemen I interviewed for this study: Earl Petznick of Phoenix, Arizona, Roy Hunsaker of Moxee, Washington, Gene Courtney of Yakima, Washington, William Howard "Bill" O'Brien of Phoenix, the El Capitan of Los San Patricios, and Colin "TC" Thorstenson, owner of the Living Legend ranch in Scottsdale, Arizona.

Over the years I have been able to inspect the special artifacts in the following museums and libraries and am grateful to museum curators, keepers and librarians for maintaining the integrity of the past. I began my quest in the national museums of Damascus and Aleppo followed by the extraordinary collections in the Topkapi and Anthropological museums in Istanbul, the Ashmolean museum and Bodleian library at Oxford where I spent several months in 1999-2000, the British museum, London museum and British library in London, the National Archaeological museum in Nicosia, Cyprus, the Israel museum in Jerusalem, the Louvre in Paris, the Metropolitan museum in New York, the Roman museum in Amman, Jordan, the Getty Research Institute in Santa Monica, California and the superb collection of the Arizona State University library.

All the photos in this book are my own.

INTRODUCTION

While strolling through the second floor gallery of the National Museum in Aleppo, Syria, a museum that contains antiquities from the 4th millennium to the 5th century BCE one morning in 1986, I came across three archaeological artifacts that caught my eye. One was the statue of a man holding a box in his hands and wearing ox horns on his head. This statue dates from the 9th century BCE and the reign of the Assyrian King Tiglath-Pilesar III (841–727 BCE) who ruled over this part of northern Syria.

The second stone carving was of a cow standing among papyrus plants from Egypt turning its head to lick its suckling calf, a delicate artistic rendering unforgettable and tender in its simplicity.

The third graceful statue I saw that morning was found in a water basin in the palace from the ancient city of Mari on the Euphrates in today's Iraq, dating from the 2nd millennium BCE. It's a woman goddess, with her long hair plaited around her neck, wearing two ox horns on her forehead. She has several necklaces and three bracelets on each hand. On her short-sleeved shirt is a scene of fish swimming in water. She carried a spherical vase in her hands symbolizing water as the gift of life. The ox horns reveal her divine quality.

Why are there so many cow and bull sculptures, I asked myself then, and what is their significance to these ancient societies where civilization is said to have originated? What, if anything, do they say to us today?

Thereafter I began to look consciously for cow and bull artifacts exhibited in museums throughout the world. I easily found examples of bull figurines, some ordinary examples used for domestic purpose like pitchers, and amulets or charms but also huge bull statues from sanctuaries like the palace of Minos at Crete. Images and carvings of the bull and cow now seemed to me to be everywhere, archaeological evidence from the treasures of all ancient civilizations. Are we too distant in time to understand if there the relevance of these highly esteemed cultural artifacts?

While watching a bull fight in southern France later it suddenly dawned on me what these statues and figurine represented. I was a spectator at a very ancient ritual celebrating the conquering of this ferocious animal for the benefits the cow provides. The bullfight was a ceremonial re-enactment of the killing of the bull in order to domesticate the cow, not a slaughter to satisfy the blood-thirsty sense of spectators but a historical dramatization of civilization's most prized food source, so important that the animals themselves were deified in every ancient society.

Now my task seemed clear. Since I could find no book on the importance of the taming of the bull and the domestication of the cow, or of the designation of them as gods and goddess, I determined to study and write about this fascinating topic, a cultural trait and food story that may even have caused humanity to survive and evolve. I soon discovered that the ceremonies of the bull cult were universal. What began as a curiosity about how the cow had sustained civilization and the bull influenced our popular culture continued as a two decades-long inquiry.

Think of our popular culture. Bull session. A bull market. Shooting the bull. A bullpen. Bull's-eye. Bullish on America. Beefeaters. Sitting Bull. The Chicago Bulls. Even the Green Bay Packers named after meat packers. The bulldog was a special breed created for bull baiting, a form of entertainment where bulls in a ring were attacked by mastiffs, declared illegal in England only in 1835. These and other idiomatic expressions, including a few verbal indiscretions, indicate how close we are to the mythology of our most globally popular animals.

Then think of beef, milk, cheese, yogurt, cream, ice cream, all dairy products, leather for clothing and shelter and on expensive car seats, baseball gloves, shoe polish, buttons, linoleum, the casing for sausages, bandages, hairs for brushes, deodorants, hydraulic brake fluid, glue, soaps, piano keys, bone meal, at one time toothbrush handles, cow's blood used in scientific laboratory tests, even cow's blood used in building materials. About 570 pounds of a 1,000-pound cow might go into useful, non-consumable products.

The devil is portrayed with horns and a cloven hoof. The second sign of the zodiac, from the time when the day enters April 20th to May 20th, is the sign of Taurus, the Bull. Signs of the bull and the cow appear to be everywhere.

Women who can afford it inject doses of bovine collagen, that fibrous, insoluble protein that helps connect tissues, into their lips and faces to avoid wrinkles. They ingest large amounts of red beef, cheese, butter and creams following certain high-protein diet plans. Vanity always overcomes health, morality, and animal activist fears. Animal rights groups never attack the killing of a cow for its collagen. (1)

Castles in Ireland built by the Normans about a thousand years ago were erected by mortaring stones together with cow blood. Outer walls were built with suitable blocks of stone and rubble and then wedged into cracks. A mixture of lime, sand and fresh oxen blood was used to fill into the inner core of the wall. The mixture congealed and hardened into a massive and almost indestructible wall. (2) Plaster covering the pole and slat huts bound with bark strips of the Maasi tribal people in central east Africa, in Kenya and Tanzania, is cow dung applied regularly as the walls decay.

But closer to the stomach, how would Wendy's, Burger King, McDonald's or any hamburger fast food outlet or summer picnic exist without beef? The McDonald's Corporation is the symbol of economic globalization. Responsible for 90 percent of new job growth in the U.S., it operates about

28,000 restaurants throughout the world, opens 2,000 new ones each year, and employs one million new workers annually. It owns more retail real estate that any other company in the world. It is the nation's largest buyer of beef, pork and potatoes and the second largest purchaser of chicken. (3)

OK, the cow is needed, but why the bull? He isn't a pretty boy and probably has breath that would knock down a heavyweight prizefighter. For those who have heard the trumpeting of a range bull, the call of the wild is a challenge to all comers and a threat to intruders. The bull represents virile strength and awesome power in a way a larger animal, like a lumbering elephant, does not. Nobody really domesticates a virile bull, but each can be sedated with drugs and castrated to remove most aggressiveness. Before domesticating the cow, as our ancestors learned quickly, it was necessary to subdue the bull, and not be gored in the process.

We often think of the long and inexorable march of civilization as manifestations of a culture, its architecture, jewelry, paintings, religion and literature. It is rare that we would think of animals, and probably not one particular animal, as crucial to a civilization. But what is amazing is that cattle, more than any other animal, and more than any other aspect of civilization with the possible exceptions of wheat, rice and maize, have been absolutely essential to the upkeep of all civilizations. A large part of the history of civilization, more than the history of gold, of guns, of exploration, of settlement, of artistic development or literary achievement, is the untold story of the cow and the bull.

The Persian Gulf, on a map looking like an enlarged human appendix, washes the shores of the sandy, parched deserts of the Arabian peninsula and southern Iran. For centuries these salty waters have borne the dhows, the familiar boats of the gulf with distinctive, slanted sails on its wind-swept back nourishing a fishing industry that supported this region prior to the discovery of petroleum. At its headwaters in Kuwait and southern Iraq the Gulf has witnessed the rise of ancient civilizations, the Sumerians, Mesopotamians and Babylonians, just as today its salty waters transport oil tankers, the industrial lifeblood that sustains modern civilization.

In 2002 while teaching on a short term assignment at Zayed University, an all-female college in the United Arab Emirates, I journeyed to Al Fujairah a city on the east coast near the Straits of Hormuz to witness an unusual ceremony in this desert emirate: Brahma bull fights. Every Friday afternoon promptly at 4:30 along the beach facing the Gulf of Oman on a football-size patch of dirt and sand, men begin gathering on the sandy slopes while about 20 Brahma bulls are off-loaded from Toyota trucks and tethered to nearby trees where they paw the ground and kick up great clouds of dust.

The difference between these Arab bull fights and other bull fights in Spain or Mexico is that the bulls are not slaughtered and they do not compete with matadors or picadors. Like entertaining contenders in the World Wide Wrestling Federation, they play fight with each other. I came to observe this ritual when I learned it has been in the Arabian peninsula for centuries, introduced by explorers and traders on the route to Asia in the 15th century. It was the Portuguese introduced bulls to the Arabs.

The master of ceremonies announced in Arabic the opening of the event with his hand-held bullhorn. A group of about twenty judges squatted on the ground at a safe distance from the upcoming action. A cluster of about fifty men sat on mounds of desert sand. A few women remained enclosed in the cars hidden from the men except those in their family, a tradition in conservative Moslem society. Soon, one bull from each side was led into the makeshift arena, one attendant on each side holding a sturdy rope around the bull's neck as they led it to the center of the field. The attendants dropped the ropes holding the bulls as the bulls bumped heads and the contest began. Often spectators scrambled for safety as the agitated bulls fought through the seating sites, oblivious of the spectators. A few spectators honked their car horns to get the bulls distracted so their vehicles would not be damaged, an insult not covered by insurance.

The bulls shoved and goaded each other alternately yielding and attacking. With heads nearly touching the ground, horns locked, they sometimes both yielded and simply blinked at each other. When the bulls became immobile attendants ran out and goaded them with sticks. As soon as one bull was judged to be the more aggressive and hence the winner, all judges rushed to separate the bulls as attendants grabbed dangled ropes, pulled them apart and led them back to protected areas near the parked trucks. A few bulls emerged with skinned bloody foreheads but without lethal injuries.

Although it might appear that this is just another spectator sport at which men congregate to celebrate masculine bonding, I believe that even this kind of bull fight, without blood, without killing and maiming, is an integral part of the ancient ritual, a remarkable association of humankind with this particular animal. Bulls are not domesticated and never will be. But what the cow yields is beyond compare.

I was warming to the study of bovines and began to read everything so I could synthesize how cows and bulls impact our lives. I was soon to discover that Portuguese descendants also have a similar tradition in America.

Halfway around the globe from the Arabian peninsula, in the agriculturally rich central valley of California clustered around the small towns of Stevinson, Gustine and Thornton, a Portuguese-American community of about 350,000, mostly immigrants from the Azores, work in local dairy farms, in food processing and the construction trades. In the social and cultural center of this community lives Dennis Borba, the only active matador born in the United States. Through May until October, but especially in June around the time of the Catholic Pentecostal season, about 20 bullfights held on Monday nights constitute a part of the religious celebration of the Portuguese-American community. (4)

But there's a twist to these bullfights. The bull is not killed. It is "velcroed." Instead of a sword or the lances of picadors, Matador Borda uses a velcro dart to hit a velcro patch on the thousand pound bull's huge shoulders. The Portuguese have a bloodless bullfight. Bullfighting where the bull is killed is outlawed in the U.S. and by state law in California. But state lawmakers have made a special provision for the Portuguese when associated with religious festivals. Not widely known nor publicized outside the Portuguese community, these social bullfights are attended by about 3,000 weekly after a series of masses and religious processions and feasts.

Matador Borba spent an eight-year apprenticeship as a novice bullfighter in Mexico joining the ranks of full-fledged bullfighters in 1987 praying solemnly before each bullfight. The danger from a Spanish or Mexican bull, though not killed in the U.S. contest, is no less real because such a bull will fight only once in the bullring. Matadors don't want to see experienced bulls. Borda journeys to Spain, Mexico, Panama and Peru during the winter months where he fights to kill the bull in a "real" bullfight and supplements his income by acting as a stuntman in movies. Most bulls are sold to rodeos or slaughterhouses after they have fought. Only the most virile and ferocious, about one in a hundred, are selectively bred.

No form of entertainment illustrates that better than a rodeo, a western competitive sporting event for cowboys and popular entertainment among enthusiastic ranching crowds. Rodeo has some killer events that would have dazzled a rowdy gladiator crowd at the Roman Coliseum, including bronco busting, trick horses and bulls, clowns, horseback riding, and lovely rodeo queens. A rodeo is part outdoor circus and part individual competitive events with cash prizes for cowboys. But the most popular event among rodeo spectators, and the most dangerous for cowboy contestants, is bull riding.

Steer wrestling and calf roping are rodeo competitions affiliated with ranching. But bull riding, as an expression of domination of an animal, is more ancient and not a typical ranching job. The bull rider rides the bull—the classic position is the man on top—to demonstrate dominance, not to get himself killed. But the bull does not want to be dominated nor ridden, and reacts instinctively by kicking, bucking and attempting to throw the unprotected rider. There are no football helmets, shoulder, hip or knee-pads for these guys. Bulls want to trample and gore riders before rescue by clowns who offer themselves as potential bait.

The entertainment value for the spectators, many of whom are rooting for the bull, is older than history. The domination of the bull, and by association the cow, has a powerful grip on the imagination. These huge animals have to be controlled as they represent a killing force. Imagine what fear and dread overtook an ancient hunter confronted by an angry bull that didn't want to become a cave dweller's steak or roast dinner. Bull riding, like bull killing in the arena, is a hunting ritual of ancient origins, part entertainment, part religious sacrifice. Our fascination is that ancient ancestors did subjugate bulls and their cows for us and we celebrate that triumph, like a military victory, over and over again.

Bull riding is the ultimate tough guy sport. Bull riding, like bull fighting, is a symbol of an ancient ritual and serves as an introduction to how these and other sports have arisen to become part of our popular culture. The Professional Bull Riders Association, for example, sponsors 28 events and has a $6.2 million tour with cable TV deals fuelled by corporate sponsorship. Each contestant rides one bull for eight seconds and receives a score based on 100. The winner is the one with the highest cumulative score for the two rides, for example 82 and 90 for a total of 177. The top 15 riders qualify for a final round where each finalist rides another bull. The name of the bulls helps convey a sense of the excitement of the ride: Trick or Treat, Evil Eyes, Dillinger, Little Man.

In time, bull riders suffer too many broken bones and ruptured body parts, just like the knees and knuckles of a professional football quarterback. Cody Custer of Wickenburg, Arizona knows. At 39 years old in 2005, one of the oldest on the bull riding circuit, his bones are brittle and it hurts to move. He has had back surgery and four broken shoulders. Waking up sore is a part of the business. In his younger days, he was among the top 20 bull riders, winning the world championship in 1992.

Ty Murray, who came in second in an event in Phoenix in 2000, returned after a year of recovery from breaking both legs in the World Championships in 1999. Norman Curry came back the same year after a broken jaw. Kagan Sirett, a junior bull rider from Montana State University, had some sense knocked into him by a bull too. He finished second in the College National Finals rodeo in June, 2000. A week later he was speared by a bull's horn, broke two bones in his forehead and two more in his face, suffered a concussion and experienced dizziness. Even NFL quarterbacks, with 300-pound tackles knocking them to the ground and falling on them, don't have that kind of day. Kagan had a speedy recovery but his performance has fallen off. Ranked 14th in the Professional Rodeo Cowboys Association ratings system, he earned $51,900 in 2000. (5)

Candy Bell is a 42 year-old thrill-seeker who has won bull-riding contests on broncos and steers, first winning her bull-riding buckle in 1998 on the women's pro circuit. In 1998 she was ranked third in the world on broncos and fifth on bulls. She paid a heavy price: a perforated lung, broken ribs, bruised chest and a horn in the back. She has a scar that runs around her eye and down one side of her face. Candy is one of six women bull riders on the gay rodeo circuit. (6) While riding, she always carries a feather blessed by a medicine man from the Gila River Indian reservation where she works for the department of environmental quality.

At home Candy Bell has a three-and-a-third acre ranch with five horses, a cow and two ducks. The trailer has two dogs, seven cats and a pygmy goat. But her real business is rodeo and her craft is riding bulls on the rodeo circuit whenever she can. But because of her age and declining health, she has decided to cut back on riding. Her performance at the 16th annual Arizona Gay Rodeo Association (AGRA) gave her the same adrenaline rush that makes her keep coming back for more excitement.

About 2,000 paying spectators help the Arizona gay rodeo event donate about $20,000 each year to local charities. The combination of rodeo and gay cultures might appear to be the ultimate cultural clash. Cowboys dressed in pink might make an onlooker wonder what event was actually going on. But there is a forum for the gay culture in rodeo and in Arizona it has become a local tradition that draws an international crowd from Europe and Canada. The idea of a gay rodeo began in Reno in 1976 and by 1982 25,000 people were paying to attend Reno's gay rodeo, complete with bull riders of both sexes and straights and gays. By 1984 the International Gay Rodeo Association (IGRA) was born and the circuit included Arizona, Colorado, Texas and California.

Besides being the world's most dangerous sport, bull riding is a special event that celebrates the subjugation of the mankind's largest fully domesticated animal. Like its older cousin bull fighting, bull riding ritually celebrates the bull's fiery resistance to domestication and mankind's dominance over the largest animal necessary for human survival, the cow. Killing the bull in bullfighting is a symbol of this human achievement, an expression of civilization's oldest ritual--hunting and killing an animal, like the American Indians did the bison.

In a contest no less dangerous and robust, Buzkashi is the rodeo of Afghanistan where one does not ask the gender preferences of contestants. Buzkashi dates from the time of Genghis Khan. It is questionable whether or not it can be called a game, but certainly it is a national pastime. Teams of men on horseback fight for control of a calf or goat carcass, often headless, weighing about 150 pounds. But in essence, it is every man for himself. The rider, usually holding the reins of the horse within his teeth, and having snatched the carcass, has to lift it close to him, fend off other riders trying to snatch it and guide his own horse to the destined circle where depositing it wins the rider the prize.

The Spanish Bull Fight

Spain has elevated the bull to a mythic and heroic status. Never really overrun by the barbarian hordes that conquered the Roman Empire in 410 that began the Dark Ages in Europe, Spain has in fact retained the ancient bull cults. Bullfights offer a peek into the ceremonial and sacrificial ancient past.

Andalusia is southern Spain's land of olive groves with quintessential cities like Cordoba, Granada and Seville, flamenco dancing and Moorish architecture like the incomparable Alhambra. (7) Andalusia is the home of bull breeding. Over 200 ranches are exclusively devoted to the breeding and rearing of fighting bulls, together producing about 3,500 bulls a year for the *corridas* of Spain.

The fighting bull bears little resemblance in physical appearance or character to a farm bull. Bulls bred for the ring are wild animals, direct descendants of the wild bulls that roamed the Iberian peninsula thousands of years ago. The genus is Black Iberian, often known as Toro de Lidia or Ganado Bravo. The fighting bull has a small head, a wide forehead, wide shoulders and thick neck. A huge back muscle extends upwards when it is agitated. It has a thick strong, glossy hide. These bulls are bred and reared on large bull ranches and you can see them grazing everywhere in Andalusia. They roam freely over thousands of acres of ranch land and have minimum contact with men until they are ready to appear in the ring. The Spanish fighting bull with a strong hide, good horns and slender legs, and is bred and selected for its fighting character. The bull character traits best represented for the fighting bull are bravery, strength, liveliness, aggressiveness, endurance. (8)

The huge neck hump of the fighting bull is a feature that even the prehistoric potters of ancient Iran artistically displayed. The Amlash bulls, displayed in the Ashmolean Museum in Oxford, England are examples of the work of these decorative potters, living between 1,350–1,000 BCE. They were especially skilled in representing these humped-back bulls, actually zebus, as pitcher pots for domestic use.

After visits to Andalusia in the south and Spain's east coast, I journeyed to Ronda, high in Spain's southern mountains, ancestral home of modern bullfighting.

Founded by the Romans, Ronda is a city where the first bullfighting arena was built. Walk into the empty arena to the far southern side and enter the museum of bullfighting under the stands. Gaze reflectively on a special painting hanging on the wall, satirically depicting Spain's obsession with the bullfight. Its theme is that life is survival and the bullfight is that reaffirmation in Spanish culture. A quick look at the hands of the two figures in the painting indicates they are children, though they have the serious look of adults. The boy is dressed in the traditional costume of the matador and she in the mantilla of the noble Spanish lady, a derivative of the aristocratic origins of the bullfight. He believes he must test his manhood from childhood in the bullring of life. Courage is his badge, not the love of the woman. He stares rather vacantly straight ahead, serious, with some trepidation, at a distant and uncertain future. He does not look at his lady. He needs to uphold the honor of the past. There could be other tests to his manhood but none more meaningful to the Spanish than a youth intending to become a bullfighter. The bull in the painting is enshrined in his heart, enveloped with roses.

She is a representation of a woman as a child in adult clothes. She has chubby hands. She is not admiring the young bullfighter. Like him she stares abstractedly into a middle distance as if waiting for someone to tell her what to do. She stands next to him yet is psychologically distant, as if she belonged in another painting. She has an indifferent flower on her undeveloped bosom, not a rose. It is as if she is impelled to stand next to him. That is her destiny. Perhaps she senses his ultimate death at the confrontation with savage animal power.

The left side of the painting is red: a symbol of his blood which must be shed, whether from the bull or the woman is unclear. The right side is black: his impending death, a final destination of which she is contributor and benefactor. Without mother, there is no man. Without man there is no conflict. The cherub in the left corner indifferently awaits the fate of the man, predestined, playing no part in the unfolding drama.

If painters have shown the conflict of the culture and the soul, writers have eulogized the forum of bullfighting. Ernest Hemingway's colorful accounts in his novels and stories describe the courage of the bull and the bullfighter and extol the tension in spectators. (9) Hemingway is a master of capturing the thrill of the bullfight and the essence of its daring artistry. What Hemingway misses, I believe, is the bullfight's ritual significance, its sacrificial nature codified in religious myths and ancient hunter qualities. It never occurs to him that what he is witnessing is not just a cultivated Spanish sport but an ancient human ritual that commemorates human survival.

Spain's obsession with the bull and its death, I believe, perpetuates this ancient cultural drama. Take any Sunday afternoon during the season that runs from March to October in one of Spain's major cities. The bull stands bleeding profusely from its shoulders. The matador has lowered the muleta, the red cape held with a stick, and gripped the long thin sword in his right hand. As the bull lurches at the cape one last time, encouraged by the movement of the cape, the matador thrusts the sword deep behind the bull's head between the shoulder blades piercing its heart. It is the final *coup de grace* and the end of a dramatic episode between man and the ferocious beast.

But the bullfighter is not the hero eloquently defeating death. He is the priest officiating at the ancient sacrifice of the bull, his cape the cloak of clerical authority. Of course a bullfight is butchery. How else are humans to survive without beef?

More than a circus or a rodeo, the bullfight is Spain's entertainment industry, alongside the flamenco, the strumming of the guitar, and gypsy songs. It is hard for foreigners to appreciate how culturally ingrained it is in the Spanish psyche. Bullrings are packed with spectators and important bullfights are televised nationally. An elderly waiter in a *tapa* bar in Cordoba lamented to me, while we both watched a local bullfight on TV, how even bullfighting had degenerated into mediocrity from its former high classical artistry.

Francisco Romero, born in Rondo in 1698, invented the rules for bullfighting. Prior to the Romero family, bullfighting was done on horseback from about the time of the Middle Ages. Prior to rules, fighting the bull had been an activity of audacious bravery and recklessness. Romero introduced the muleta, the red serge cloth held by a maneuverable stick. His son Juan introduced the team of picadores, toreros, or other ring attendants, and the two matadors.

The most idolized bullfighter of modern times was Manolete (a.k.a. Manuel Rodriquez Sanchez), born in Linares, Cordoba in 1917, a man who revolutionized bullfighting. From 1938 when he began fighting bulls until his early death in 1947 at age 30, gored by the bull known as Islero, he was the Spanish hero par excellence. While other bullfighters took small steps backwards when the bull charged, Manolete stood his ground and never moved from his original stance. Of course, his unflinching daring got him killed. But his courage and early death earned him heroic status in Spain, the acknowledged master of the matador in the 20th century.

Pablo Picasso as a Spaniard held a personal passion for the bullfight and painted a series of abstract oils on canvas. In 1933, in a painting with flaming, vivid colors and twisted bodies in intense conflict, *Bullfight: Death of a Toreador,* in the Picasso Museum in Paris shows both a toreador and his horse dying in agony from a bull's goring. *Bullfight* is the expressive image of a bull lanced by a picador who manages to gore the picador's white horse while blood pores from the wound. The bold clashes of color, fierce lines, and the obscured background of the spectators, reveal Picasso's symbolism of the frightening power of the bull, and perhaps the enslavement of the Spanish by the dictator Franco.

Canadian historian Shubert claims that the Spanish bullfight is an expression of " commercialized mass leisure." (10) But it is much more. The bullfight is the symbol of mankind's ancient struggle for survival—the taming of wild animals for food, shelter and clothing. The bull uniquely represents that ancient struggle and the bullfight is a lingering symbol of man's triumph over the ferocity of animal nature. The bullfight is the ancient hunter enactment of bull domination, overlaid with the religious ritual of animal sacrifice.

The flamboyant lifestyle of the Spanish bullfighter is equally exhibited in flamenco, a dance closely resembling the relationship with bullfighting. The body lines of the bullfighter and the flamenco dancer have similarities. When bullfighters have a successful bullfight they often retire to a cafe where a flamenco dance is held in their honor. The struggle between the strength of the bull and the cunning, daring and intelligence of the bullfighter are on display in the bullrings of Spain weekly during the season. The static steps of the flamenco dancers portray the dance of death, of love, of life. Matching the artistry of a love dance with the dance of death in the bullring is as old as ritual itself. The only real question is who gets sacrificed.

In contemporary times, this means that a doctor has to be in attendance.

Dr. Vasquez waits in the shadows of the corridor leading to Plaza de Toros in Mexico City. It is a bullfight day and he is on duty. He rushes into action when a matador or anyone associated with the bullfight is injured. He is passionate about the bullfight, what it represents and is happy to sacrifice time away from his family to perform this professional, almost sacred duty. He is, according to one matador, "his guardian angel," as he is to scores of bullfighters. He is adored, depended on and yet feared. Matadors will not speak to him even while standing next to him in the corridor leading to the bullring, because they believe it is bad luck, an ill omen, even if they also believe that he could save their lives if they are gored that day. Mexico has 150 bull rings and all of them are equipped with as many as three surgery tables ready to perform emergency room service to injured participants, as often only seconds separate an injury from a fatality. About 4% of matadors will die in the bullring.

In a moment of distraction, in a flash of an impulse from the bull, when least expected, the bull lunges or dips his horns and flips the matador, who has been toying with him, over his head, and if the matador is lucky, only gored in the leg. The wound received by a matador from a bull has had the force of a ton exerted on it, though the bull weighs less than a half that. The ashen young fighter is then carried out of the ring and onto the surgical table where his five-inch leg wound is stitched shut and the exuding blood flow stopped. Before the end of the afternoon he is back in the ring to finish off the bull with a sword lunge behind the shoulder blades. The bull is accorded respect for its fighting spirit and allowed to die fighting. There is no surgeon for the bull.

It appears to be cruelty to animals. But among devotees there is no contradiction between kindness and cruelty. The bull, like the prizefighter or the matador, is raised to inflict injury in a contest. Some claim that for the fight to be fair the bull should also have a sword. Matadors, and bullfight surgeons like Dr. Vasquez say that the bull already has two swords, both on its head.

The spectacle of killing the bull had a more carnival atmosphere even into the last century, as an extended quote from *The Illustrated London News* of 1849 demonstrates.

"This spectacle was given at Madrid and appears to have afforded unbounded gratification to many thousand spectators," begins the article. The story continues by describing how a dozen dogs dispatched a deer as a part of the preliminaries, and then six large Spanish dogs mastered a white bear.

"All attention was now fixed upon the last and chief part—the combat between the bull and the tiger. The former, a noble animal, and with formidable horns, was introduced into the arena on one side, a moment or two before the tiger, which was let in on the opposite side—making a bound from his cage, when the door as opened...On seeing the bull, he moved slowly towards him, crouching down as if seeking an opportunity for a spring. The bull did not see the tiger at first, but when he did, he made straight for him; but they were still a pretty good distance apart when the tiger suddenly turned and made off amidst the derisive shouts of "bravo toro." The latter, finding no enemy to attack, remained staring about, while the tiger ran slowly around the arena close to the railing as if looking for a point to escape from it. The bull was roused from his immobility by assistants who went round waving colored handkerchiefs outside the barrier and in this way was got near to the tiger, whom, seeing in motion, he again ran at. And this time the tiger, as the bull neared him, tried to spring upon him as close quarters; but the bull received him on one of his horns, and threw him over his head, inflicting a severe wound under the lower jaw, which rendered him helpless. The tiger made off to the other side of the arena and stretched himself by the barrier and nothing could induce him to combat again. The bull was drawn towards him again but appeared to disdain to touch him again."

After this description the writer tells us that dogs were let in, after the bull was let out, and made a swift kill of the tiger. Spectators were allowed to make off with tiger body parts. (11)

Pamplona and the Running of the Bulls

There are several modern equivalents of bull domination and expressions of courage in the face of this dangerous animal, most in Spain. Take an imaginary trip to Pamplona, the central Spanish town in the northern Pyrenees, reputedly named after Pompey.

Since 1591, the San Fermin festival, known as *enciero*, has let bulls run through the narrow streets. The bulls allowed to run are chosen to fight in the late afternoon bullfights. The festival is named after the patron saint of the city and the runners called *sanfermines*. Each morning during the eight day festival six fighting bulls are released to run through the town to the bullring. It takes about 4 minutes for the bulls to run the 825 yards from their starting point near the Church of St. Fermin, past city hall, across the Plaza Consistorial, and the next three blocks to the bullring. About 1,500 hundred young men participate, clothed in white and festooned with red berets, sashes and scarves, testing their manhood before the eyes of admiring women. The observance attracts adventurers from throughout the world, most of them pretend matadors (the average age of the runners is 28) who dart through the narrow streets avoiding the bull's pent-up anger and testosterone charges. In the past century, 13 people have been killed, but no bulls.

In 2000 in Pamplona, Spain, 206 young men were injured, 22 seriously, and six hospitalized. In 1995, a young American, Matthew Tassio, 22, from Glen Ellyn, Illinois, tripped and fell onto the cobblestones and was then fatally gored to death when a young bull ran its long horn through his midsection piercing his liver and a main heart artery. He was gored when he stood up after falling rather than rolling into a ball on the ground as more experienced runners would have. It was his first trip overseas. He had just graduated from the University of Illinois in electrical engineering and was getting ready to start work at Motorola. He clearly was unaware of the danger of this animal or the history of its long struggle to survive against those who try to tame, corral or dominate it. College did not prepare him for the most elemental survival or hunting skills a Cro-Magnon man would have possessed.

In 2001 rain kept the cobblestone streets of Pamplona wet and slick. The running bulls slipped and in the noise and confusion did an unusual thing: they ran off in different directions rather than following the normal route leading to the bull ring, catching even seasoned runners off guard. Jennifer Smith, 29, a tourist from New Jersey, took a horn in the thigh. Six others were badly gored on the first day.

Ironically, Spanish law forbids running with the bulls. But the law has never been enforced and no one tries to suppress spontaneous actions. The galloping morning herd always wins in this contest and a few unlucky men escape serious hurt while those who do manage to evade the lethal horns have a story for their children. Because of the drinking that occurs prior to the run, the high alcoholic content of the runners, fatigue from all-night partying, participants often stumble over each other regardless of whether bulls are near.

But many of Spain's towns run the bulls, and my experience was no less exciting than Hemingway's descriptions. (12) My role of spectator occurred one summer afternoon in the small fishing village of Javea midway between Valencia and Alicante on Spain's eastern coast. I arrived early in the afternoon in the downtown square of Javea on the eastern coast of Spain eager to watch all the preliminary activities. I wandered near the improvised corral where the young bulls were penned and noted their restlessness and pent-up energy. I then watched as attendants slowly constructed portable scaffolding fences around the perimeter of the square and located a site behind one of these fences to stand to watch safely behind the metal fence. There was no way I intended to play boy matador in the streets of this village.

The town band, young men and women clothed in jeans and yellow jersey tops, paraded through the square beating drums and clapping in unison to an unfamiliar and occasionally discordant tune. The band was followed by small groups of other youths, some soccer teams, clothed in red and green jerseys chanting a captivating song. It was a gay moment and the square was filled with laughter and excitement. Apparently the entire population of the town was in attendance.

There was a huge whoop from the crowd when the bulls were released. They ran at everyone who moved and they quickly spread through the square and nearby streets, scattering the young men darting in front of them. Agile youths leaped onto a high metal fence as any bull got close to them or they slipped behind barricades or palm trees avoiding the charge. One young participant dressed in sneakers, shorts and a large red shirt like a matador's cape, foolishly courageous, tried to engage a bull to run after him. When the bull charged, he quickly scrambled behind a wooden pillar. Other young men, flushed with wine, ran barefoot alongside the bulls out of their weak eyesight. If there are members of the Humane Society or its counterpart, they are unobtrusive and not vocal. Observers and participants alike are close enough to feel the hot breath of the young bulls and sense their fear and nervousness. Both bulls and runners exhaust themselves from the athletic frenzy, though runners are fuelled as much with alcohol as by exercise.

By the time I had learned the rituals associated with bulls, I then had to learn how many different kinds of bovines there were. So what are all the names of a bull and cow and how are they categorized? Here are the short descriptions.

The male is first a bull calf that matures into a bull. If castrated, the bull becomes a steer and evolves into an ox. Males raised for meat or for draught work are generally castrated to make them more docile on the range. Mature bulls weigh between 1,000–4,000 pounds. The female is born a heifer calf, becomes a heifer and then a cow. Mature cows weigh between 800–2,400 pounds.

Bos Taurus, or *Bos Indicus* for humped Asian breeds like the zebu, are classified as bovine farm animals raised chiefly for milk or meat or used for draft work. The short-horned *Bos Longifrons*, named after its long forehead, arose in Neolithic times in western Europe, and survive today in the Jersey cow and the Brown Swiss in the Alpine foothills. (13) We might still refer to cows endearingly as "bossy," after *bos*, the Latin term for a cow. "Neat" cattle originated from an Icelandic word *naut* which the English changed to "neat."

There are many cousins of the ox, such as the anoa, the pigmy buffalo of the Celebes, the Indian bubalis, the bibos that include the gaur, zebu (or Brahman, *Bos Indicus*), sanga and yak. The bison, yak, gaur, banteng of Indonesia and all forms of domestic cattle (*Bos Taurus*) will interbreed and produce fertile female offspring. Domestic cattle and bison rarely produce fertile male offspring.

Certain sub-species of cattle are raised specifically either for meat or milk production, for example Angus, Galloway or Herefords for beef, and Guernsey, Jersey or Holstein-Friesian breeds for milk. A related cousin, the yak, is used almost exclusively by Tibetan and western Chinese herdsmen as a food, clothing, shelter and commodity source. There are over 150 million water buffalo or Asian cattle, and 97% are in Asia, 50% in India alone. (14)

There are at least 277 cattle breeds throughout the world with varying types depending on whether they are cultivated for draft purposes, milk or meat. There are 52 breeds in the United States whose population is over 100 million. (15) Only bison or buffalo are native to the North American continent. All cattle have evolved from three breeds, the Urus and the Celtic shorthorn in Europe, and the Brahmin in Asia. (16)

The Roman word for money, *pecunia*, came from *pecus*, the Latin word for cattle. Today in most parts of Africa cattle are the sign of the owner's wealth. There are about 1.5 billion cattle on earth, an amount not just because of natural selection but also breeding choice, constituting about a quarter of the landmass on earth. All the earth's cattle weigh more than all humans, each mature cow ideally at about 1,100 pounds. Cattle and livestock eat 70% of all U.S. grain. The Canaanites about 4,000 years ago worshipped the bull god Baal, as we shall see in the Bible Chapter, and the Hebrews adopted their bull worship until belief in the Yahweh emerged.

A papal bull in Roman Catholicism is an official document promulgated by the Pope. The old Roman name for a seal (bulla) was traditionally attached to important documents. Since the 12th century, the bull has been a term used to designate this kind of letter with a red-ink imprint from the Pope including pronouncements such as dogmas or the canonization of saints. The more common English word designating a public notice from an authoritative source, bulletin, derives from the same Latin word.

Holy Bull was neither holy nor even a bull. The great Holy Bull was a 3-year old champion race horse, Horse of the Year in 1994, inducted into thoroughbred racing's Hall of Fame in Saratoga Springs, New York August 6, 2001. Holy Bull was a gray, front-running colt winning five Grade I races in 1994 and 13 of 16 starts over his racing career. Holy Bull had a leg injury in 1995 that ended his racing days but allowed him to enter his second career as stud. (17)

If civilization has made advances, the domestication of the cow was the one great achievement that still sustains civilization on every continent except Antarctica. The cow could survive, as the American bison has over thousands of years on the Great Plains, without humans. But unlike the American bison hunted by Plains Indians, the great leap forward was the cow's domestication. Its deification was a major step in acknowledging that momentous accomplishment.

Ancient and modern civilizations throughout the world are inextricably linked to the cow as food and cultural imperative. It is not just globalization that has converged countries and cultures into a mass unit of trade and commerce. The cow has synergistically linked all humans to the natural world and to the survival of both. Bull fighting, bull riding, the beef industry and cow diseases have shocked the world's attention to the global importance of beef products and loudly proclaimed this synergy. (18)

I synthesize the ancient and modern story of this one animal in this book to show its history in literature, art, myth and religion and its links to human survival from diverse geographies and cultures, and through a few interviews with cattlemen and cowboys, to reveal a cultural commentary on the evolution of cattle's special and intimate relationship with humans. The pieces of our cultural history are embedded in us all and include the history of our food and our animals with the cow at the top of that food chain.

CHAPTER ONE
TAMING AND DOMESTICATION

Domesticating Wild Things

Large domesticated mammals were critical in supplying needed provisions and transport to ancient peoples. The major five animals over time domesticated for human consumption and transportation include the cow, sheep, goat, pig and horse. Minor domesticated animals include the camel, donkey, reindeer, yak, alpaca, llama and various bird species including the chicken.

Other animals figure in the pantheon of gods from ancient civilizations, like the lion, the serpent, the scorpion, even the dung beetle in Egypt. Camels played a similar necessary role in civilizations where the land and the presence or absence of water and vegetation for grazing dictated animal survival. In mountainous terrain or in areas of thin vegetation, sheep and goats survive better than cattle as do camels in deserts or semi-arid plateaus. Camel skins became the tents of the nomadic tribes and camels their transport vehicles. The two-humped bactrian camel, hairier and less resistant to water than its desert cousin, the Arabian dromedary, was the main luggage carrier in the northern climes across Anatolia, modern Turkey, and South Asia. The equivalent of bull riding in the Middle East is camel racing an entertainment (and betting sport) I observed in Dubai.

But though many ancient civilizations have had various forms of animal worship in the belief that they too enjoy a form of immortality, humankind has not attached the same importance to any other animal as it has to cattle. We naturally assume special relationships between humans and animals: a boy and his dog, a rider and horse, a shepherd and sheep. But none approximates the universal and integral bond as that between humans everywhere and cattle.

There are very few sheep or ram stone carvings visible in any museums, though this animal was highly prized for its food and wool for clothing. The lion, for example, figures prominently in all civilizations in the Middle East and Africa, as does the tiger in Asia. But neither was ever domesticated. The elephant was tamed but not domesticated in Asia because of its strength for heavy labor, but its meat is not regularly eaten nor its milk consumed. No one drinks the milk of a pig as they do goat's milk from which some preferred cheeses like feta come. The pig, whose genetic structure closely resembles humans, may one day be cultivated for its organic parts for human use. Because of religious prohibitions among Jews and Muslims pork is forbidden meat. Humans seem to have hunted deer for as long as hunting has existed and venison has been and continues to be a staple protein source throughout American and Europe. But the elusive and fleet-footed deer has never been domesticated as livestock.

Domesticating certain animals like sheep and dogs did not usually include the possibility of human death as did the taming of the cow with unpredictable and ferocious bulls nearby. The indomitable strength of the bull, its courage when it rushes headlong into adversaries and intruders who poach on its territory, its belligerence when roused, exciting admiration and fear, its awesome reproductive powers make it one of the most enviable and potent forces and symbols for the generative impulse in all of nature.

But small animals, fowls like turkeys, ducks and chickens, also had their domestic usefulness as accessible food whose feathers made comfortable pillows and bedding. The dog emerged from the domestication of the wolf because of its proven benefit as a hunting companion and as a guard for provisions and livestock. Canine evolution as pets occurred because dogs had a genetic propensity and a social structure to follow an identified pack leader and man became a de facto herd leader through subjugation and domination. Even domestic dogs, if allowed to run together freely, will quickly and instinctively identify a leader and run wild as a hunting pack.

Large mammals like the cow were prime candidates for domestication because they had a relatively rapid growth rate and a food source that did not compete with humans. Because cows congregate in herds, once the bull can be killed or isolated, cows will band together more docilely and acquiesce to human domination. By contrast, solitary animals, like bighorn sheep or deer (though not reindeer), cannot be herded and don't tolerate the presence of other animals of their species, particularly other territorial males.

Urbanized culture seems to have disassociated our bonding with the animals from which we get our food and clothing and turned our animal husbandry to dogs and cats mostly which give us no tangible food, clothing or shelter, only unquestioning loyalty and devotion.

Let's take a journey to a distant past as I did in the mid-1990s to south-central France, to where Cro-Magnon peoples, the first *Homo Sapiens*, roamed the plains and valleys of Europe, when art was first invented and when the first artistic paintings depicted wild bulls. (1)

Paleolithic Cave Art and Extinct Aurochs

The solitary flashlight of the guide penetrated the absolute darkness and shone on the painting of the shaggy bison on the cave wall. Everyone in the tour was silent and in respectful awe at the exquisite detail and still vivid colors. For this cave painting at Niaux, and many others in southwestern France, was over 20,000 years old. These magnificent examples of pre-historic art are unique illustrations of human artistic expression. (2)

Southwestern France and northern Spain and the paintings found there from 30,000 to 17,000 years ago constitute the first known examples of human art. The caves may also be the sites for initiation into religious ceremonies for the young, possibly the first schools into the mysteries and dangers of the hunt and religious education of the time. (3) Among the representations are multiple paintings of aurochs, the extinct wild bull of Europe.

Marcelino de Sautuola, a Spanish archaeologist working in the latter part of the 19th century, noticed some paintings of the roof of a cave he was exploring in 1897 in Altamira near Spain's northern coast. They appeared in the dim, artificial light of the cave to be in excellent condition. There was widespread skepticism among the general public about the alleged antiquity of these paintings when first reported. However, the verification of their age to the Magdalenian Period (15,000–12,000 BCE) gave scientific authenticity to this remarkable cave art, especially when similar cave paintings were discovered in the Dordogne Valley in southern France about 40 years later. And what do the paintings depict with such artistic accuracy and vivid colors? The European Bison in sleeping positions, stretching out, leaping, galloping across a landscape with surprising realism. Bulges in the rock have been used by the anonymous artists cleverly to illustrate bison body shapes.

Hunting the bull was probably not a first food choice of ancient humans as its suspicious and unpredictable nature, its fierce long horns and its large size is discouragement enough unless the clan was starving. Smaller, more highly maneuverable animals like rabbits and foxes were less likely to kill a hunter. Yet the cave wall paintings from this Paleolithic age are of large animals, not the drawings of the daily meat rations of *Cro-Magnon* people who roamed around the Mediterranean.

But why did these ancient artists paint the bison more than any other animal? Why not the huge and terrifying bull elephant, lions, tigers and bears, predators that can claw or stomp a hunter to death and eat one for lunch. The bison, bull and cow are not carnivores but grazers of the plains, the grass-munchers of antiquity. They are not naturally aggressive unless threatened or injured, perfect animals for hunting, large enough to provide a meal for the clan yet dangerous enough to kill members of the hunting party. Yet it is improbable that these ancient bison were domesticated at the time despite the artistic renderings of their formidable prowess. Perhaps at the time of these Paleolithic drawings ancient peoples wished they had tamed the bison and hoped for the day such animals were in their power to control.

Imagine further entering a stream into a cave in France, paddling a canoe upstream, getting out, walking for awhile, then crawling for about two hours before finally reaching a terminal chamber about a mile underground. Here, about 14,000 years ago, ancient ancestors sculpted from clay taken from the cave floor a pair of bison about two feet long, some of the world's first ceramic art.

Deep in the recesses of the cave in a section known as the End Chamber is a most unusual etching. It is dubbed the Sorcerer. The ancient artist has drawn the image of a figure that is half human and half bison. This figure seems to clasp a woman, represented with legs, hips and a triangle of pubic hair. Why is the artist creating such a half-human, half-bison form? A mythical creature, the Minotaur, half bull and half human, would occur again in Greek fable. Was this a male or female initiation rite, or just a playful artist's creativity?

Picture yourself along limestone cliffs near the Ardeche River in southern France north of Avignon, an area long popular with cave explorers. In 1994 three spelunker specialists found chambers deep underground with walls covered with paintings and drawings of lions, mammoths, horses, bears, bison and rhinos, even an owl, that have been reliably dated to 35,000 years ago. (6) This discovery radically pushed back in time the world's the oldest human art by another 15,000 years and highlighted the bison as an animal worthy of representation. While we appreciate the beauty of these animal depictions, we may never know their true symbolic significance or whether or not they were used in initiation rituals. But it is clear that these early humans had a dignified respect for their environment, for the world they inhabited, and for the variety of animals that provided hunters with such an abundance of food sources and subject matter, including bison, for the world's oldest known art.

Deifying the cow was a natural acknowledgement of an economy that placed cattle at the apex of wealth. The cult of the bull was a prehistoric practice associated with religious ceremonies that extended from eastern Europe to the Indus valley in what is today Pakistan. Often paired with the goddess of fertility, the symbol of the bull was a phallus carved in stone or represented in clay that stood for virility and the spirit of generation. Phallic bull symbols exist in India today in some temples dedicated to Shiva and throughout ancient Middle Eastern cultures. Other animals and birds have also achieved the status of deities in various religious beliefs, like the falcon, the snake, the sacrificial lamb. But only the cow and bull have won divine status based on their life-giving sustenance, fertility and strength and not as just as a symbol for a particular force in nature.

Our relationship with the cow is one of the most long-lasting associations of humankind with any animal, paralleling our love of gold and precious gems. The sheep and goat may have been the first domesticated followed by the zebu and ox, dog and pig, horse and camel. But regardless of the order, location of domestication and the imprecision of dates, the cow has been domesticated about 10,000 years. Animal domestication, and its corollary plant domestication, made a nomadic society sedentary, released it from the fluctuations of the environment, and with tool improvement made food production more predictable, thus invariably increasing the size of the community. Settlements began along river embankments as farmers displaced hunter-gatherers, and job specialization and trade mark the beginnings of urbanization that led to the invention of writing, an obvious benefit to trade and property rights and an enormous profit to literature. Animal and plant domestication is the most important step in human history and advancement prior to the invention of writing. (7)

Once the cow became divine and the bull supreme in the earliest civilizations, subsequent civilizations copied and embellished this profound cultural attachment. Diodorus, a 1st century BCE Roman historian writes: "For the bullock was of greater service than any other animal to the discoverers of the blessings of grain, such as in planting the seed and in other agricultural assistance of benefit to everyone." (8)

Were these paintings of bison and other oversized vicious animals from 30,000+ years ago used in initiation ceremonies inducting youth into the skills of the hunt? The encouragement of hunting as training for youth is noted by ancient authors beginning with Xenophon (c.430–355 BCE) whose treatise *On Hunting* describes the importance of the socialization of the hunt on the young, an activity that promotes athleticism, virtue and civic responsibility. (9) Hunting makes a young man courageous in the eyes of his peers and universally allows him to enter the world of the adult males.

Moreover, the art and science of hunting is depicted in the night sky. Orion, whose father was Poseidon the god of the sea, is pictorially represented as a hunter, and is the most recognizable constellation near the equator in the winter sky. In Greek myth Orion was a hunter who, after proving his prowess by ridding the island of Chios of wild beasts, in a drunken state reputedly violated princess Merope, daughter of Oenopion who blinded him for his offense. The multiple versions of the myth say that Artemis, the sister goddess of Apollo, killed him with her arrows.

The ancient Greeks had an artful way of capturing conflicting human desires and passions in their myths and Orion is no exception. The hunter is the pursuer of women and the hunt is the chase and speaks volumes about man as the hunter of a mate and the captivator of even goddess women who see him as an idealized form of manhood. Whoever brought about the death of Orion, Artemis gave way to grief and placed him immortally in the heavens where he still stalks the night sky with his sword at his belt, not far from Taurus the bull that he appears to be pursuing. (10) It is noteworthy that the hunter is stalking, not a lion, a tiger, or an elephant, but a bull.

Archaeological findings are represented in a wide variety of artistic renderings, in basalt, limestone and marble carvings, wall paintings, in ordinary vessels for daily use such as pitchers and cups, in pottery, silverware, seals, and a host of figures for household and temple adoration. From the earliest written records we know that cow skins were used for clothing, belts and shoes. Horns and bones were used for buttons, handles, and musical instruments. (11) The ordinariness of the articles for daily use, combined with the enormous stone carvings for worship and adoration, reveal that all ancient civilizations held this particular animal in veneration. Until Christianity was introduced as the new Roman state religion under Constantine in 325 CE, every culture had some form of religious belief in the bull or cow or both from as early as 7,000 years ago. Coming of age for the bull meant that it became divine.

Domestication Origins

The ancient city of Aleppo lies below the foothills of northern Syria in a fertile valley settled thousands of years ago by farmers, herdsmen and traders who crossed from the fertile crescent in the river valleys of the Tigris and Euphrates rivers to reach the Mediterranean coast. The meaning of Aleppo is unknown but Arabic speakers I questioned believe it derives from the Arabic word *halaba*, which means to milk. *Halaba al shahbaa* means "he milked the light-colored cow" and is often referred to as an act Abraham performed. (12) If this derivation is correct, one of the first most ancient and continuously inhabited cities was originally named after the milking of a cow. Also if true, since Aleppo is at least 10,000 years in settlement, the city's name shows the antiquity of the domestication of the cow.

According to scholars, the domestication of the cow occurred separately in Africa and in Anatolia in what is today eastern Turkey and northern Iraq. (13) Let us imagine that it is the beginning of spring when green shoots thicken in the plain, water flows more freely in the streams, the sun appears higher in the sky, the air has a sweeter fragrance, and songbirds chant a familiar chorus. The earth swells and claims the fruitful seed, rearing to life all teeming seed germs. Fields unlock their bosoms to the warm west winds and the avocado-green blades face the warming sun, fearless of the north wind's driving rains. It's the vernal equinox, and although no one is there to record it, an historic moment is about to change humanity for all time.

Imagine further a small encampment of a dozen or so extended families living on ground wheat and grains, berries from the mountain bushes, venison (chickens would not be domesticated for another 4,000 years), goat's milk, perhaps a strong mead fermented from sorghum silage. A band of related men, uncles and nephews, after a ritual hunt in which they manage to kill two bulls in a herd, realize that the cows can be herded into a cul-de-sac below the cliff near where the stream flows where they can be guarded from escape until a barrier is erected.

In such a place, under such probable circumstances, originated the greatest accomplishment of animal husbandry as a method of supplying food to a population. Afterwards and for all time, prior to writing, mathematics, symphonies, philosophy, sculpture, architecture, politics, drama, the harnessing of electricity, this would be the epitome of the intimacy and domination of the animal kingdom. Taming the bull would signal a high watermark in civilization, not only in animal husbandry, food, clothing, shelter, transportation and draft work, but in art, religion and myth, more significant than "inventing" fire.

Evidence from mitochondrial DNA (mtDNA) demonstrates that cattle were domesticated independently on the Indian subcontinent and in western Eurasia, in the region around what is today northern Iraq. (14) Mitochondrial DNA, more abundant in a cell and easier to extract than nuclear DNA, is transmitted only through the female. Genetics and the application of DNA, in addition to its multiple human applications, have enormously enhanced our understanding of the domestication of the cow. It's unknown whether the Indian cow, the zebu, the *Bos Indicus* the original cow, the aurochs, the *Bos Primigenius*, the extinct cow of Europe found in cave paintings, were the originator of all modern American and European cattle over 7,000 years ago.

Scientists extracted mtDNA from extinct aurochs bones in England and compared them with modern cattle. (15) Analyzing differences between the DNA sequences shows how closely related common ancestors are. The accepted theory now is that modern cattle, *Bos Taurus*, are the result of two distinct domestications based partly on the anatomical differentiations of European cattle and Middle Eastern cattle compared to *Bos Indicus*, the Indian cow or zebu. The extinct aurochs constituted a separate branch of the evolutionary ladder. Scientists surveyed cattle in 20 different locations to determine the evolutionary relationships and molecular biogeography of different cattle populations from Africa, Europe and Asia and found a marked distinction between humpless (*taurine*) and humped (zebu) cattle, providing strong support for a separate origin for domesticated zebu cattle at an estimated divergence time between the two subspecies of 610,000–850,000 years. (16)

Here is some of the scientific evidence gleaned from archaeology and art to illustrate some of what we know about this early period of Neolithic humankind's ancient relationship with bovine domestication.

Around 10,000 BCE, in a period known as the Neolithic, in the region of Mesopotamia, today's Iraq, southern Turkey, eastern Syria and western Iran, herds of wild cattle, wild boar, antelopes, gazelles and even elephants roamed freely in the marshes, lagoons and jungle areas. Sheep and goats roamed the hills in the north. Neolithic peoples began domesticating a few of the wild animals as they began settled agricultural societies in order to rely less on hunting as primary food sources. The systematic management of herds made more sense than herding and killing them. Whole communities could simply move with the herds, or begin to corral them as necessary when a group wanted to remain in one location, usually near a water source. By 7,000 BCE, in an excavated village called Maghzaliya in southern Turkey, 40 percent of the bones discovered were of sheep and goats, while 60 percent were from wild animals indicating that hunting still fulfilled the largest food needs of the village. (17) Domesticated cattle appear by the early 7th millennium BCE in Turkey and in the preceding century in Greece. Bones of domesticated cattle have been found in Thessaly dating to 8,500 BCE. Long horn cattle were domesticated first by 6,400 BCE, but the smaller short horn cattle not until about 3,000 BCE. (18)

Other scientists have collected evidence from limestone carvings, mural paintings, cylinder seals, revealing how pastoral communities milked their cattle. (19) Once captured or domesticated, cattle were used as draught animals as they still are in many parts of the world until large horses, donkeys and camels replaced them. Methods for harnessing cattle to be used for draft work occurred much earlier in time, around 3,200 BCE, and spread to northwestern Europe. (20) For example, a wooden model of a team of oxen yoked to pull a plough I saw in the British Museum in 1999 dates from 2,350–2,000 BCE in Egypt, so we know that cattle have been not only domesticated but used as draught animals for at least four thousand years.

By 5,000 BCE shards of pottery vessels reveal aspects of daily life including highly stylized bull's head and horns, the kind that can easily kill a careless hunter. These appear to be a common theme in art of the period as wild cattle were still widespread in the region. Soon, horns would be symbols of divinity in ancient Mesopotamia. (21)

The Hittite city of Catal Huyuk in southeastern Turkey is the largest ancient human settlement uncovered. It flourished between 6,500–5,650 BCE. Houses were made of dried mud bricks and were one-story high with only one room and sleeping platforms built into the walls. People entered them by climbing onto the roof on wood ladders and climbing down into their residences through a hole in the roof. As ladders were retracted at night, enemies or wild animals were prevented from entering. Woven mats covered the floor. The principal trade was obsidian, or volcanic glass, from which these early peoples built mirrors and daggers. The community had many shrines, special rooms where the walls were painted with religious scenes and decorated with the heads of bulls. Primitive wall paintings show men touching the bull's horns as if to receive some potency from them. Human skulls have been discovered under the bullheads suggesting sacrifices to the bull deities, and an artifact c.6000 BCE showing the birth of a bull god. (22)

The world's oldest farming communities sprung from the upper Euphrates River along the foothills of what is today southern Turkey. Archaeologist husband and wife team Robert and Linda Braidwood found fossilized fragments of cloth dating from 7,000 BCE not only pushing back farther in time the introduction and use of textiles but providing evidence that flax had been domesticated as the fabric of choice. (23) These discoveries confirm the link between settled agriculture and the rudiments of civilization. By about 6,000 BCE cattle had been domesticated, a form of irrigation existed, and all the staple crops were cultivated thus creating settled agricultural and pastoral communities. It is probable, though not confirmed, that a primitive plough was in use and that it was drawn by cattle. (24)

We can link the study of cattle domestication with religion from a very early date, roughly the 6th century BCE. The people who lived In the mountainous plateaus between northern Iran and northern India at the time of the priest-prophet Zoroaster about 2,600 years ago were pastoral herders whose economic foundation was cattle. The Zoroastrian texts the *Avesta* testify to this early society's reliance on cattle. (25) Like tribes in east Africa this ancient society depended on the cow as we do today for its milk and milk products, meat, leather, dung for fuel, and draft leverage. Its urine was used as a purifying agent.

A religion that believed in a modified version of monotheism (a god of light and a god of darkness), which preached that individuals must choose between good and evil, which believed in the soul and its immortality, of a savior, of a personal relationship with God, and that individuals would be judged in the afterlife by the level of their righteousness in life, a religion that heavily influenced Greek philosophy, Judaism and Christianity, also had a migratory society built around the cow and its need for pasturage and water. For Zoroaster the cow was a sacred animal, and in the *Avesta* he sees the soul of the cow as a prototype for all animal creation.

> *The soul of the cow lamented to you: For whom have you determined me? Who fashioned me? Wrath and violence, Harm, Daring and Brutality (each) have*

bound me! I have no other pastor than you—so appear to me with good husbandry! (26)

It may well be that this bovine lament is an allegory for a sedentary form of peaceful livestock life contrasted with lawless nomads who raided agricultural settlements and used cattle for sacrifices. The lament can also be viewed as a struggle to establish a religious form of life built upon respect for all life, as is still true in contemporary Indian society.

The first civilization, the culture of a people who had established writing and literature, sculpture and metallurgy, decorative art, and the refinement and good taste that are the benefits of a stable agriculture and pastoral life that produces wealth to sustain the pursuits of leisure is also the first to reverence the cow.

Sumer: The First Cow Culture

Eden is a Sumerian word which means "grassland to the south," a watery, flowering paradise during the time of this civilization's height before the Tigris and Euphrates changed courses, the salinity of the soil depleted agriculture and killed livestock, and the desert reclaimed the fields and the cities. Sand and desert storms long ago swept the region clean of vegetation begetting a lunar landscape. But in its day Sumer was the first and largest permanent settlement of any people, containing two large temple complexes, with art that demonstrates its sophisticated social, economic, religious, and political organization and thriving artistic life. The Sumerians were the first people to live continuously in one settled place with a degree of security and leisurely wellbeing over hundreds of years, cultivating crops and livestock with cattle as the most important. (27)

Sumer is humankind's earliest civilization, beginning about 7,000 BCE and reaching its zenith about 5,500 years ago. It is significant for this study because it is civilization's origin of writing, the chief ingredient of any civilization, a system that evolved into cuneiform, or writing with a wedge-shaped stylus. In fact, it has been the challenge and triumph of philology to decipher this unique language. The first pictographic texts emerged in the Early Dynastic Period, from 3,400–3,200 BCE, and the cuneiform text appeared between 2,900-2,350 BCE and with it a flowering of literary texts, proverbs, hymns, narratives about boundary disputes and cultic figures like Gilgamesh.

The Epic of Gilgamesh, one of the first heroic legends in the world's oldest literature dating from c.3,000 BCE, contains the stories of the creation of the world, the flood, and the titanic conflict of good with evil. It too has bovine similes and metaphors. The greatest of all gods in the Sumerian hierarchy was Ishtar who had the power of life and guardianship of all living things. Ishtar the great mother goddess was the giver of herds and flocks and often is shown with horns on her headdress, a feature later to be incorporated into Egyptian mythology with Hathor, the cow goddess who had a similar headdress. Found at Lagash, an ancient Sumerian city located between the Tigris and Euphrates rivers in southern Iraq near the modern village of Telloh, is a representation of Ishtar with a sacred cattle house. It is also the first known civilization to feature the bull. One of the most striking sculptures of naturalistic art is that of a bull in silver kneeling and with its front hoofs holding up a vase.

The cities of Sumer, Ur, Eridu, Uruk (biblical Erech and modern Warka) and the Sumerian religious capitol, Nippur, finally ceased to exist as a place of habitation altogether by the time of the Arab conquest of the region and were abandoned to desert winds and the inexorable encroachment of sand. This civilization flourished in the lower delta of the Tigris and Euphrates rivers near the Persian Gulf. Its main city was known in the Bible as Ur of the Chaldees, said to be the city of Abraham's birth. Marauding bands from the north eventually overran its limited defenses and the high civilization moved north to Nineveh and Babylon. All Middle East civilizations, as culled from cultural artifacts, have a reverence for the cow as a goddess, for bull deities, for sacred temple herds, for sacrifice to other gods or goddesses, and obviously for the milk the cow produced. (28)

Ironically, when the coalition military forces of the West came to Kuwait and southern Iraq in 1991 during the Gulf War and again in 2003, they fought at a location that was the heart of Middle Eastern civilization's origins. Here, in the ziggurat or step-pyramid temple cities of Sumer now lying submerged under sand, were developed the wheel, law, science, astronomy, writing and literature, inventions and developments that are the heritage of modern life. Civilization means urban life and almost always includes large ceremonial buildings dedicated to religion.

Within a short time period as we compress centuries, cattle became associated with deities. Once Sumer decided to make the cow and bull sacred, subsequent civilizations borrowed this cultural and religious trait. Weather controls the agricultural environment and hence food sources and living conditions so it is not unusual that the awesomeness of the weather should also be associated with the powerful and forbidding animals like bulls. The storm god of the Sumerians was Enlil, which became under the Akkadians Ada, and under the Canaanites Hadad-Baal, or Baal-Hadad (where Baal means lord). Baal is frequently shown brandishing lightning bolts and standing on his cult animal, the bull, as portrayed on an Assyrian relief in the Louvre in Paris. In the National Museum in Damascus in Syria is a Syrian sculpture of a figure of Baal-Hadad with the head of a bull and a dagger at his belt, with earrings dangling from his bull ears. Here, for the first time in recorded history that we know, the bull becomes a part of religious motif and remains thus through all later civilizations including the early Hebrews.

Canaanites, precursors of the Hebrews, elevated storms, lightning and the weather to the status of gods. (29) A basalt stele discovered by accident in the fields around a village just north of Aleppo in Syria and dated to the 8th century BCE, is that of Baal Hadad, the weather god riding an ox. He holds the reins in his left hand and a spear in his right hand. The stele is covered with Hittite hieroglyphics. By the second millennium BCE bulls and oxen were routinely offered to the gods for sacrifice. A mosaic in the Aleppo museum in northern Syria shows a priest conducting such a ceremony.

A restored Sumerian lyre from Ur, dating from 2,600 BCE on display in the British Museum in London, is decorated with a pure gold bull's head colored with bright blue tufts of hair both on its head and under its chin and capped with ivory horns. It has bright blue eyes, the same blue of the tiles of the temple shrines in Sumer. A similar lyre, also with a decorated bull's head appears on the Standard of Ur, also from 2,600 BCE. So was the bull venerated as the patron divinity of music? Did the bull appear on all musical instruments as a reminder to the musicians of where their patronage lay? Or was the image of a bull simply a matter of artistic discretion that the artist could choose? Clearly, the prevalence of cow and bull images from this earliest civilization is an indication of elevated esteem.

A Sumerian cylinder seal from the 4th millennium BCE in the Ashmolean Museum in Oxford shows a community of cows, thought to be part of a temple herd clustering around reed houses, residences still found in the reed marshes around southern Iraq. Some of the huts appear to have cattle living in them.

An invasion of Sumer occurred in 2003 when, within weeks of the fall of Babylon, the extraordinarily rich cultural heritage of Mesopotamian civilization in the Baghdad Museum was sacked as organized thieves looted scores of its treasures. The immensity and antiquity of the cultural significance was over-shadowed by the quantity of jewelry, sculpture, ceramics and metalwork stolen, over 10,580 pieces at the last confirmation, all irreplaceable art objects, national treasures to the world and not just Iraq. The art objects we possess from Sumer had resplendent depictions in gold, lapis lazuli and carnelian, many of them in cylinder seals so lavish and yet so miniature that they are less than one and one-half inch in height.

The systematic and mechanized plundering of the archaeological sites in 2003 deepened the artistic loss as clandestine and illicit excavators using trucks, earth-movers and armed guards dug trenches and gouged the earth for antiquities. It was the worst desecration of unprotected cultural sites in memory, perhaps in all time. The cow we will always have with us. But examples of artistry depicting the cow and other aspects of Sumerian life may be lost forever, squirreled away in private residences around the globe.

The Cattle of Ancient Mesopotamia

Mesopotamia, as the Greeks called the land between the Tigris and Euphrates rivers, is the place where civilized life flowered, architecture originated, and art and metallurgy had their fountainhead. Mesopotamians built houses and molded pottery from dried mud but had to import stone, metal and wood that were locally unavailable. They had easy access to water and from a perpetual flowing reserve they founded irrigation systems that gave them plentiful harvests of wheat, fruit and vegetables to feed a growing population of people and livestock. Agricultural wealth provided the leisure to promote literary recordings, the stories of creation, floods, of the Garden of Eden, of arks, of the tower of Babel, all had their sources in the literature of Mesopotamia and provided source material to biblical scribes. Luxury items on display in museums throughout the world are an indication of the wealth of this ancient civilization founded on agricultural abundance and mercantile trade with neighboring communities.

Cattle were the backbone of this labor-intensive economy. The marsh Arabs of southern Iraq, the Madan peoples, who live in the swamps where the rivers flatten out into a broad delta, live almost identically to their ancestors by fishing, cutting marsh reeds for the construction of boats and houses, and by cultivating the water buffalo. Their reed houses, some as long as a hundred feet, were depicted the same way on seal impressions from Sumer over 5,000 years ago. Located in the marsh area in southern Iraq around the town of Basra, these people inhabit an area that experienced war between Iraq and Iran in the 1980s, the Gulf War in 1991 and the invasion of Iraq in 2003. The marsh Arabs care for their animals, recognize them instantly as if they were members of the family and are distressed when the buffalos are ill. (30)

The milk of the water buffalo supplies modern Iraq with its dairy products. The marsh Arabs nurture the water buffalo exclusively for its milk and do not work them for draught purposes. Its meat is consumed only after years of milk production, though the natives will occasionally kill a male calf for a wedding feast or other festive occasion. Females will calf after about four years and thereafter about every 12 months. Over a lifetime, a dam will yield about 15 calves. The buffalo graze most of the day on the large reed plants ubiquitous in the swampy region that the locals use to build boats and huts.

These marsh Arabs suffered the obliteration of their culture under Iraq's Saddam Hussein in the last years of the 20th century. Saddam suppressed their rebellions, murdered Shiite clerics and bombed marshland villages with impunity. In a six-month period in 1992 Hussein's battalions drained the wetlands turning them into an inhospitable desert, described by the United Nations as an environmental disaster, and displacing 100,000 to 250,000 marsh Arabs, many fleeing to Iran.

From King Ashurnasipal's well at Nimrud about 2,650 BCE, and displayed in the Hall of Ivories in Baghdad, has come a small ivory head of what is likely a water buffalo. So the water buffalo, like cattle, have been in this region of the world for at least 5,000 years. However, it's possible that the swamp buffalo was introduced into southern Iraq from the Indus river valley civilization as long ago as 4,600 BCE. (31)

Cities may crumble, art get pilfered or destroyed, but cattle culture endures. The Sumerian culture was in turn absorbed by the Akkadian, the Babylonian, the Persian, the Canaanite and yet the role of the bull continued assuming unparalleled prominence in successive generations. Babylon at its height had the divine bull at the apex of its ceremonies.

Babylon: Expanding the Role of the Bull

Astarte placed upon her own head a bull's head as an emblem of kingship. (32)

Babylon, if not the whole contiguous culture of the rich Mesopotamian valley, had a moral prejudice introduced by the Hebrews and by John, the author of *Revelations* who wrote:

What city is like unto this great city...that great city that was cloaked in fine linen and purple and scarlet and decked with gold and precious stones and pearls! Babylon, the Great, the Mother of Harlots and the Abomination of the Earth. (33)

This passage alluded to the morality of Rome for contemporary readers but also hearkened back to the Hebrew captivity after the Persian destruction of Jerusalem and the temple in 586 BCE and the subjugation of the people and their exile in Babylon. But the passage is a moral judgment upon a civilization and not descriptive of its principal cultural or religious elements.

Situated in the lower Euphrates river drainage in what is today central Iraq, Babylon was the geographic and cultural heartland of Mesopotamia extending from roughly 7,000 BCE, through the civilizations of the Sumerians, Akkadians (2,350–2,000 BCE), and Babylonians (c.2,000–500 BCE). As people were able to grow food instead of having to travel long distances in search of it, they were able to pursue trade and business. The domestication of cattle and other livestock, occurring about the same time or just after the agricultural revolution and birth of settled villages, contributed to the development of these burgeoning city-states in Mesopotamia and other nearby civilizations along the Indus river in today's Pakistan and the Nile in Egypt. (34).

With the Sumerians, who invented writing and literature, and the Babylonians, who improved math and law with the code of Hammurabi, invented the calendar, gave us measures of keeping time by divisions of 60: 60 minutes in an hour, 60 seconds in a minute, conveniently divisible by two into a month and by 5 to equal 12, the length of a day and a night, and who perfected star study, we enter into more sophisticated and refined urban life. Schools, and the leisure required for youth to study at all, would not have been possible unless Babylonian society had developed a solid agricultural base with working cattle permitting urban settlements and an expanded leisure class.

Thirty-five hundred years ago oxen were routinely used as draft animals for sowing seeds. A clay seal impression from 1,730–1,155 BCE gives an excellent depiction of a Babylonian ploughman and seed-sower. Drawn by two hump-backed oxen, a central figure places seeds in a funnel that trickle down through a seed-drill into the plough furrow from a seed bag slung over his shoulder, while a bearded ploughman works the wooden plough with both hands. A golden bowl from Ras Shamra (14th century BCE) residing in the National Museum in Aleppo shows a hunting scene with various wild animals. Prominently displayed are two bulls with heads and horns lowered, as if to gore anyone approaching, and with tongues hanging from their mouths.

The young certainly would have been apprenticed in agricultural chores, tending animals like cows and sheep and helping irrigate fields and the sowing, reaping and harvesting of crops. The settled agricultural communities presented new survival challenges. Many of the young in adapting to these skills would have to learn the cycle of planting and harvesting crops and this may have included learning about how to read and calculate the signs in the heavens for the changing of the seasons. (35)

Babylonians had a large number of festivals throughout the year, none more festive than the New Year celebrations that occurred during the month of the spring or vernal equinox that lasted about a week. There were ceremonial prayers and purifications, the reciting of epic stories, and rituals involving the demeaning of the king to show that he must bow before and venerate the great god Marduk. In the evening of this day, a white bull was sacrificed.

Here are a few of the surviving monuments from Babylon showing the ubiquity and importance of the bull cult. Each appears to be a simple yet elegant testimonial to the semi-deification of the bull, but assembled here they represent how these people revered this one animal more than any other and how their lives were so thoroughly dependent on its bounty, as we are today but without the accompanying cultural symbols.

- Babylon was capitol of Nebuchadnezzar II who built it in 604–562 BCE Its famous sculpture is the gate of Ishtar, the goddess of the Sumerians. Covered with colored glazed tiles, it is decorated with bulls and dragons along the walls in brick relief. (36)

- A black basalt memorial stone, with inscriptions describing the restoration of the walls of Babylon, depicts the Assyrian "sacred tree," or tree of life and knowledge. A king stands before an altar with a horned crown resting on it, and behind the king looking on, an attentive bull. (37)

- One of the earliest representations of a ziggurat comes from a stone relief from the palace of Assurbanipal at Nineveh who lived in the 7th century BCE. Atop the five-storied ziggurat is the horned shrine, the symbol of the high eminence of the bull cult.

- Seals from this same period portray a variety of contest scenes in which heroes, kings, or famous athletes overcome protagonist animals, always shown as life-sized compared to the size of the man so the contest seemed fair. The beast contestant may be mythical, but more often is a representation of a real animal like a lion or bull. The contestant usually is bearded, nude, and commonly holds the defeated bull by one hind leg, dangling him upside down to show his dominance and victory and the bull's defeat, the modern equivalent of the bullfight. (38)

- The mighty palace of the Assyrian King, Sargon II, who reigned from 721–705 BCE, in the ancient city of Khorsabad contained extensive treasures. A huge pair of man-headed winged bulls, about 20 feet tall by three feet wide, guarded the entrance to the palace. These enormous bull bodies, probably portrayed with the head of Sargon himself, stand in the first floor of the British Museum. Were these enormous stone bull figures at the gates of the palace and citadel guardians against misfortune, as the description notes? Or does the bull represent the power of nature that the king commands? The bull is so paramount and commanding that artists would not spend time on such a vast project unless the king approved and the representation of a bull was not symbolically at the core of a cultural imperative.

Collapsing millennia and centuries into tiny blips has the disadvantage of skipping over important events, figures and periods of relevant human development. However, at the expense of appearing unsystematic, this brief summary synthesizes selected bits of archaeological history to show how the cow and the bull in the dawn of civilization captivated human imagination and molded it into the useful, the ritualistic and the religious. Unknown artists left artistic remnants of a dynamic cultural context, a world in which the cow and the bull had a colossal and abiding influence. And though such ancient domestication efforts and cultural incorporation appear far removed from modern society, there are people whose practices with modern herds parallels these ancient methods of survival. The reindeer people in Siberia are one such group.

The Reindeer Peoples: Surviving with the Herd

The Chukchi in northeastern Siberia, the Nenets and Khanty in north central Siberia, and the Sami, or Lapp people in northern Scandinavia, are native peoples who base their economies upon the herding of reindeer, an example of contemporary groups who mirror ancient peoples who first domesticated cattle. Reindeer in arctic Asia are slightly smaller than caribou, their North American and Greenland cousins, and belong to the same species of cloven-hoofed animal.

The Chukchi, a hardy, adapted people who have never been subjugated number about 15,000 and inhabit the inhospitable tundra of northeast Siberia across from the Bering Strait from Alaska, are related genetically to all American native peoples. About 15,000 years ago Chukchi ancestors crossed the Bering Strait, then frozen solid prior to the retreat of the last Ice Age, and populated North and South America. Today, taking their nourishment and subsistence from the reindeer, they live without electricity or any modern conveniences exactly as their progenitors did 5,000 years ago. So total is this dependence that it is difficult to say whether or not the reindeer might have tamed humans. If Chukchi were to die out the reindeer would still survive; but if the reindeer were all to die, the Chukchi would become extinct.

Their lives and livelihood throughout the harsh winters, food, clothing, shelter, and transportation, revolve around the migratory patterns of the reindeer, numbering about 2.4 million in the Arctic. These hardy animals migrate annually to calving domains and follow the grazing areas of lichen which they paw to eat through the powdery winter snow. Each Chukchi family will have a herd of about a 1,000 reindeer and make a living from the sale of meat, hides and antlers. Reindeer hides supply beautiful, light and warm clothes for survival in the severe cold, provide bedding and, when sewn together with sinew, become the winter coverings of the large round tents, quickly disassembled when the herds begin to migrate to find new sources of lichen growing beneath the snow cover. Sleds of tents and supplies are pulled by teams of reindeer who are lassoed and put into service for the occasion.

The Chukchi are contemporary evidence of the intimacy of a people with a domesticated animal and show how a similar domestication of cattle thousands of years ago changed forever the lives of the people who cultivated them, a mutual dependency which has persisted through the centuries as we shall see in the following chapters.

CHAPTER TWO
THE COW AND BULL IN SCRIPTURE

The Sea stood on twelve bulls, three facing north, three facing west, three facing south and three facing east. The Sea rested on top of them, and their hindquarters were toward the center. The Book of Kings (1)

Footprints in moist ocean-side sand will disappear almost as quickly as the traveler has passed. Cultural footprints, however, leave a more indelible mark. The relationship of humankind with cattle and its divine status recurs repeatedly in the Bible, like the mythical view of the world with bulls holding up the sea in the quotation above.

There are 84 references to cattle in the Bible and gospels, 157 citations for bulls, one or more appearing in: *Genesis, Exodus, Leviticus, Numbers, Deuteronomy, Joshua, Judges, Samuel, Kings, Chronicles, Nehemiah, Job, Psalms, Isaiah, Jeremiah, Ezekiel, Daniel, Joel, Jonah, Habakkuk, Hosea, Matthew, John, The Acts of the Apostles, Revelation,* and Paul's *Letter to the Hebrews.* The number of references, however, is not as significant as those few statements that give us an insight into the relevance of the bull or cattle in Hebrew culture. Most are only occasional references to the slaughter of bulls or cattle (*Joshua*: 6, *1 Samuel*: 14), but there are also bull sacrifices made with goats and sheep, and how the sweet aroma of the cattle sacrifice is pleasing to the Lord. In the time of Abraham cattle are spoken of as a sign of wealth. Abraham is blessed with sheep and cattle and this is pleasing in God's eyes. (*Gen.* 32) And just who was Abraham?

When the late author Tad Szulc retraced the steps of Abraham he was shown a book containing reproductions of a fresco painted in Mari in northern Syria dating from the early second millennium BCE, a fresco reputedly of Abraham. What Szulc observed in that fresco was a decidedly un-mythical, ordinary man with brown skin, semitic-looking, with a small black beard, wearing a black cap with a white headband. But in his lap, conspicuously, was the two-horned head of a sacrificial bull. (2) Was Abraham originally a follower of the bull cult as was the rest of the Mesopotamian and Canaanite culture? One answer is yes, because he lived prior to Mosaic monotheism and the declaration of the Torah or Law. Abraham was not a follower of Moses because, based on the inaccuracy of biblical calculations, he reputedly lived at least 500 years prior to Moses, and, like all Hebrews prior to the founding of Jerusalem, was profoundly influenced by the culture of his birth, Ur of the Sumerians, and by Mesopotamian and Canaanite culture that had the bull cult. Prior to his attempted sacrifice of his son, Abraham sacrificed a bullock, not a lamb, when angels came to visit him. "And Abraham was very rich in cattle, silver and gold." (3)

God speaks to selected individuals and tell them specifically what to you with selected bulls. "That same night the Lord said to him, 'Take the second bull from your father's herd, the one seven years old. Tear down your father's altar to Baal and cut down the Asherah pole (a symbol of the goddess Asherah) beside it.'" (4) All this is very Canaanite and not Hebrew.

But in *Exodus* we find a curious and over-looked allusion to cattle. Just as every firstborn son in Pharaoh's Egypt is to die so is every firstborn cattle in the realm was to die (5), and this is not an incidental phrase but part of the law: "As it is also written in the law, we will bring the firstborn of our sons and of our cattle, of our herds and of our flocks to the house of our God to the priest ministering there." (6)

The scribe(s) of *Exodus* are also steeped in the tradition of cattle and bull sacrifice and describes accurately how this ceremony is to be conducted. "Take some of the bull's blood and put it on the horns of the altar with your finger, and pour out the rest of it at the base of the altar." (7) This same ceremony is described again in *Leviticus*: "The priest shall then put some of the blood on the horns of the altar of fragrant incense that is before the Lord in the Tent of Meeting. The rest of the bull's blood he shall pour out at the base of the altar of burnt offering at the entrance to the Tent of Meeting." (8)

At this time in Hebrew history bull sacrifice is clearly a part of the culture and even Moses takes part, as the following passage in *Leviticus* makes clear. "Moses slaughtered the bull and took some of the blood, and with his finger he put it on all the horns of the altar to purify the altar. He poured out the rest of the blood at the base of the altar. So he consecrated it to make atonement for it." (9) Why would Moses, who killed his friends and relatives for worshipping a golden calf participate in a ceremony of slaughtering a bull for sacrifice? To whose god was he thus making propitiation and for what purpose? The *Exodus* scribe is unconcerned that the Hebrews at various times in their history worshipped Baal like the Canaanites and then later repudiated the practice.

Later in the history of the Hebrews sacrificing a bull becomes an unrighteous act. According to Isaiah, "Whoever sacrifices a bull is like one who kills a man, and whoever offers a lamb, like one who breaks a dog's neck; whoever makes a grain offering is like one who presents pig's blood, and whoever burns memorial incense, like one who worships an idol. They have chosen their own ways, and their souls delight in their abominations." (10) This is indeed degrading since the days when it was not only expedient but essential that bulls be sacrificed, as the example of Moses demonstrates.

When Solomon dedicated the temple, he sacrificed 22,000 cattle and 120,000 sheep and goats. (11)

In John's gospel, Jesus finds the moneychangers in the temple precincts. But there are also men there selling cattle and he drives them and the cattle away with a whip made of cords. (12) Even after the death of Jesus the disciples found bull sacrifice popular among the Greeks living in the region. "The priest of Zeus, whose temple was just outside the city, brought bulls and wreaths to the city gates because he and the crowd wanted to offer sacrifices to them." (13) But for Paul the blood of animals can no longer be an expiation for sins. "The blood of goats and bulls and the ashes of a heifer sprinkled on those who are ceremonially unclean sanctify them so that they are outwardly clean...because it is impossible for the blood of bulls and goats to take away sins." (14)

How, after so many centuries when God sanctioned and even commissioned the sacrifice of bulls did it become religiously unacceptable for both Jews and Christians? Did God change his mind, or did men misinterpret the message? According to Paul, the blood of bulls can still "sanctify" those who are "unclean," though it is unclear whether this is because it has been a longstanding Jewish tradition or a new Christian conviction. From the conflicting biblical references, one could argue that sacrificing bulls is good, evil, or un-necessary.

I have tried to recapture here what the bull, prevalently mentioned in the Bible, meant to the Hebrews and their civilized predecessors. Based on my own extensive travel and living experiences in the Middle East and inquiries, let me retrace part of that literary and artistic history to show the continuity of the cult of the bull. The recent discoveries of the Ebla and Ugarit archives have shed new light on these ancient cultures from which Hebrews borrowed heavily and reveal how pre-eminent the bull was in all Middle Eastern and Egyptian cultures.

What I have deduced from literary evidence is that the Hebrews lived essentially in a Canaanite culture that over the centuries became overlapped with Mesopotamian influences grounded in the Babylonian exile, and then absorbed Egyptian influences. The Hebrews sought a unique national identity through a monotheistic religion founded on the Torah. But biblical literature is replete with a powerful and permanent traces from these other cultures, most prominently the bull cult. We can perceive this cattle influence if we turn to the earliest biblical scriptures.

The domestication of the cow in the Middle East brought meat, milk, bone, leather, hides, traction for planting and reaping and dung for fertilization, all essential elements in a civilization's survival. The cow's presence among its owners signified wealth and *Genesis* describes the conflict between large herd owners, Abraham and Lot, and Jacob's family in

Goshen. (15) During David's reign, overseers were appointed to supervise these herds. Young calves were considered a special delicacy for festive occasions. The bull, on the other hand, was the mythical, religious symbol of fertility and power, and it and the calf are cited innumerable times throughout scripture. The golden calf in *Exodus* 32 is just one extended narrative of this ubiquitous religious influence. The principal Middle Eastern deity in Babylonia, Canaan and Egypt was a bull and all ancient literature makes copious and analogous reference to its divine intervention and protective powers.

For example, from at least 5,000–3,000 years ago bull animal sacrifice was a rite practiced by all ancient civilizations. The biblical gods of the Hebrews, El and Yahweh, represent the force, virility and strength of a bull. "The Mighty One" of Jacob is also translated as "the Bull of Jacob."(16) A similar phrase appears in ancient Egyptian literature referring to Akhenaten: "The living Horus: Mighty Bull, beloved of Aten." (17) Golden calves, such as the one mentioned in *Exodus*, were erected in Dan and Bethel. (18)

Sacrificial biblical cattle had to have certain requirements and these are clearly enunciated. (19) These sacrificial restrictions may have changed over time and place, but a universal prescription is that the yearling calf, or fattened bull for festive celebrations, cannot have been used as a draught animal. (20) Sometimes, several kinds of animals, heifers, she-goats, rams, turtledoves and pigeons, as God told Abraham, were all sacrificed together. (21) By the time of the establishment of the monarchy under David and Solomon whole herds were sacrificed. (22) The ritual of bull sacrifice had been firmly established by the Hebrews in the first millennium BCE, but like much of the rest of the culture the ritual was borrowed in whole or part from the Canaanites, Edomites, Babylonians and Egyptians.

Let's take one example of an existing hill of sacrifice from the Edomites who inhabited the area of what is today southern Jordan. I journeyed to Petra in Jordan the city of the ancient Nabateans and Edomites in the spring of 2002 climbing to such a hill of sacrifice to gain a direct experience of where bulls were likely sacrificed.

The ascent of Jebel Attuf in Petra is physically demanding but affords the adventurer stunning views of the wadis and the surrounding mountains. The elevation is 3,414 feet and is reached after a couple hour hike through flights of steps carved in the rocks, sometimes existing as ramps etched into the face of the red rock quarry through narrow valleys ringed by precipitous rock walls. At the top stand two obelisks, one 19 feet and the other 23 feet high, probably representing the Edomite divinities Dushara and Al Uzza. (23) It was at the summit of this mountain and open air sanctuary where the religious celebrations occurred and where the Hill or High Place of Sacrifice is located. Tradition has it that Aaron, brother of Moses, sacrificed here. It was at such locations that the early Canaanites and Edomites performed sacrificial ceremonies, undoubtedly with bulls.

The faithful gathered at this site in a large rectangular-hewed section of the rock, about 48 feet by 21 feet, surrounded on three sides by stone benches. A rock altar, known as a *mensa sacra*, or sacred table, is a perfect rectangle just six inches high where supposedly non-blood offerings were placed. I rather believe, because of its smallness, it was not a regular table but a ritual table where large animals like bulls were slain. A small drain for the entire rectangle or courtyard is at the northeast corner where liquids, blood among them, drained steeply down the hillside. I stared down the sculpted edge of the drainage where the blood would have run off in a torrent during ceremonies and wondered how the spectators would have reacted emotionally to this sacrifice atop this hill with the awesome view in all directions.

An altar and carved pools for collecting blood and washing faces faced south at the edge of the mountain. The rock-carved altar is reached by ascending three steps. In the top is a square hole where the non-human representation of the divinity such as an ovoid rock form known as a baetyl was likely placed during the ceremonies. Standing before the altar you can see towards the south a panorama of the wadis or narrow mountain passages below, and in the distance to the southeast the ruins of the Roman city and the amphitheatre. The officiating priest or priestess certainly had a commanding view. To the immediate left of the altar is a rock basin reached by four steps where other sacrifices, possibly of small animals were made. A three-foot deep rectangular basin carved into the rock might have contained water for purification of the officials or cleansing the animals (or humans) to be sacrificed.

The most extensive Hebrew ritual is associated with the Passover, celebrated as a Seder meal in Jewish tradition, in which a yearly lamb without blemish is sacrificed. This firmly-established ritual by the time of Josiah c.1,250 BCE is derivative of a herding and animal husbandry nomadic desert past emanating in the south or east of Israel and not from the settled agricultural societies which bred cattle in the north. (24) The shift in animal sacrifice indicates a shift in cultural traditions. The early history of the Hebrews is influenced by Canaanite culture and Israel in the north, whereas later history falls under the influence of the Edomites and Judah in the south. In the New Testament it is the blood of the lamb that is empowered to cleanse sins and expiate guilt. In the Torah or Old Testament, it is the blood of the cow that expiates guilt and penalties for crimes. *Deuteronomy* has an odd expiation ceremony involving a heifer, not a lamb. The following quote from *Exodus* gives an insight into this mentality.

If the corpse of a slain man is found lying in the open on the land which the Lord, your God, is giving you to occupy, and it is not known who killed him, your elders and judges shall go out and measure the distances to the cities that are in the neighborhood of the corpse. When it is established which city is nearest the corpse, the elders of that city shall take a heifer that has never been put to work as a draft animal under a yoke, and bringing it down to the wadi with an ever flowing stream at a place that has not been ploughed or sown, they shall cut the heifer's throat there in the wadi...Then all the elders of that city nearest the corpse shall wash their hands over the heifer whose throat was cut in the wadi and shall declare, 'Our hands did not shed this blood, and our eyes did not see the deed.' Thus they shall be absolved from the guilt of bloodshed. (25)

The cow has significance beyond food, beyond the giving of milk, and beyond draught work in the fields. The cow for the Hebrews was the ultimate sacrifice for expiating the sins of the community. Notice that the prescription above does not urge the community to seek justice, to start the process of finding the killer. The more important act is to cleanse the community of associated guilt for the death of an unknown man. The murderer may never be found but the community will be spared any divine or human retribution from families if the dead man's identity becomes known. As MacBeth said to the imaginary ghost of Banquo, "Thou canst not say I did it. Never shake thy gory locks at me." But what if one of the elders or judges had in jealous rage or imperious anger killed that fellow citizen? Would the cutting of a cow's throat avenge the dead man's untimely death, or capitulate to the perpetration of an injustice by not seeking the killer? Knowing this prescription, might not a future killer simply drag the man he has killed to a place farther away from his town so a distant village would have to sacrifice a cow to be expiated?

A cow should not be asked, in its own unwanted death, to perform the deed of absolution for something it had nothing to do with, nor substitute for justice by neglecting inquiries into human fatalities. Yet by calling on the sacrifice of a cow the Hebrews were accentuating a social condition that is far removed from contemporary society: that cow sacrifice is more important than justice, even in the search for a killer.

Canaanite Civilization

Canaan roughly corresponds to the territory known as the Levant, equivalent to the modern states of Israel, southern Turkey, Lebanon and northwestern Syria. The biblical people who were variously known as Ammonites, Moabites, Israelites and Phoenicians were all ethnically Canaanites. (26) The Phoenicians, whose name *phoinikes* comes from the murex of Mediterranean mollusks that yield the immensely popular purple dye, the symbol of royalty, took to the sea after the withdrawal of the

Egyptians about 1,200 BCE to trade and expand their economy establishing colonies like Kition in Cyprus and Carthage in Tunisia. Everything understood by the artistic and literary accomplishments of the Phoenicians pre-existed in the Canaanite culture.

One of their chief and enduring contributions to civilization was the reduction of the approximately 800 symbols used in cuneiform writing to consonants where vowels were understood to occur between them and thence to the manageable alphabet we have today of 26 characters. We commonly associate the development of the alphabet (from the Greek *Alpha* and *Beta*) with the Phoenicians, from which we get the phonetic lettering. The first letter of the modern alphabet, A, is shaped like the image of a bull's head. (27)

The story of the founding of Carthage by the Phoenicians concerns a cow. According to classical legend, and in particular Virgil's epic Roman poem the *Aeneid*, settlers from Tyre in Phoenicia under Elissa, or Dido according to Virgil, Pygmalion's sister and great niece of King Ahab's wife Jezebel, came to North Africa with 80 maidens and the priest of Astarte who had been picked up in Cyprus. Upon arrival in Carthage they bargained to purchase as much land as an ox hide could cover. They cut the hide into very thin strips and hence were able to acquire more than enough land. This same legend, in a slightly modified format, would appear centuries later in a Danish myth.

Canaan was the Promised land where Hebrews were to settle and over whose local inhabitants they were to prevail. However, it is possible that the Hebrews were themselves a cultural subgroup of Canaanites. The discovery of Canaanite writings is relevant to this conclusion. When the ancient city of Ebla in what is today northern Syria a few miles southwest of Aleppo was first discovered in 1974, over 15,000 cuneiform clay tablets were found revealing a rich urban kingdom of the third millennium, the Canaanites, originating roughly from the early bronze age about 2,500 BCE. To date, the discovery of the archives at Ebla has been the largest cache of material uncovered from the early Bronze Age, showing that Ebla was a bustling trade center of foodstuffs, textiles, metals and wood products. (28) The tablets of documentation contain domestic and international political conditions, accounts of dynastic rulers, military campaigns and business transactions. Its king administered a vast agricultural bureaucracy and well-regulated social system and was elected by an oligarchy of elders and not a religious or priestly caste. Ebla was a secular state with a religious base. It was not, as Egypt was about the same time, a dynasty whose king was the divine head of a priestly order.

Ebla's wealth was sustained by extensive farming, livestock breeding, and commercial industries. Agriculture consisted of the growing of several kinds of wheat and barley. Sheep were bred for food and for the wool used in textiles. But cattle breeding was intensely cultivated. One Ebla commercial or governmental tablet lists 417 bull calves, indicating an extensive herd. A few of these cattle were destined for religious sacrifice, but the majority were employed in the fields and for transporting carts with messengers. The textile industry, weaving wool and flax, employed the largest number of workers. But the precious metals and gems industry, workers in gold and silver especially, also had an impressive numbers of workers. In one four year period, three tons of silver were mined and worked. Ebla was also a center for lapis lazuli, copper, tin and bronze.

Since the tablets come from the archives of the state economy and record business transactions, we know that the value of an ox was about 30 times that of a sheep, slightly less than the value of one gold dagger. A cow, the second highest valued commodity in the empire after a gold dagger, was valued only one-third that of a mercenary soldier. (29) Oxen were occasionally slain in religious offerings. One text, without elaborating on the form of the ritual sacrifice in a religious feast, describes in an accountant's laconic style:

2 oxen for the goddess Hapatu
2 oxen for the god Astabi
2 oxen for the god Nidakulof Arugatu
2 oxen for the god Rasap (30)

The Ebla archives are a rich store of pre-biblical literary traditions dating from nearly 4,000 years ago, far prior to any known biblical writing. The archival tablets have not, as of this writing, been fully translated and analyzed by scholars. But it is clear that oxen are an integral part of the agriculture and mythology.

Let's advance ahead chronologically about a thousand years to the late Bronze Age about 1,500 BCE to the apex of the Canaanite civilization that flourished along the eastern Mediterranean prior to the establishment of Hebrew cities in the highlands around the Jordan valley. The cult of Baal, the storm god who had the bull as his animal symbol, thrived and was the chief deity. As Enlil was the Sumerian storm god, the Canaanite Baal the storm and bull god was to have an enormous influence on biblical scribes. The cow and the bull evolved over centuries in several neighboring civilizations from a principal food source and draught animal into a powerful mythical image of the divine. The following section describes this relationship.

The Canaanite Ugarit Texts and Bull Cult

The Ugarit cuneiform texts, first discovered in 1929 in the tell of Ras Shamra along the Syrian coast, are clay tablets which date from 1,400 to 1,200 BCE, at least 500 years prior to the writing of the Torah or Pentateuch and represent the culture of the Canaanites. Based on the Ugarit texts, scholars have identified about 1,500 scriptural cross-references to literary phrases and ideas in the Torah, the prophets, the New Testament, the four evangelists, the Acts of the Apostles, and the letters of Paul and Revelations. Since these texts were found between two temples it is likely they were used in temple ceremonies. (31)

Ugarit was then the capitol of a flourishing trading culture and the texts reflect the extensiveness of the trading partners, including the Egyptians and Babylonians. But more importantly, these texts contain the ritual chants, incantations and myths of the Canaanites, a culture that incorporated the circulating myths that included the bull and cow as primal deities. Bull images are prevalent throughout these texts and portray Baal as a bull and as having bulls sacrificed to him.

However, the meanings attached to the words we use to describe what these texts are may deviate from their cultural context. Words like "religious texts," "myths," "legends," "cult" "literary convention" are contemporary understandings and may not actually represent what the texts were in the middle and late bronze ages. (32) It may be impossible to enter comprehensively into an ancient culture's worldview simply by reading its literature, especially about the quarrels gods have among themselves, a pattern replicated in Greek stories and epic adventures like the *Iliad* of gods and goddesses.

But if we view the Canaanite texts from Ugarit as formative in the succeeding literature of the Middle East, including scriptural writings of the Bible especially the *Psalms*, it is possible to see a duplicated pattern of images, word phrases, allegories and references to biblical stories drawn from more ancient literary sources. When lowly humans cannot exercise control and dominion over nature, then superhuman gods need to be called into action to assist the cause of fertility, regeneration, and rebirth in the vegetative earth and in human progeny. Enter the cow and bull.

Like the biblical *Song of Songs*, one poetic Canaanite text describes in detail the love of Ba'ul or Baal with 'Anatu his consort and his sister. Since 'Anatu is a virgin and unable to bear offspring because she is the sister of Baal, she proposes to Baal that she enter a cow so that the cow will bear offspring for her. Here is that text where the within-parentheses letters and phrases constitute scholarly guesses about the probable meaning of the missing sections.

The (wom)b of the cow can bear (for me),
the ox for the Virgin 'Anatu,
yes, the beautiful (beast) for the Wanton Widow of the Nations!
And Ba'ul the Almighty answered:
'Surely I can mount (you) like our Creator,
like the old generation that created (us)!
Ba'ul strode forward with a filled 'hand',
*   the god Haddu filled his 'finger'.*
The orifice of the Virgin 'Anatu was deflowered,
*   yes, the orifice of the most graceful of the sisters of Ba'ul...*
A f (luttering of   ) fell upon the ox,
*   a trembling of (   ) fell upon the ox.*
She flowed and started to convulse
*   (with) happiness, with delight, the pangs (of birth).*
The cow, the cow (of the Virgin 'Anatu),
*   bore a bull to Ba'ul*
yes, a wild ox to the Rider on the Clouds.
*   The c(ow) embraced (her son),*
the cow embraced (her young),
*   and covered him with her udder...*
She addressed herself to Ba'ul:
*   'Receive divinely inspired news, Ba'ul,*
yes, receive good news, child of Daganu!
*   For a bull is born to Ba'ul*
yes, a wild ox to the Rider on the Clouds!'
*   Ba'ul the Almighty rejoiced. (33)*

The birth of the bull god by the chief god Baal by means of his sister transforming herself into a cow is an elevation of the cow to divine status in Middle Eastern mythology 3,500 years ago, a theme with echoes from Sumerian, Babylonian origins and Greek and Egyptian traditions. The creation of divine animals is an act commemorating the participation of divine beings in the initial creation of all living things.

Anatu, the cow of Ba'u was eager,
*   Anatu was eager to bear.*
Her womb had not known pregnancy
*   (nor) her breasts nursing. (34)*

Sometimes, a suppliant, such as one wanting to be cured of unseemly language, uses the generic name of Cow as a petition for the goddesses 'Anatu.

Cow! Take away the word (spoken) in negligence,
*   the stammering from my mouth,*
*   the negligence from my tongue (35)*

A ritual incantation evokes similar expressions of sacrificing cows to the deities:

> *On the seventeenth,*
> *the king shall wash himself clean.*
> *The sanctuary of Ilu, a cow,*
> *A cow for Ba'luma.*
> *A cow for the Lad,*
> *two ewes and a cow for the two Lasses.*
> *The house of the officiant shall sacrifice.* (36)

> *The bull, o Ba'ul, we shall consecrate,*
> *the vows to Ba'ul we shall fulfill.* (37)

> *Raise your hands to heaven,*
> *sacrifice to the Bull, your father Ilu.* (38)

Without knowledge of genetics and basic life science principles, the Canaanites created a poetic myth assimilated from previous civilizations that helped explain their worldview for the propagation of cattle as a main food and life-sustaining force. The images, incantations, and ritual chants prevalent throughout these Ugarit texts testify to bovine cultural ubiquity. These rituals and cults were first accepted and then overthrown by the Hebrews when they adopted monotheism, as a passage in *Judges* notes.

> *But once the rest of that generation were gathered to their fathers, and a later generation arose that did not know the Lord, or what he had done for Israel, the Israelites offended the Lord by serving the Baals. They followed the other gods of the various nations around them.* (39)

I pointed out in the Introduction a sculpture in the Damascus museum that shows a seated figure with a bull's head with a dagger at his waist and is commonly thought to be Baal, or Baal-Hadad, the god of the Canaanites. Baal is often shown standing on a bull or with a bull's head. The bull pedestal is very common among the gods of the ancient world, especially the ancient near eastern weather gods. Baal can fight like a bull, has bull's horns and is probably related to a bull. (40) From the period of Tiglath-Pilesar III (d. 728 BCE) in the museum in Aleppo I stared at the life-sized statue of a male holding a box in his hands, wearing a long dress and a decorated band, and on his head ox horns protruding from his headdress, symbols of divinity and its accompanying power. (41) Two basalt stones found at Hazor in northern Israel show Baal astride a bull holding weather symbols like lightning bolts. Baal is usually shown carrying a plant, a symbol of life-giving properties. The conclusion among scholars is that Baal is related to royal and cosmic forces, often shown with Egyptian headdresses or wearing Egyptian kilts. One sculptured figure from about 1,500 BCE shows a pharaoh with a bull in relief on his high headdress.

The various representations of the bull, showing a god or royal figure standing on a bull means that the bull has not only been domesticated for a few thousand years by this time in history, but that it has become adopted into mythology and deified and that kings became associated with this divine power. The deification of the cow, and the transmigration of souls in and out of a cow or bull, is a part of Persian religion under the god Mithras and of Hindu culture, as we shall see in other chapters.

Baal, as storm god bringing water and shown astride a bull, represented for the Canaanites the forces in nature associated with fertility, vegetative and animal, both crucially dependent on a society's survival. In a semiarid climate, the arrival of the rains was essential for the harvesting of crops and a god who could be invoked to deliver the rains was savior indeed. Indeed, Yahweh, the God of the Hebrews, assumes this divine deliverance of rain.

> For the land which you are to enter and occupy is not like the land of Egypt from which you have come, where you would sow your seed and then water it by hand, as in a vegetable garden. No, the land into which you are crossing for conquest is a land of hills and valleys that drinks in rain from the heavens, a land which the Lord, your God, looks after; his eyes are upon it continually from the beginning of the year to the end. (42)

So Yahweh has assumed the same powers of Baal of the Canaanites because rain or water is desperately critical to crops. The Hebrew Bible contains 76 references to Baal. Clearly, the Canaanite cult of the bull god had embedded itself deep in the Hebrew culture if it merited so many attacks. (43) But the most telling and conspicuous repudiation of the cult of Baal occurs when Elias summons the prophets of Baal to a showdown on Mount Carmel, a prophet's duel, himself as a prophet of the Lord and 450 prophets of Baal, to see which god will produce the desired results in the offering of a sacrifice. (44) In this account, not surprisingly, Yahweh emerges as superior to Baal in responding to invocation. The sacrificing of bullocks also figures prominently. Because the episode highlights the intended subjugation of the bull cult of Baal, the full story needs re-telling.

> And he (Elias) said: I have not troubled Israel, but thou and thy father's house, who have forsaken the commandments of the Lord, and have followed Baalism. Nevertheless, send now, and gather unto me all Israel, unto Mount Carmel, and the prophets of Baal four hundred and fifty, and the prophets of the grove four hundred, who eat at Jezabel's table. Achab sent to all the children of Israel, and gathered together the prophets unto Mount Carmel.
> And Elias coming to all the people, said: How long do you halt between two sides? If the Lord be God, follow him: but if Baal, then follow him. And the people did not answer him a word. And Elias said again to the people: I only remain a

prophet of the Lord: but the prophets of Baal are four hundred and fifty men. Let two bullocks be given us, and let them choose one bullock for themselves and cut it in pieces and lay it on wood, but put no fire under: and I will dress the other bullock, and lay it on the wood and put no fire under it. Call ye on the names of your gods, and I will call on the name of my Lord: and the God that shall answer by fire, let him be God. And all the people answering said: a very good proposal.

Then Elias said to the prophets of Baal: Choose you one bullock, and dress it first, because you are many: and call on the names of your gods, but put no fire under. And they took the bullock that he gave them, and dressed it: and they called on the name of Baal from morning even till noon, saying, O Baal, hear us. But there was no voice, nor any that answered: and they leaped over the altar that they had made. And when it was now noon, Elias jested at them, saying: Cry with a louder voice: for he is a god, and perhaps he is talking, or is in an inn, or on a journey, or perhaps he is asleep, and must be awaked. So they cried with a loud voice, and cut themselves after their manner with knives and lancets, till they were all covered with blood.

And after midday was past, and while they were prophesying, the time was come of offering sacrifice, and there was no voice heard, nor did anyone answer, nor regard them as they prayed. Elias said to all the people: Come ye unto me. And the people coming near unto him, he repaired the altar of the Lord, that was broken down, and he took twelve stones according to the number of tribes of the sons of Jacob, to whom the word of the Lord came, saying: Israel shall be thy name. And he built with the stones an altar to the name of the Lord: and he made a trench for water, of the breadth of two furrows around the altar. And he laid the wood in order, and cut the bullock in pieces, and laid it upon the wood. And he said: Fill four buckets with water, and pour it upon the burnt offering, and upon the wood. And again he said: Do the same a second time. And when they had done it the second time, he said: Do the same also a third time. And they did so a third time. And the water ran round about the altar, and the trench was filled with water.

And when it was now time to offer the holocaust, Elias the prophet came near and said: O Lord God of Abraham, and Isaac, and Israel, show this day that thou art the God of Israel, and I they servant, and that according to thy commandment I have done all these things. Hear me O Lord, hear me: that this people may learn, that thou art the Lord God, and that thou hast turned their heart again. Then the fire of the Lord fell, and consumed the holocaust, and the wood, and the stones, and the dust, and licked up the water that was in the trench. And when all the people saw this, they fell on their faces, and they said: The Lord he is God, the Lord he is God."

And Elias said to them: Take the prophets of Baal, and let not one of them escape. And when they had taken them, Elias brought them down to the torrent of Cison, and killed them there.

Apart from the obvious statement of supremacy among the gods and that the Israelites are declaring Yahweh their own god and not inheriting a god from a neighboring community, this story emphasizes the bullock as the favored animal of sacrifice. The symbolism in this story is redolent with irony. Baal, as the storm god with his preferred animal symbol of the bull, is also the symbol of lightning. Yet it is lightning that ignites the chosen bullock sacrifice of Elias. "Cappadocian cylinder seal impressions show the Storm god (Ba'al) with his lightning and his animal attribute, the bull." (45)

In a minor inconsistency in the account, Elias instructs the people to pour the water on the "burnt" offering, but it hasn't been burnt yet. It is, in fact, not a burnt offering because no fire (lightning) has yet come.

"After many days the word of the Lord came to Elias in the third year, saying: Go and show thyself to Achab, that I may give rain upon the face of the earth." (46) There had been a famine and it was time for a storm. In this rendition, Baal is trumped at his own prerogative. The message is that he is not the real storm god any more and Israel has selected a new God, but one who also assumes the powers associated with the former gods. Some scholars believe that the entire Deuteronomic history of the Bible is a polemical attack against the cult of Baal. (47)

By the time of the Hebrew prophet Jeremiah, in a time of famine and drought, Yahweh has definitely replaced Baal as the main rain god. *The Book of Jeremiah* has a flood of references to drought and famine. For example: "Are there any among the graven things of the Gentiles that can send rain? Or can the heavens give showers? Art not thou the Lord our God, whom we have looked for? For thou hast made all these things." (48). By the 14th century BCE the Canaanite Baal had replaced the Mesopotamian EL, a name that would survive as the name of God in the creation stories in *Genesis*.

In order to totally disassociate themselves from their neighbors and establish cultural and religious independence, the Hebrews had to disenfranchise the Canaanite gods and reliance on Baal as a source of divine power, to eliminate Canaanite gods, even resorting to killing the prophets of Baal and to vigorously assert the primacy of the one God, Yahweh, for their transformation as a separate people and nation to be complete.

There are multiple examples of this new prerogative. *The Book of Hosea*, a word that means deliverance or salvation, is a prophecy that teaches that the God of Israel is the only true god. God loves his people, his chosen ones, despite their infidelities. The superiority of the God of the Israelites over all heathen gods is aptly demonstrated in the *Book of Daniel*. *Psalm* 68: 16-17 are hymns of praise for Yahweh but are very similar to earlier Canaanite paeans written for the god dweller on mountains, the storm god, Baal. Mount Carmel was one such mountain sacred to the Canaanites. Baal and the female deities Ashereh and Astarte are named. (49) The term "beelzebub" appears as a synonym for Satan or the devil in later scripture.

That same night the Lord said to him (Gideon): 'Take the seven-year-old spare bullock and destroy your father's altar to Baal and cut down the sacred pole that is by it. You shall build, instead, the proper kind of altar to the Lord, your God, on top of this stronghold. Then take the spare bullock and offer it as a holocaust on the wood from the sacred pole you have cut down." (Judges 6:25)

Male deities begin their ascendancy about three thousand years ago when Canaanites turned towards Syria and Phoenicia for cultural and religious inspiration and away from Egypt. The bull, the principal attribute of the weather god slowly emerges, and the lion, the symbol of the Egyptian god Amun declines. "According to the Old Testament, worship of Yahweh that included the use of one or more bull calf images was the most distinctive element of the state cult in northern Israel." (50) Yet near the close of the Iron Age (900–800 BCE) images of bulls play no significant role in the specialty crafts of the age in Israel. The Egyptian Amun and Canaanite Baal bull cult had both been supplanted in favor of Yahweh. By this time there are no defined boundaries separating the religions of Israel, Judah and Phoenicia. Israel and Judah both believed that other deities besides Yahweh existed, but were subordinate to him and not in competition for supremacy. (51)

Yet, although the level of divinity of the cow or bull ebbed according to the religion of the era, archaeology reveals that by about 1,200 BCE in the highlands of today's Israel cattle were integral to daily agricultural life. A close analysis of animal bones at late Bronze Age highland settlement sites, a relatively large number of cattle bones have been found indicating extensive field farming with a plough. (52) Thus, apart from the religious significance of the bull cult among the Canaanites and Hebrews, there were practical, pastoral reasons for cultivating cattle during the Bronze Age. Egyptian rulers like Thutmoses III and his successor Amenhotep II were aware of the cattle grazing in the region and received large cattle tributes to feed hungry armies. Similar texts report the Assyrian military acquisition of cattle on marches. Ashurnarsipal II, Shalmaneser III, Tilath-Pileser III and Sennacherib during the Babylonian occupations all exploited the successful breeding grounds and cattle riches of biblical lands on their campaigns. Egyptians by 1,200 BCE had villages in Canaanite territory, like the settlement at Tell es-Saidiyeh in the agriculturally rich east central Jordan river valley, which served as a military outpost as well as major trading site, providing wine from the vineyards at nearby Gilead, olive oil and cattle for consumption in Egypt.

Exodus: *The Golden Calf*

One of Michelangelo's masterpieces is the gigantic statue of Moses displayed in Rome's church of San Pietro in Vincoli (St. Peter in Chains) where he is shown as a virile liberator with curling locks and a flowing beard with ringlets cascading down to his mid-chest. But on the top of his head, where the casual observer might miss the irony, are two short horns. A horned Moses? Can Michelangelo, that keen observer of human nature and unsurpassed sculptor of heroic figures, have actually portrayed the Hebrew hero as a semi-bull god in keeping with Mesopotamian and Canaanite tradition? Moses reminded me of the Bronze statuette from the time of Tiglath Pilesar I saw in the museum in Aleppo. It is a fascinating speculation, a conjunction of divinities and heroic figures, but clearly Michelangelo knew his subject and his history and he portrays Moses as a cultic bull hero.

Although *Exodus* describes the problems Israelites led by Moses encountered on their journey to a new land, the scribe(s) reveal in little ways how thoroughly imbued he was with the urbanized and sophisticated cultures of the Mesopotamians, Hittites, Canaanites and Egyptians who all worshipped the bull or cow. In worshipping a golden calf the Israelites were simply following contemporary religious tradition. But the *Exodus* theme, like the Isaiah rebellion and slaying of the priests of Baal, is that the Hebrew people felt compelled to divorce themselves from these influences and assert religious independence.

There are scholars who claim that Moses had been a high Egyptian official during the reign of Akhenaten, a pharaoh who established belief in one god, Aten. (53) Akhenaten moved the capital from Thebes to Amarna to escape the enmity and influence of the Egyptian priests whose duties and stipends revolved around the sponsorship of multiple gods. After Akenhaten's death the priests reinstated the allegiance of multiple gods then ordered his name effaced from engravings, walls and columns. It is surmised in this scenario that Moses, carrying the belief in one god, escaped Egypt from this religious reversal and fled with his believers to migrate to a new country, Palestine, but at that time still under Egyptian control. A Canaanite myth has Baal using a club to subdue the sea god, reminiscent of Moses using a rod to part the Red Sea. Many argue that Moses may be only a heroic legend and not a historical person at all. (54)

In the spring of 2002 I traveled to Mount Nebo in Jordan to the hill site overlooking the Jordan valley, the river, and the Dead Sea. On a clear day it is a place of natural beauty, with gentle breezes combing your hair, and, because of its historical significance, a place of inspiration. Mount Nebo is where Moses reputedly died and from which he gazed into the land that was not his to enter. "This is the land which I swore unto Abraham, unto Isaac, and unto Jacob. I have caused thee to see it with your eyes, but you shall not go over." (55) For here Moses would have been able to see across to the Dead Sea, and slightly to the north Jericho and the Jordan valley and south to the Negev desert. To climb this mountain he would have had to come from the valley below.

A commemorative stele is erected noting this ("and no man knows of his sepulcher unto this day") where a 3rd century CE Franciscan church was

first erected. Did Moses carry the monotheism born in Egypt with him to the
new land? Did he and his followers also practice the worship of the calf and
bull as the Canaanites did? Archaeological evidence is beginning to fill in
many gaps. But it is clear that whichever peoples entered this agriculturally
rich land along the Jordan river valley borrowed an existing cow culture and
all its attendant religious traditions. That those people erected an altar and
made a semblance of a god they had known in Egypt should not be
surprising.

Or is it possible that the story of *Exodus* and Aaron represents the
overthrowing of Baal from the Canaanite tradition, and Moses the new
worship of the one god from Egypt, Aten? Here is the story in *Exodus*.

> *When the people became aware of Moses' delay in coming down from the
> mountain, they gathered around Aaron and said to him, 'Come, make us a god who
> will be our leader; as for the man Moses who brought us out of the land of Egypt,
> we do not know what has happened to him.' Aaron replied, 'Have your wives and
> sons and daughters take off the golden earring they are wearing, and bring them to
> me.' So all the people took off their earrings and brought them to Aaron, who
> accepted their offering, and fashioning this gold with an engraving tool, made a
> golden calf. Then they cried out, 'This is your god O Israel, who brought you out of
> the land of Egypt.' On seeing this, Aaron built an altar before the calf and
> exclaimed, 'Tomorrow is a feast of the Lord.' Early the next day the people offered
> holocausts and brought peace offerings. Then they sat down to eat and drink, and
> rose up to revel.*
>
> *With that, the Lord said to Moses, "Go down at once to your people, whom
> you have brought out of the land of Egypt, for they have become depraved. They
> have soon turned away from the way I pointed out to them, making for themselves a
> molten calf and worshipping it, sacrificing to it and crying out, 'This is your god, O
> Israel, who brought you out of the land of Egypt!' I see how stiff-necked this people
> is, continued the Lord to Moses. 'Let me alone, then, that my wrath may blaze up
> against them to consume them. Then I will make of you a great nation.'* (56)

Having talked God out of being wrathful to his own people he had just
led out of Egypt, Moses then returns to find the Israelites reveling and
adoring the golden calf. He then asks who will be on the side of the Lord,
and all the Levites rally to him. He then commands them to put on their
swords and "slay your own kinsmen, your friends and neighbors," a
decidedly capital punishment, cruel and unusual, and very contrary to the
prohibition in the commandments. *Exodus* tells us, "The Levites carried out
the command of Moses and that day there fell about three thousand of the
people." (57) Clearly, this is a Hebrew death knell to the followers of the
bull and calf cult and a repudiation of previous beliefs adopted from the
Canaanites. Here, too, is the first indication that God and his chief ministers
can sanction the killing of thousands for the sake of religious political
correctness. (58) And what was the proximate cause of this human disaster?

The image of a calf. The killing of "kinsmen, neighbors and friends" represents the death of the old religious belief in the cow and bull.

Despite these misgivings about cattle, it is clear from further biblical entries that bull sacrifices and analogies continued, as in *Numbers* the following scene occurs: "Then Balaam said to Balac, 'Build me seven altars, and prepare seven bullocks and seven rams for me here.' So he did as Balaam had ordered, offering a bullock and a ram on each altar." Or in the phrase: "It is God who brought him out of Egypt, a wild bull of towering height." (59) Clearly, the sacrificing of bulls and the prohibition against cows did not cease among the Hebrews as did the adoration of cattle as a divine image from the Canaanites. There are countless other examples including the following:

> And finding out a device he made two golden calves and said to them: Go ye up no more to Jerusalem: Behold thy gods, O Israel, who brought thee out of the land of Egypt...for the people went to adore the calf as far as Dan. (60)

> And going up to the altar, he did in like manner in Bethel, to sacrifice to the calves which he had made." (61)

> They exchanged their glory for the image of a grass-eating bullock. (62)

These quotations illustrate the prevalence of a cattle culture and bull cult among Hebrew people and the rebelling against the divine status of the cow and bull culturally inherited in order to establish a separate monotheistic and nationalistic society. Biblical scribes may denounce the bull and cow as gods and goddess but the habitual use of cattle in food production and animal sacrifice did not cease as these scriptural quotations testify.

Nor did bull sacrifice automatically die out when Christianity became the prescribed religion in Europe. A certain Rev. Dr. MacLean obtained presbytery records as late as 1656 from parishes in Scotland that showed men still sacrificing bulls. "The Minister being inquired be his brethren of the main enormities of the parishes of Lochcarrone and Appilcross declares some of his parishioners to be superstitious, especially the men of Auchnascallach, finding among other abominable and Heathen practices that the people in that place were accustomed to sacrifice bulls at a certain time upon the 25th of August, which day is dedicate, as they conceive, to St. Mourie, as they call him." (63)

The Christian era too has its peculiar moments with cows associated with religious activities. Impruneta is a village in Tuscany a few miles south of Florence, Italy producing terra cotta. In the church in the village flanking the high altar are chapels made of terra cotta. The chapel on the right reputedly contains a part of the true cross on which Jesus was crucified. The chapel on the left supposedly contains an icon of the Virgin Mary painted by

St. Luke the evangelist (though I have seen a similar one exhibited in a monastery in Cyprus). It was buried during the time of Roman persecutions by Christians and unearthed by an ox when the foundations were being excavated. Thus it can be claimed that a cow was partly responsible for the preservation of significant religious relics.

Like all the surviving literature of the Middle East, the collected books of the Bible contain the main elements of the culture of the times. Even a cursory analysis reveals that the bull and the cow figured prominently in the mentality, culture and religious practices of the Hebrews as they first adopted and then repudiated the bull cult, and even abandoned the cow as a food source and reverted to the pastoral lamb. Christians, adopting the Hebrew religious culture, made this the Paschal Lamb.

54

Figure 1: The Evangelist and the Bull From the Bronze Doors in Florence on the Baptistery of the Duomo, The Cathedral of Santa Maria del Flore by Ghiberti

CHAPTER THREE
EGYPT AND THE CULT OF THE COW

Nefertari was the queen of Rameses II, bearing him at least six or possibly seven children. Her radiant and lovely figure has been universally associated with Egyptian grace, femininity and beauty. Though her mummy has never been found, the countless depictions of her on wall paintings from the reign of Rameses II and her impressive tomb in the Valley of the Queens in the necropolis of Thebes reveal a lissome, slim figure with a gracious and beautiful countenance dressed elegantly, an image forever etched as the envy of women and the captivation of men. Her tomb, though small, has the most highly aesthetic paintings in the Valley of the Queens. One of the most interesting scenes located in a small room in a recess is the scene with seven sacred cows, brown and reds mostly with some speckled, and the bull with four steering oars representing the power of the world. (1)

Let this simple scene of cows displayed in a burial crypt, depicting both pastoral simplicity and religious devotion, serve as an example of ancient Egypt's unusual relationship with the bull and the cow. Soon after the domestication of cattle, at one of those indeterminate periods in prehistory, both the bull and the cow became divine in Egypt, assumed sacred attributes and sustained the timely devotions and temple donations of Egyptians for thousands of years. In fact, the cult of the bull was readily adopted by the Ptolemaic Greeks and lasted until the 4th century CE when the famous temple in Alexandria sacred to Serapis, who combined the divine qualities of the Apis bull and the god Osiris, was demolished by Christian zealots after the death of the emperor Constantius, son of Constantine who Christianized the Roman empire. (2)

Early Bronze Age Egyptians domesticated the oxen from the long-horned wild ox, the *Bos Primigenius*. But by 4,500 BCE humpless, short horns entered from the Middle East into the Nile valley and cross-breeding between short and long horns began. (3) The Narmer Palette dating from 3,000–2,800 BCE depicts the first king, Menes, who united both lower and upper Egypt, as a bull attacking the walled fortress of city.

The Hekanakhte Papers from Dynasty 11 (2,009–1,998 BCE) reveal an account written by a scribe that lists several bulls and young cows he had handed over to a steward with instructions to sell them when the price reached a certain level. The scribe describes how produce is to be distributed for the bull's feed and how to treat the 11 cows recently calved. Speaking through his scribe as an astute businessman, this ancient Egyptian noted:

"Be very energetic in cultivating the land; take great care. My seed corn should be looked after; all my property should be looked after. I consider you responsible for it." (4) He was also very instructive in conclusion about his mother and the care that should be taken with his concubine: "May Hathor (the cow goddess) gladden you for me."

During the reign of Cheops (2,584–2,566 BC), whose pyramid stands on the sandy plain of Giza, the largest and most photographed of ancient Egyptian burial places, long horned cattle are depicted on wall paintings on a funerary monument indicating that provisions from the previous season will be sustained. A similar depiction from a shell cylinder dating from roughly the same time period during the height of the Akkadian civilization in Mesopotamia, shows an oxen drawing a plough fitted with a large seeder.

In polytheistic religions many deities became grouped together, collected from previous cultures with divine qualities intact, combined with local attributes or given new ones. Gods and goddesses were not always neatly defined nor characterized with particular powers. Over time their abilities might change, they might acquire a new name, or the distinctive virtues of lesser deities. "The Egyptians are themselves distinguished from the rest of mankind by the singularity of their institutions and their manners," remarked Herodotus. "It would be indeed difficult to enumerate their religious ceremonies, all of which they practice with superstitious exactness." (6) *The Papyrus of Ani* proclaims. "Not shall I be repulsed by the bull of terror, I shall come daily into the house of the Double Lion-god, I shall come forth from it into the house of Isis, I shall see holy things hidden" (7) The satirical Roman poet Juvenal mused satirically about what monsters were being adored in demented Egypt.

We think of religious belief as a defined legal and ethical set of principles for justifying behavior. But Egyptian religion is different, filled with complicated cosmology and a pantheon of divinities. Despite the technical mastery, sophisticated artistry and refinement of its skilled craftsmen, goldsmiths and sculptors, (8) its formal religion remains inconsistent and illogical. (9) We'll never know the degree to which individual Egyptians believed in the multiple cosmic gods and goddesses represented through animal caricatures. But we do know that religious practices were highly formalized around natural occurrences like harvest festivals, that royalty was closely associated with divinities and religious rituals, and that a large cadre of state priests and priestesses performed sacred functions in innumerable temples and shrines. By about 1,000 BCE, the priesthood ruled the kingdom and the temples owned half a million head of cattle. (10)

Similarly, if we believe in the superiority of our preferred religious belief, it's relatively easy to presume on the doctrinal errors, the apparent naivety, the strangeness of the gods and rituals, the worship of animals and stone statues. How can anyone, we might say, believe that God inhabits a bull or cow like Apis or Hathor? For that matter, how can anyone believe God inhabits the replica of a ship, a burning bush, a temple precinct, a marble statue, a piece of unleavened bread? If God is all-powerful then any concept of what, who, or where he inhabits is conceivable and permissible. Even if God chooses to live in a cow. We may disparage the beliefs of our cultural ancestors, worshippers of Enlil (Sumer), El (Canaan, Hebrew), Hathor, Apis (Egypt), Zeus (Greek), Diana or Jupiter (Roman), but an honest appraisal indicates that ancient mythological motifs, after having been reinterpreted and carefully sifted, have been blended into new cultural and religious preferences. That God manifests himself in, and actually inhabits animate and inanimate objects, is just one of those derivative beliefs. It is difficult to separate the pantheon of gods and goddesses from elemental natural forces and from the animal figures they choose to enter as living embodiments of their power.

At one time or another in ancient Egypt nearly every animal, cows and bulls, goats, serpents, crocodiles, hawks, all assumed some divine power and became the object of a totem and worship. Herodotus writes: "The superstition of the Egyptians is conspicuous in various circumstances, but in this in particular: notwithstanding the vicinity of their country to Africa, the number of beasts is comparatively small, but all of them, both those which are wild and those which are domestic, are regarded as sacred." (11) Taweret, the pregnant hippopotamus, probably borrowed from the Akkadian pantheon of deities, protected women in childbirth. In Akkadian legend, Tiawat was a rebellious female crocodile monster that presided over the swirling waters of the Euphrates that often flooded until it was eventually harnessed and diverted by a vast system of dikes and irrigation canals. Taweret, also a goddess of a mighty river, clearly performed a similar function for the Egyptians.

Anyone who even accidentally killed an ibis, symbolic of Thoth, the god of wisdom, or a hawk, the symbol of Horus, son of Isis and Osiris, was put to death. The bull was especially sacred because it represented creative sexual power. Osiris himself was the supreme representation of this natural force and Apis, the bull god, the animal representation of this awesome power. Together Osiris and Apis became Serapis worshipped until Roman times and the obliteration of all pagan images, temples, shrines by Christian zealots.

Understanding nature's creative forces transmitted through sexual activity is not any easier to grasp today than it was for the ancient Egyptians who lacked microscopic observations and a scientific vocabulary. We may attribute scientific knowledge to the banishing of superstition about human and animal sexual powers. For ancient people, deities, through their embodiment in ordinary animals, intervened and governed these reproductive mysteries. Nudity in Egypt was often associated with divine powers. (12) Contemporary science, in the minds of some, may have replaced religion in the knowledge of the most profound of human mysteries, the regeneration of new organisms within a species. The profound mystery is only partially understood whether we see regeneration through sex as a religious, scientific, philosophical or biological enigma. And those who have not heard the stentorian bellowing of a bull mating cannot begin to appreciate sexual dynamism in nature.

Throughout the high culture of Egypt, there were two main cattle deities relevant to our theme here: one was a bull, Apis, and the other the cow goddess Hathor.

All Egyptians sacrifice bulls without blemish and calves. The females are sacred to Isis and may not be used for this purpose. The Egyptians venerate cows far beyond all other animals. (13)

Apis the Bull God

I wandered slowly through the mezquita in Cordoba, Spain, the cathedral church built literally inside the enormous mosque erected by the Moors, which in turn was constructed over the site of a 6th century Visigoth Christian church. In the center of this fascinating structure I found an improbable marble carving. A black bull writhes from the talon of an eagle embedded in its back. It is improbable because an eagle is unlikely to attempt to kill or capture, much less eat a bull, and doubly improbable because the carving serves as the pediment for the cathedral's pulpit. There is no biblical reference to such a scene or symbol. In fact, the black bull represents Apis, the divine bull of ancient Egypt often portrayed with the figure of an eagle on its back. The unknown 16th century sculptor placed this carving symbolically under the pulpit thus consciously or unconsciously merging symbols of divinity from Egypt, transmitted through the Romans, into a Christian liturgical environment.

Here, in the center of Catholic Spain, before the establishment of the rules for bullfighting and the art and entertainment of mass audiences, we can see the presence of a bull as a symbol in a Christian religious context. Such a scene may seem peculiar in any church, but perhaps not in Spain where the tradition of the bull remains proud and conspicuous. This one sculpture is an example of the continuity of religious symbols passing from one religion to another through centuries.

Apis was the Egyptian god associated with fertility. (14) Apis became incarnate in a specially chosen bull after the god Ptah impregnated a virgin cow with the power of his lightning. Ptah was the patron god of arts and crafts, including writing, who usually carried out the commands of Thoth, the god of wisdom and a primeval, creative force symbolized with the head of an ibex. Apis would in time become associated with the judgment god of the Egyptian living and dead, Osiris. A divine Apis bull was eaten at the end of the old year by the pharaoh in order to obtain the bull's strength and power. After its death Apis was enthroned in a sanctuary in Memphis in a temple dedicated to Ptah. Black in color with a white blaze on its forehead with an outline of a vulture or eagle on its back, Apis was one of the great tourist attractions of ancient Egypt. Women exposing their genitals in the presence of Apis the divine bull is a reflection of the ancient fertility ritual associated with the bull.

From the 1st Dynasty nearly 5,000 years ago Apis evolved to become the sacred bull worshipped at Memphis, but buried after death at Saqqara until late in the Roman era. According to Herodotus, the bull calf must have identifiable markings: black skin, a white star on its forehead, the figure of an eagle on its back, a divided tail and an insect like a beetle under its tongue. Apis is often shown on stone carvings in black and white with these special markings.

The sacred Apis bull was kept apart, like a vestal virgin, in a pampered state in special stalls and at special sites known as *Apieons* at Memphis. When a sacred bull died, its successor calf was installed at the same *Apieon*. The deceased bull was then embalmed, just like a pharaoh, and buried in underground galleries. This place was known as a *Serapeum*, named after the god Serapis whose original temple stood on the site. These catacombs for deceased Apis bulls were expanded during the reign of Rameses II, the pharaoh believed by some biblical scholars to have released the Hebrews from captivity.

All Egypt mourned when an Apis bull died. His burial ceremony dates from the reign of Amenhotep (c.1,390–1,352 BCE) when animal cults reached a zenith. (15) A ritual performed by priests that included mummification lasted 70 days and a papyrus dating from the 2nd century BCE discovered in 1821 and now in Vienna describes this ceremony. The priests were supposed to wear their hair long for 70 days, abstain from bathing, wear special costumes, wail loudly each day, fast totally for four days and abstain from milk and meat the remaining 66 days. An elaborate embalming occurred on an alabaster platform in a special chamber in the temple precinct of Ptah. The bull's body fluids and innards were collected and buried with the bull, mummified similar to human embalming practices. (16) The bull was ritualistically washed, dried, anointed, wrapped in linen and buried in a special crypt. One section of this papyrus indicates how complex this ceremony was for just the embalming of the tongue.

He begins the embalming of the tongue. They anoint it with the warm medicament. He cuts up a cloth that measures three palms in length and 6 digits in width. He swathes it at its front with 3 wrappings while the cloth is soaked in the warm medicament. He pulls up his tongue. He places the cloth under it. He pulls the edges of the cloth up until his tongue reaches in front of it upwards. He makes go...the left one to the right, the right one to the left, the (bandage) in the midst goes upon them. (17)

At one time Apis bull chambers were dug below burial chapels in the tombs of pharaohs. But a son of Rameses changed the routine so that Apis bulls were buried in chambers that lined the long burial corridors. In his day this complex of bull burial chambers had become extensive, winding through 1,150 feet of corridors. Mummies of sacred Apis bulls have been found in several sites in Egypt buried in large stone tombs and these can be seen in depictions in the British Museum in London. One sacred bull is depicted on a sandstone column showing the burial of a bull in the reign of Ptolemy II (284–246 CE). The writing on the carving notes that the bull was aged 20 years, 8 months and 14 days old when it died. Diodorus Siculus was a Sicilian and contemporary of Caesar who wrote in Greek over a period of thirty years a history of the world that constituted 40 volumes, about half of which survive. Here is his description of an Apis ceremony.

We must add to what has been said above an account of the circumstances surrounding the sacred bull called Apis. For whenever one has died and has been buried in splendor, the priests concerned with these matters seek out a young bull whose bodily markings are similar to those of its predecessor. When they find it, the people put away their mourning, and the priests whose duty it is to conduct the bull calf first to Nilopolis, where they keep it forty days, then they put it aboard a state barge with a gilded stall and convey it as a deity to the temple of Hephaestos in Memphis. And during the said forty days, only women may look upon it, and these stand before it and pull up their clothes to reveal the generative parts of their bodies. But at all other times it is absolutely forbidden for women to enter the presence of this divinity. (18)

"They esteem bulls as sacred to Epaphus (the Greek name for Apis)," writes Herodotus, "if they can but discover a single black hair in his body, he is deemed impure; for this purpose a priest is particularly appointed, who examines the animal as it stands. Its tongue is also drawn out, and he observes whether it is free from those blemishes that are specified in their sacred books. The tail also undergoes examination. If in all of these instances the bull appears to be unblemished, the priest fastens the byblus round his horns. He then applies a preparation of earth that receives the impression of his seal and the animal is led away. This seal is of so great importance that to sacrifice a beast which has it not is deemed a capital offence." (19)

Buchis bulls have a similar history but are associated traditionally with RA, the sun god, and Montu, an ancient deity of Thebes. The Buchis bull was also conceived by a god in a virgin cow and, after elaborate ceremonies, was likewise entombed with special cemeteries, interred in a stone sarcophagus and placed on a raised platform in a vault. In the foreground in the crypt stood tables and stands with pots holding offerings, often with a stela of the reigning pharaoh or monarch. (20) This cult flourished for about 600 years until the reign of the Roman emperor Diocletian (284–305 CE)

The Mnevis bull was representative of the herald of RA and also had distinguishing external features. A bull with these similar markings had to be located as its replacement when one of them died. Tombs of Mnevis bulls have been discovered dating from the time of Rameses II (1,279–1,213 BCE) and Rameses VII (1,136–1,129 BCE) that were covered with roofed vaults of limestone in caverns below ground level. The walls were inscribed with paintings just as in human burial chambers.

What joyful harvests spring, what heave'nly sign
Invites the plough, and weds to elms the vine;
How rear'd, Maecenas, flocks and castle thrive
And what experience stores the frugal hive... (21)

Cyrus the Great founded the Persian Empire in the 6th century BCE and unified the disparate tribes and nations of the Middle East. He is mentioned in *Chronicles, Isaiah, Ezra* and *Daniel*. After conquering the Babylonians, Cyrus rehabilitated the Jews by setting them free from captivity to return to their homeland and in gratitude they eulogized him more as their liberator than the conqueror he was. When he died in 529 BCE, his son, Cambyses, named for his grandfather, dispatched his ambitious brother Smerdis so he could rule without opposition and, inheriting the whole of the Persian Empire except Egypt, invaded it and sacked Memphis.

On one occasion while Cambyses was in Memphis the Apis bull appeared and everyone in the city donned choicest apparel and began to rejoice. (23) Cambyses thought the Egyptians were celebrating because he had lost many troops fighting them in the desert. The city magistrates responded that when the bull deity appeared to them it was their custom to celebrate with a solemn festival. Cambyses thought they were lying and condemned them to death, a huge cultural and linguistic misunderstanding. Then he called the priests and asked why the celebration and received the same response. Cambyses asked that the Apis bull be brought to him and when it was he flew into a rage and raising his dagger to stab it wounded it in the thigh. Believing he had been mocked, Cambyses had the priests scourged and ordered that no further solemnities of Apis be observed. Thereafter the cult of the Apis bull fell into disrepute until partially revived during the rule of the Ptolemys, named for the successors of the rulers Alexander the Great installed after his conquest of Alexandria in the 4th century BCE.

The acerbic, skeptical and witty Voltaire's (1694–1778) short story, *The White Bull*, pillories superstitious beliefs. (24) A fictional Egyptian Princess, Amasidia, grieves for her lost love until she recognizes him in a field tended by an old woman sorcerer. He has been changed into a white bull and eventually it is revealed that he is Nebuchadnezzar (c.605–562 BCE), a real Babylonian king who defeated the Egyptians in Syria becoming the undisputed master of all of western Asia. Pharaoh Necho defeated Nebuchadnezzar years later. Ever the military man jealous of his conquests, Nebuchadnezzar then turned to crush the revolt in Judah and destroyed Jerusalem in 586 BCE. These historical events are turned upside down as Voltaire seeks to de-mythologize real characters, making fun of prophets clad in ragged clothes and elevating reputed biblical enemies to a divine status. Nebuchadnezzar as a white bull becomes a divine Apis bull in Egypt where he is adored, even though in reality he was Egypt's dreaded enemy conqueror. He is later reincarnated as a mere mortal king. Voltaire's clever prose is still easy reading and ingeniously satirical. The last line of *The White Bull*, chanted by the Babylonians, is "Long live our great king, who is no longer an ox."

Hathor The Cow Goddess

The form of the goddess may undergo transfigurations from time to time, like Istar (Babylonian), Isis (Egyptian), Pallas Athene, Aphrodite (Greek), but devotees will never tire of trying to comprehend female divine significance because in all mythology she represents the mysterious, unknown creative force. (25) Sentimentally touching and pious tributes to Hathor are everywhere in Egyptian texts, sculptures and wall paintings. Whatever we may think about divine power, Egyptians have a way of humanizing the sacred with a special human touch making the divine palatably earthy. In such a context, investing a cow or a bull with divine qualities somehow doesn't seem so bizarre.

The cow was so crucial to Egypt's survival that it was raised to the level of divinity: the popular, much-beloved, majestic cult goddess Hathor. Why the form of a cow? The answer may never be resolved, but the cow in Egyptian lore was so beloved because it represented, along with the cobra, an essential life force of the all-powerful and awesome magnificence of the sun god, Amun-Re. I believe the Egyptians adopted the bovine deity from the Canaanites who held Baal at the apex of their divine hierarchy. Hathor's representation assumed strategic and primary importance among all Egyptian goddesses.

Hathor's worship is of great antiquity and probably originated in the 4th millennium BCE. Hathor means "estate of Horus." Horus was the son of Osiris and Isis, the two principal creation deities. Hathor is represented in all paintings as a human figure, sometimes with the head of a falcon but always with a replica of the divine sun-disk between her horns. She is portrayed as a falcon in hieroglyphics, the representation also of the god Horus, enclosed within a rectangular box, or cartouche, a symbol of regal and divine status.

Archaeologist Alain Zivie and his team dug deeper in the tomb of Netjerwymes, a trusted diplomat and courier of Rameses the Great. The team found an unusual carved masonry. (26) It was the stone image of a giant Hathor image with a sun disk on her head signifying divine royalty. Standing beneath her chin was Rameses himself, depicted as a god, welcoming Netjerwymes into the afterlife. Hathor is the goddess of the dead promising resurrection to the worthy.

Hathor became associated with the goddess of the sky, Nut, and of women, fertility, and love while acquiring the symbol of a cow with a solar disk on her head, the symbol of the sun god, Ra. She became identified with the attributes of other Egyptian gods and goddesses, with the characteristics of a mother, of the sky, the sun and royalty, as a goddess of the dead and music and dance, assuming the qualities of mother and consort to a variety of bull deities, to the heavens in general, and other local gods. Her symbol was a sistrum, an ancient percussion instrument made with a thin metal frame and a wooden handle with numerous metal rods and loops that vibrate when shaken. Grateful hymns were composed in her honor and her benevolence was exalted in festive life. (27)

Her emblem was cow horns mounted on sticks. She is frequently shown in wall paintings as full-faced, a rarity in Egyptian art since the full face represents power, and as a woman wearing the solar disc between two cow's horns. (28) She is also associated with Isis, the female companion of Osiris, the judge of the dead. The once favored cow began to take on the trappings of other deities in the Egyptian pantheon and thus she was petitioned for an ever-widening variety of requests once associated with other goddesses.

Hatshepsut (c.1,420–1,482 BCE), daughter of Thutmose I, Egypt's only female Egyptian ruler until Cleopatra, managed to rule by out-witting and out-maneuvering her husband and brother. During her peaceful reign she was a major builder of temples and shrines who encouraged economically productive mining efforts in the Sinai. At the time, Hathor was worshipped throughout the Egyptian zone of influence including the Sinai, and her cult extended from Nubia (today's Sudan) to Syria. (29) A small temple to her was found south of the Dead Sea between Israel and Jordan in a copper mining region. Hatshepsut's three-tiered temple-tomb, a popular tourist attraction, contains a Hathor shrine on the second tier, a sanctuary testifying to the royal adoption of her cow goddess's divinity and of her protective mantle throughout the kingdom.

The deceased were said to pray to her, goddess of that part of the sky where the sun rose and set, as a source of celestial food. A recitation notes: "Look, Hathor will lay her hand on her with an amulet of health."(30) In the Metropolitan Museum is a huge head of Hathor dating from 1,417–1,379 BCE with five glossy finishes revealing a warm and technically proportionate balance.

Many votive offerings found in temples and in local shrines were addressed to her, the most numerous coming from the New Kingdom temples at Thebes and from the late Ptolemaic period, testimonials to the long chapter of the goddess' success in satisfying the cultic life. The most frequent request to the goddess was for human fertility. Chapter 162 in *The Book of the Dead* notes that if a certain spell is invoked over a cow of gold, that the petitioner will become "a lord of the phallus...shining with rays...a lord of transformations." (31) The "cow of gold" would become, *mirabile dictu*, a "golden calf" in *Exodus* 32. Innumerable wooden phalluses as votive offerings scattered by suppliants seeking Hathor's intercession were found at her shrine at Deir el-Bahri in the necropolis near Thebes at the turn of the 20th century.

Besides human sexuality, Egyptians turned to Hathor for human and agricultural fertility blessings. A psalm from the temple of Medinet Habu contains the following invocation:

> You honor the great god in Medinet Habu,
> And the wives (praise) Hathor,
> the mistress of the fort of the West,
> In order that the goddess has your wives
> Give birth to boys and girls,
> In order that they become not sterile
> And you not barren. (32)

The "fort of the West," the direction of the setting sun, is the Egyptian gateway to the afterlife.

Hathor is not a love goddess like Aphrodite or Cupid, making people sick with passion. Rather, she is the goddess who brings fructification to life: in happy hearts, in successful childbirth, in plentiful harvests, in a pleasant transition to the afterlife. Her image is found in royal tombs, like that of Seti I, and is present as the patroness of festivals honoring the dead, like that conducted at the necropolis at Thebes, as the dead were invited to take part in the joyous celebrations.

The Papyrus of Ani, first acquired by the British Museum in 1888, is the best example of ancient Egyptian religious documents known as *The Book of the Dead* dating from 1,500–1,400 BCE. Ani was the High Priest of the religious revenues of the temples of Abydos and Thebes. The extensive documentation of his high standing, the glorification of his duties and his publicity of his tributes to the gods in this papyrus are meant to ensure his survival in the hereafter, so he can become "one of the shining beings." The text's authoritative status makes it representative of funeral texts of the era and is a remarkably preserved source of religious beliefs among Egyptians. "I am provided," says Ani through the voice of Osiris, the judge of the living and the dead in the *Papyrus of Ani*, "I am intelligence provided, I have made the way to the place in which RA and Hathor are." (33) In a later section we can find: "My face is RA. My two eyes are Hathor." (34)

But even in inscriptions on the walls of catacombs, petitioners make it plain that they want to identify with the god not just receive her blessings. One proclaims, "I am Hathor, I have appeared as Hathor, who is descended from the primeval age, the queen of the All." Egyptians wore cow effigies of Hathor around their necks to ward off evil, as today believers will wear crosses as pendants or necklaces to reveal their beliefs. (35) Representations from her temple at Dendera illustrate the close relationships of the pharaohs with Hathor. She is shown suckling the children of pharaoh, licking the hand of the queen, as if to give her power, laying her hand on the shoulder of pharaoh as if to indicate support for his authority. Pharaoh responds by making tributes to her service as he is mythically also her son.

One of the more endearing representations in sculpture and wall paintings is of pharaohs sucking sacred milk directly from the goddess's udder. The future king Amenemhet III is shown in a scene from a gold pectoral sucking from Hathor's udder while an attendant gently strokes the goddess's neck. Likewise, in a small outer sanctuary of Hatshepsut's burial tomb, a phrase on the doorway reads: "The doorway of Hatshepsut who is imbued with the vitality of Hathor, the supreme one of Thebes." In this reverential burial sanctuary, where Hathor's reddish-brown tint can still be seen, Hatshepsut crouches to drink milk from the sacred udder. Hatshepsut's nephew, Thutmose III, has a similar life-sized statue of Hathor carved in a shrine in his burial suite, also with the goddess sucking the king.

Kheruef was a high official in the court of Amenhotep III and his position of authority entitled him to his own private tomb in the Theban necropolis. Amenhotep III (d.1,372 BCE), the father of Akhenaten who was the founder of monotheism, ruled during the age of Egypt's greatest splendor when architecture and sculptor entered a period of extreme elaboration. Amenhotep greatly extended the buildings at Thebes, Karnak and Luxor. A wall carving in Khereuf's tomb reveals a beautiful young woman symbolizing the Hathor goddess seated next to Amenhotep, while the Queen stands behind them. Hathor is crowned with a horned headdress, symbolizing the cow's horns, with the solar disk symbolizing the sun god RA between the horns. The sacred asp, the uraeus, the symbol of sovereignty, is positioned at the front of her headdress on her forehead. The various scenes in Kheruef's tomb describe the celebration of Sed, an important festival to which foreign dignitaries are always invited, and was only conducted after a pharaoh had ruled for 30 years. One wall scene shows a young female dancer, a monkey, a pintail duck and a leaping bull-calf. This is just Hathor's kind of celebration where dancing and music venerate her presence and her attributes. Singers and dancers would chant hymns to Hathor's glory as the king would need her powers for the remainder of his reign. Here is what the celebrants sang:

O Hathor, you are exalted
in the hair of Re, in the hair of Re,
For to you has been given
the sky, deep night and stars.
Great is her Majesty
when she is propitiated.
Adoration of Gold
when she shines forth in the sky.
To you belongs everything
in the sky while Re is in it...
There is no god who does what you dislike
when you appear in your glory. (36)

It may seem simplistic that people would actually pray to the likenesses of a cow and expect such an animal to grant special favors. But animals are not the gods any more than statues in Catholic churches are the divinities or saints they are supposed to represent. Hathor was the symbolic representation of the god of creation, a life force. The cow and the bull were chosen because of their supreme importance in the life of the Egyptian people associated with divine and nurturing properties.

It is commonplace to see Hathor acting like a mother suckling her young or standing close to her suppliants, while at the same time shown with the symbols of divinity such as the sun disk or the power of the life-sustaining forces of the cow. The glow or halo we observe in early western paintings of the saints showing a shining radiance around their heads associated with a holy identity is just like Hathor's representations in her temple with the glow of the sun revealing her solar character surrounding her head. The halo, derived from Egyptian and Sumerian symbols of sanctity, represents power from the sun god.

Cheops or Khufu, who flourished about 2,680 BCE and is famous as the builder of the largest and famous pyramid at Giza, had ordered the temples at Giza closed. His son, Mycerinus commanded that the temple be opened. The people rejoiced as they were able to offer sacrifice again at the shrine of their gods. Revered by the people for his exemplary conduct, Mycerinus grieved when his daughter, an only child, died. Wishing to honor her memory, he enclosed her body in a purple cloth and placed it in a wooden heifer encasement ornamented with gold, not burying her but placing her in a hall in a palace. (37) Rich aromatics were burnt daily before her and every night her shrine was illuminated. A coffin in the form of a cow in this Egyptian context does not appear to be frivolous.

You would not have been able to engage an Egyptian in a dialogue about life's truths without getting into Egyptian metaphors. There was no search for an eternal, universal truth or even life's truth. Those would later become Greek conceptions as reason rather than myths and storytelling became tools for understanding the universe and people's role in it. An Egyptian would tell you that the sky was supported by four pillars and that these four pillars were very likely the four legs of a cow. (38) Moreover, the sun, moon and stars moved along the belly of the cow. Alternatively, the pillars could also be the arms and legs of Nut, the sky goddess. But these seeming contradictions and elaborate use of symbols and animal incarnations to convey visual clarity are approaches to understand the mysteries of divine beings. Although Hindu theology accepts cow spirits as potential human reincarnations, our appreciation of the cow for its beef is like a comparable cultural belief. Language and imagery, together with science, still strain to capture what divinity and the creative power of the universe is.

For example, the Egyptian hieroglyphic symbol of the KA is a bull. It would be unwise to translate this as "soul," although its meaning comes close to the Egyptian idea of a spirit inhabiting the body and surviving it into eternity. The related belief by the Egyptians was the BA, closely resembling our term for personality, or alter ego, or such phrases as "the shadow of our former selves." For the pharaohs the KA is eternal and is symbolized by the virile power of the bull derived from the mother goddess. The pharaoh is often called "a mighty bull," an epithet that reveals the regenerative power and strength granted him by divinities and a phrase found often in the Bible. The KA is also associated with a nutritive power, which is why food offerings, given to the KA of the owner of the burial chamber, are frequently placed in tombs with the deceased. It was the function of the BA to escort the KA to the afterlife and abode of the eternally living. What happened to the BA afterwards—it is after all only a guide and not immortal—is uncertain.

Osiris was the Egyptian quintessential divinity, part mortal and part god, who rose from the dead, judges the living and the dead, keeps accounts of the actions of all people, and decides who merits eternal life. All the kings and pharaohs of Egypt delighted in being referred to as a "mighty bull." Apis was thought to be the son of Osiris as well as Ptah. (39) How the bull came to be associated with Osiris is unclear but the relation of the bull to agriculture and the harvests is close enough in antiquity to welcome divine status.

The Egyptians were unfamiliar with the philosophical and theological distinctions of Jewish, Christian and Muslim scholars about gods. The concept of the Trinity, of Father, Son and Holy Spirit, of three persons in one god, comes close to an Egyptian understanding of gods inhabiting one space like the sky, like Nut, Horus, and Hathor, or her mixed association with Isis, Ptah, or Thoth, separate but all also having distinct, divine attributes. Egyptian pictorial representations in hieroglyphics of ordinary objects, animals and domestic life conveys a simple quality to religious properties and divine awareness, with the bull and the cow at the apex of the pyramid of deities.

CHAPTER FOUR
GREEK BULL MYTHS

What is it that leads to a renaissance of intellectual and artistic activity? What particular group of geniuses do we need to assemble to merit a flowering of culture, an aesthetic outburst such as occurred in Greece in the 3rd century BCE? The question for all generations is not only how a civilization blossoms, but how its flowering can be sustained. We encounter throughout life a variety of experiences in language and literature, the fine and performing arts, music, athletics, participation in government, and architecture. All of these art forms the Greeks who lived in Athens three centuries before Christ had in abundance. And they had bull cults.

Unrivalled masterpieces of Mycenaean art were discovered last century dating from about 1,400 BCE, golden treasures depicting the popularity of bull sport from an age when metal-working and engraving reached a high art form. Two of these prizes are twin cups of thinly beaten silver portraying the snaring and taming of bulls. On one cup a bull is snared in a net of heavy rope. The bull has an open mouth and one can almost feel the hot breath from its flared nostrils as it struggles angrily to free itself. On another panel of the same cup, a second bull gallops away, terrified of a similar fate. A third scene shows a bull charging an ancient cowboy who snags it courageously by the horns. (1)

The companion cup shows an alternate method of capturing the bull whereby hunters deploy a dummy female cow. The artist shows this with a raised tail of a decoy cow indicating its readiness for mounting. In the first scene the bull noses the cow's tail. The second scene shows the bull and cow turning their heads towards each other reminiscent of an amorous conversation. In the third scene, the hunter takes advantage of this dalliance to lasso the hind leg of the bull shown with uplifted head bellowing in a roaring rage. Such exquisite examples of simple yet refined and unparalleled artistry were not to be duplicated again for centuries. Bull catching and taming, portrayed as sport, are here ennobled with magnificent engravings over 3,400 years ago.

The Greeks had no bull god but held the bull sacred nonetheless. Hera, wife of Zeus and Queen of Olympus, is sometimes referred to as "cow-eyed," perhaps prefiguring a more ancient cow goddess. However, Pan was the cloven-hoofed, sensuous, horned god of the shepherds. His attendants were satyrs and fauns and his magic flute could be heard throughout bucolic environments and his cry brought *panic* to herds within earshot.

The Greeks had a common language, a common intellectual life, a passion for sports and athletics and a love for beauty through art and architecture admired and copied in American monumental structures. Their rituals and religious beliefs, recounted in their literature, had been until recently the core of western educational curriculum since the Enlightenment. The fables Greek creative minds invented included bull cults and feats of daring with bulls that revealed their awe and admiration for this enormous animal and its virility and potentially destructive power. (2) Before Superman, Batman, Spider Man, when children's comics became popular movies with special effects, there were Greek heroic myths, tales and adventures of larger-than-life heroes with divine or semi-divine ancestry whose powers were beyond ordinary human capacity. Many myths were about bulls.

The river god Achelous, while dining one evening with Theseus recounted stories of exploits with Hercules, of how Hercules had strangled him as a serpent so he quickly assumed the form of a bull. Theseus is left with only one horn having lost one in a struggle with Hercules who grasped his neck with one arm and dragging his bullhead down to the ground threw him onto the sand then tore one horn from his head. The river maids Naiades took the horn, consecrated it and filled it with fragrant flowers. The goddess Plenty adopted it and called it a cornucopia, the horn of plenty. We might think this a real bull story, but the emblems of our culture, the abundant cornucopia filled with fruits and flowers decorate our tables and kitchens today, evidence of the adoption of ancient fables as contemporary cultural items.

Despite the fact that it has no consonants, Io may appear to be a strange name for a princess—Ionia, the Ionian Sea, and Ionic columns and the Bosporus are named after her—but her name is not as strange as what became of her. Io was the daughter of the river god Inarchus and because of her beauty Zeus (Jove or Jupiter in Roman tradition) made her his mistress. In order to hide her from his wife Hera (Juno to the Romans), he transformed her into a heifer, a cow that hasn't calved. Hera was too suspicious and enlisted the help of Argus, who had a hundred eyes in his head, to keep watch over her and report to Hera who she was if she became a woman again. Not to be outdone, Zeus told Hermes (Mercury among Romans), son of Zeus, messenger of the gods, depicted with a broad-brimmed hat and winged feet, to go to Argos and rescue his mistress Io lowing in the form of a heifer. When Hermes related stories of how he had invented musical instruments like the flute and lyre, Argos fell asleep and Hermes cut off his head. Hera sent a gadfly to torment Io and only called off her revenge tactics when Zeus agreed not to see her again. Domestic deceptions do not always end so peaceably. In this literary context, the commingling of physical characteristics among gods and people does not seem so unusual. The fact that a heifer was used as a lover's ruse clarifies this unified vision of nature and life while magnifying the importance of the cow.

Hesiod, whose writings are some of the oldest in the Greek language, wrote in the 8th century BCE. His classic *Works and Days* describes the bucolic life of a farmer and a prescription about the use of ox as a draft animal. "For draught and yoking together, nine-year old oxen are best, because being past the mischievous and frolicsome age, they are not likely to break the pole and leave the plowing in the middle." Hesiod writes in another place in the same work: "The cattle should be kept in good condition, and ready for work, when the migratory crane's cry let us know of winter's advance and the prospect of wet weather. Everything should be in readiness for this; and it will not do to rely on borrowing a yoke of oxen from a neighbor at the busy time." (3)

Let's trace this literary legacy about cows and bulls through a few key myths to see how this one animal graced the Greeks' distinctive and glorious civilization.

The Rape of Europa

Though exact details vary depending on the literary sources, the story about the myth of Europa, from which the continent Europe is named, goes like this. It was said that Zeus, always fancying lovely maidens, whether mortal or immortal, beheld Europa as she was picking flowers by the seashore with other maidens. He appeared to her in the shape of a bull and the maidens were unafraid of him. With his charming manner, and by laying before her in an enticing manner, he got her to mount his back and forthwith carried her over the sea to Crete where she bore him mighty sons, including King Minos. Some see the enchantment of Europa to Zeus, even in his bull shape, to be comparable to that of Dionysius, as vase drawings often depict her carrying vines laden with grapes or with fishes or flowers in her hands. (4) The charming and elegant Roman poet Horace (65-8 BCE) composed a delightful poem about this myth:

> *Thus Europa trusted her snowy body*
> *To the crafty bull, and though once undaunted*
> *Paled at billows teeming with monsters and the*
> *   Perils around her.*
>
> *Oh, if anyone in my wrath will give me*
> *Now that cursed bull, I shall try to wound him*

With the sword, and break off the horns of that once
Greatly loved monster. (5)

The Uffizi gallery in Florence contains many of the richest art treasures in the world collected by the vast holdings of the Medici family who ruled northern Italy in the Renaissance. I revisited the gallery again in Florence to retrace the Renaissance painters who enlivened this story because there are several southern rooms in the Uffizi filled with paintings just of the rape of Europa as this Greek myth captured the imagination of Renaissance artists as much as any heroic adventure in history. What I found were panels and whole rooms depicting Europa's plight. Eleven painters from 1480 to 1859, Peter Paul Rubens (1577-1640) is the most famous, depicted Europa and the bull god that are now displayed on large canvas paintings. I had previously stared in wonder at a similar wall painting of the Rape of Europa by an unknown artist in the Museum in Naples, a painting that once decorated the walls of a Roman villa in Pompeii before its extinction by the eruption of the volcano Vesuvius in 79 CE.

In a continuation of the same myth, Cadmus, a Phoenician prince and brother to Europa goes in search of his sister, ordered by his father King Agenor not to return home without her. He searches diligently but is unable to locate her. Hesitant to return home without her, Cadmus rushes to consult the oracle at Delphi. The oracle instructs him that a cow will meet him as his leaves the sanctuary and tells him that he is to build a city on the spot where the cow lays down. He is led to the site where Thebes now stands. Realizing that this is to be his city, Cadmus sacrifices the cow as an offering to the gods.

The Greeks may not have had a bull god, but Greek gods did often assume the shape and characteristics of a bull, in effect sanctifying it. This profound religious concept, that divinity and supernatural beings can inhabit animals at will, resonates throughout literary legends and has insinuated itself into religious persuasions as well. The talking serpent in the *Book of Genesis* or Balaam's talking ass in *Numbers* are two biblical examples.

The Adventures of Hercules

Hercules is certainly one of the most ancient deities of Egypt, and, as they themselves affirm, is one of the twelve who were produced from the eight gods. (6)

The rock of Gibraltar was visited by Hercules in his quest to locate the Red Oxen of Geryon, the monster with three bodies. During this strenuous contest Hercules tossed a rock across the straits, known as Cuenta in Morocco, or Abya by the Greeks. These twin promontories are the Pillars of Hercules, the permanent guardians of the Mediterranean.

Hercules (or Greek Herakles) was born the child of Zeus, as were so many Greek heroic figures, and a mortal woman, Alkmene, whom Zeus seduced even though she was the virtuous wife of another man. Hera, Zeus' main squeeze, was extremely jealous of the offspring of her husband by other women, supernatural or mortal. As a result she mercilessly hounded Hercules throughout his journeys and labors. The literary sources of Hercules' adventures date from the plays of Euripedes in the 5th century BCE, and continued with the poet Ovid during Rome's Augustan Age. The oral traditions of these fables pre-date Homer by hundreds of years.

Hugely popular throughout the ancient world, Hercules has been known as the Superman of the Greeks, a figure of unbelievable strength like the biblical Samson modeled after him. (7) As a child, Hercules strangled two snakes that tried to crawl into this crib. A man of strong passions and excitable anger, he killed his wife and children in a mad rage induced by Hera. As a result, the oracle of Delphi said that he must be punished. Hera turned him over to Eurystheus King of Mycenae who thereafter commanded him to perform his extraordinary labors. All of his exploits required courage and brute strength, but unlike the deeds of Theseus, none required intelligence. His skill was in thinking of ways to kill the monsters before him, just like David slaying Goliath. (8)

Four of the twelve mythical labors of Hercules involve the killing or abduction of cattle and bulls. Other labors entail Hercules in, well, Herculean tasks with monsters like the three-bodied Geryon and the nine-headed Hydra, seizing the girdle of the Amazon, Hippolyte, and fetching for Juno's wedding golden apples from Atlas, her father, a Titan who holds up the heavens.

A second bull task was the capture a wild bull from Crete. An ivory carving surviving from the 2nd century BCE shows Hercules wrestling with a prancing bull, not yet totally subdued. Our hero holds his body against the bull's flanks and restrains its head with both his hands. He releases the Cretan bull, however, so that Theseus can kill it and so exercise his manhood and destiny. A similar image would be adopted for the Mithras cult of bull slaying, a legend discussed in the following chapter.

The third labor was to go to the legendary ends of the earth, which for sailors of the day meant the western limits of the Mediterranean, the Pillars of Hercules, there to subdue in Spain the triple-bodied herdsman, Geryon, and to claim all his superb red cattle. After formidable struggles, including encounters with the two-headed, snake-tailed dog Orthos, brother of Cerberus, the guard dog at the entrance to Hades and the Underworld, Hercules returns with the cattle.

A fourth labor was to find the giant Cacus who has stolen the oxen and to return them. Cacus had hidden them in a cave, but cattle, being cattle, low, and when our hero hears them he enters the cave, slays the giant Cacus and leads the herd home. In the last of six of the labors, Hercules is asked to clean out the Augean stables in a single day. The stables belonged to King Aureias who owned vast herds of cattle, not horses as popularly assumed. Since the stables had never been cleaned in 30 years, the dung was assumed to be several feet thick and therefore impossible to clean in a day. Hercules rose to the occasion and diverted two rivers into the stables thus washing them clean quickly.

It's possible to deduce virtuous analogies from these tales, of the unwilling acceptance of kingly demands, of the dangers of straying beyond the local geography, of pursuing adventure no matter where it may lead. But no matter how absurd the adventures seem, Hercules is shown to be the epitome of male audacity, nimble on his feet but not in his head, fond of food, wine and women, the ancient version of a wrestling exhibitor of World Wrestling bouts, the perfect fantasy male hero. The fact that Hercules had to engage bulls in his adventures serves to make the point that these animals, worshipped as gods in other parts of the Middle East, are still perceived as man-killing animals.

Presently, the palace gates were flung open and at the head of the procession were led out the bulls for sacrifice, beautiful creatures, four and four together. They were to be offered to Zeus and to any other gods that the Persian priests might name. (9)

Herculaneum was the Roman town named after Hercules showing how deeply the Romans had borrowed Greek myths and the reverence accorded this one legendary hero. Together with Pompeii, it was buried in the eruption of Mount Vesuvius in 79 CE. (10) Vesuvius, the volcanic mountain commanding the horizon from the Bay of Naples, erupted on August 26th, 79 CE burying Herculaneum, Pompeii and the countryside under 60 feet of volcanic ash and mud together with most of its residents. Herculaneum was covered with lava flow and lay dormant under 50 feet of lava for 1,800 years. Herculaneum, like its companion city Pompeii, was the smaller but wealthier Roman town as the rich residents lived on the edge of the sea. Pompeii was at the foot of the mountain and its artistic treasures still reveal impressions of Roman aesthetic achievements, as well as plaster reproductions of the fallen victims, an extraordinary if unintended donation to posterity. Serious though not systematic excavations only began in 1709, as archaeology was then only a pleasant pastime engaged in by the aristocracy.

Herculaneum, the ancient town that was five hundred years old before its liquidation, is a microcosm of Roman life inhabited by a people who enriched their lives with varied aesthetic sensitivity. Since Herculaneum numbered among its few hundred population a few wealthy merchants and traders, we can reasonably assume that all Roman cities, influenced by the high standards of Greek civilization, were similarly decorated with paintings, frescoes, mosaics and sculptures of heroic figures from natural life and mythology two thousand years ago.

Where could you go today, walk into anyone's house and observe colorful wall decorations, or into public baths and see incredible inlaid mosaics, or to the basilica and pause in awe at the marvelous bronze statutes? Only a half-mile wide by a half mile long, the excavated part of old Herculaneum (the modern city of Erculaneum sits atop it) represents the artistic sophistication of Roman life at the height of the empire under Titus. The intact town retains the trappings of urban life: public sidewalks with stone benches where citizens could sit and watch the world go by, cobbled stone roads for wheeled traffic, public baths, basilicas for public meetings, theaters for entertainment, bakery and cereal shops, beverage counters for hot and cold drinks, schools, swimming pools and temples. Unlike Pompeii that used its streets for runoff of waste products, Herculaneum had a city sewage system.

But, on the walls of nearly every residence were painted murals of mythological figures, birds, hunting and domestic life scenes. Pompeii red, the predominant color background in many frescoes, cannot be reproduced because the berries that only grew on Vesuvius's hillsides that served as the color base were destroyed in that volcanic explosion. A white altar in a building (the Eumachia) facing the forum in Vesuvius depicting in sculpted white marble the sacrifice of a bull customarily an imperial cult, and may even represent the sacrifice held to inaugurate this temple and its compound. (11)

But the love of beauty and color in these excavated towns is ubiquitous and perfectly preserved after two thousand years. The columns, built first with brick then covered with a layer of limestone, a process faster than simply carving an entire column in limestone, were also painted with this unique red. It might be impossible to locate a town of comparable size anywhere in the world today and find a similar amount of sculpture, paintings, mosaics and architectural design to match Herculaneum's singular achievements.

Theseus and the Minotaur

About 475-470 BCE in a tomb on the island of Skysos, the bones of a very large man were found and immediately assumed to be those of Theseus, the legendary figure who, like Hercules, had to perform heroic deeds and feats of superhuman strength.

It is difficult to imagine a hero, especially an ancient Greek hero, not having to fight a bull at some stage in his career. Theseus, son of Aegus, King of Athens, had to fight two bulls. The first seemed easy, but nothing is simple where a goddess is involved. Medea had placed Theseus' father, Aegeus, King of Athens, under a spell. In order to get rid of Theseus when he arrived on the scene, Medea challenged him to kill the Bull of Marathon that had been ravishing the countryside and breathing fire from its nostrils, the same bull Hercules had left for him on Crete. Theseus subdued it and brought it back to Athens where it was sacrificed to the city patroness, the goddess Athena.

The second challenge of Theseus was not so easy: overcoming the Minotaur, the huge bull inside the maze at Crete. Queen Pasiphae, wife of King Minos of Knossos, fell in love with a glorious white bull that magically one day rose from the sea. The artist Daedalus, in order to gratify her passion for the bull, made a hollow wooden cow and covered it with a real cowhide. The Queen hid inside the makeshift cow and the bull came to her, though how he was induced is unexplained. The issue of this coitus was the Minotaur, a hybrid creature with the body of a man and the head of a bull, a monster so grotesque King Minos had it hidden in the Labyrinth, a cavern full of deceiving passages seemingly without exit.

Theseus arrived in Crete, was feted by Minos, King of Crete, whose daughter, Ariadne fell in love with Theseus. For daring to fall in love with a princess, Theseus was sentenced to be one of the victims of the Minotaur. Extracting a secret promise between them that he agree to take her back to Athens and wed her, Ariadne manages to give him a ball of thread to take into the twisting labyrinth built by Daedalus and ties one end to a point at the entrance of the cave. When Theseus meets the Minotaur (a word coming from a combination of Minos, the place, and Taurus meaning bull), this huge monster shaped half like a bull and half like a man, Theseus realizes it is more terrifying than anything he has previously encountered.

The Minotaur was fed periodically by a tribute from Athens of 12 youths and maidens as food, so it lives off human meat. But with his bare hands Theseus manages to fight off the monster's attack and break its neck. At the end of his rope, so to speak, but certainly not his thread, he finds his way back to the entrance of the labyrinth with the aid of Ariadne's creative plan to lead him out of the maze. A red Athenian drinking cup in the British Museum dating from 440–420 BCE shows Theseus dragging the Minotaur from the labyrinth in the center of the design, and around the edges of the cup are displayed a few of his other heroic adventures.

This legend may indicate more about the wiles of women in the defense of their men than the athletic struggle of men against enormous odds and wily, seemingly unbeatable beasts. Grappling with uncommon beasts is a common theme of all mythical heroes and of grade B action movies, as it appeals to men's courage in overcoming fear about physical challenges.

In 1933 Pablo Picasso, and other surrealist painters in his generation, entered an intermediate stage in his artistic career that blended his intense erotic inspirations with classical mythical themes. He made several paintings and drawing of sexual imagery and savagery when he combined the myth of the Minotaur and bullfights with sexual encounters. In 1933 he designed a cover for a surrealist journal and in the same year drew with ink on paper *Minotaur and the Nude*, combining the forces of sexual and animal passion. In *Minotaur Raping a Woman, 1933* Picasso showed centaurs and a Minotaur assaulting young women apparently in rapture. Similarly, in *Bacchanal with Minotaur, 1933*, he displays a subdued woman caressing centaurs and a Minotaur while engaging in bacchanalian delights. A 1935 etching *Minotauromachy* appears to show the Minotaur as a subject of pity as a dead, half-naked woman is led away by a horse and a girl lights the way with a candle. (12)

One of the great masterworks of the 20th century in art is Picasso's *Guernica*, based upon the destruction of the Basque town by German bombers in the service of Spanish fascists under Franco on April 26, 1937. It was painted using only black, white and grey tones. Although screaming women to the left and right of the canvas call attention to the agony and horror of war, prefiguring the coming of World War II, the center depicts the death of a horse from the bullfights. Conspicuous in the left corner is a horned bull with the head of a Minotaur bellowing its displeasure.

Picasso's genius was in inventing new forms of the human body to challenge perceptions of reality. But his creative merging of explicit sexuality with ancient Greek mythology of the bull revives a powerful theme and gives a visual range of images from the genius of his imagination.

The Bull Cult in the Minoan Civilization at Crete

To what green altar, O mysterious priest,
Lead'st thou that heifer lowering at the skies,
And all her silken flanks with garlands drest? (13)

In the latter part of the 19th and early 20th century, the discoveries of the palace of Knossos in Crete revealed a culture rich in artistic beauty, an ancient civilization that had a bull cult, as the story of the Minotaur and Theseus partly shows. The civilization of Minos had flourished during the period of the Middle Bronze Age at least 3,500 years ago, about a thousand years prior to the apex of ancient Greek civilization and two thousand years prior to the height of the Roman civilization. In artistic splendor and architectural grace it is unrivalled in the ancient world except by that of Egypt. (14)

Heinrich Schliemann, the first archaeologist to excavate the fabled fortress palace of Mycenae, the home of Agamemnon and the palace from which Helen was kidnapped prompting the Trojan War, found scores of golden goblets, vessels of silver, faience and alabaster artifacts. Found in the Fourth Grave was a silver "cow's head" (later to be recognized as a bull) with two long golden horns. It has a splendidly ornamented golden sun, of two and a fifth inches in diameter, on its forehead. There were also found two cow heads of very thin gold plate...which have a double axe between the horns."(15) Further excavations at the Palace of Knossos at the turn of the 20th century by Sir Arthur Evans, later Curator of the Ashmoleum Museum at Oxford University, demonstrated the persistence of the cultic bull belief. One of the more stunning findings by Evans in 1900 was a great relief in painted stucco of a charging bull adorning the wall of the north portico. The bull was subsequently found on other frescos, wall paintings and seals.

But the most unusual find came shortly after these sensational discoveries: a remarkable fresco depicting a young man in the act of somersaulting over the back of a charging bull with a young girl waiting with outstretched arms as if trying to catch him, or at least steady him from his leap. This delicate, yet unusual painting has had wide appeal far beyond the professional archaeological community. Both the acrobat and his female companion, of thin waist, trim and athletic, were attired in skimpy outfits resembling a kind of toreador costume. Similar frescoes and seals show the same kind of scene, the slim and youthful figure of a bull-leaper.

We can imagine the spectator thrill of gathering on a holiday in the local amphitheater to watch the popular contest of men and women perform death-defying acrobatic feats with charging bulls. A daring hunter leaps on and then somersaults over a bull while it laps water from a pool, or wrestles the animal's head sideways. A clay cylinder from Cappadocia (in central Turkey) dated from 2,400 BCE shows a similar scene of athletic daring.

The bull was not killed and nobody was carrying a weapon. Was this a real sport? A religious initiation rite? A test of courage for selected athletes? Cowboys and steer-wrestlers who roam the American west and people familiar with bulls say that such feats would be impossible as the bull charges with its head sideways with a horn down so as to gore the person, making horns difficult to grab, like trying to clutch a sword thrust. No one, not even a matador, has attempted this feat to demonstrate whether or not it is a possible. On the other hand, it is possible that the bull could have been trained, like a circus animal, to run with its head held high making such a potentially serious and lethal game for the athletes conceivable.

A piece of ivory sculpture also from the palace at Knossos shows an elongated acrobat, assumed to be a bull-leaper in a semi-flying position. It exemplifies a fluid, natural grace that both describes the sport and shows off the skill of the artist. The bull may represent the masculine side of Minoan culture, while the artistic renderings of slender, delicate female figures and bird illustrations, the feminine side.

On one side of the palace, Evans found the remains of what he called "Horns of Consecration," which had apparently been mounted on the roof of the palace so all could see when approaching from the southern road. Evans concluded that the bull had not been worshipped as a god but was considered a favorite animal of the earth god so its representations were predominately displayed. Hence, the bull was often sacrificed to the god.

In the early 20th century prior to the digs at Knossos, two golden cups, as I noted at the beginning of this chapter, had been found south of Sparta on the Peloponnesian peninsula dating from about 1,500 BCE. Because these cups had engravings of bull scenes, new interest was aroused when the bull figures and paintings had been unearthed on Crete. They were recognized as Minoan in style and depict the capturing of a bull by trapping it with nets. One side shows young men attempting to trap the bull in a wood. The bull had been driven between two trees where a net had been stretched. One scene shows the bull enmeshed in the net, but another scene has it evading capture, throwing one of the hunters to the ground on his back while the other tries to grab its horns and wrestle it down. The life-size bullhead found at Knossos, of which numerous replicas have been made, captures the spirit of the bull cult throughout the Mediterranean world over 4,000 years ago. A rhyton, a vessel in the shape of a bull's head used for pouring libations, dates from about 2,200 BCE. It is formed in black steatite with jasper mucosa and rock-crystal eyes and a mother-of-pearl profile. (16)

Odysseus and His Odyssey

Odysseus, the legendary surviving Greek hero of the battle of Troy who reputedly took 20 years to reach his home in Ithaca, all the while experiencing adventures suitable for a young person's imagination, arrived with his ship and crew one fine day within sight of an island of where the god Apollo kept his herd of cattle. Odysseus had been warned by the goddess Circe and by the old soothsayer Teiresias that if he ever wanted to see his homeland again that he should avoid touching any of Apollo's cattle. He explained this to his seamen, but they were tired from the long sea voyage and wanted to make anchor in one of the harbors. Fearing a mutiny if he did not comply, Odysseus reluctantly agreed to spend the night there. But he made his men swear that they would not touch any of the cattle.

That night a fearsome storm blew in from the south, keeping the men from sailing anywhere and it raged for a month. As food was exhausted and Odysseus temporarily away, they rounded up a few cattle from the herd. They believed that if they sacrificed these cattle in honor of the gods that the gods would be appeased. Odysseus soon returned to the odor of meat roasting and could do nothing. As soon as the men had eaten the wind died down and they could set sail. But Odysseus knew the gods would not be appeased. As soon as they were at sea, a gale sprang up and the ship was smashed and struck by lightning. All the sailors were lost except Odysseus who clung to the mast until he was washed up on the shores of another island that just happened to be the residence of the beautiful nymph Calypso, who made Odysseus her lover for seven years, as he had no way of leaving the island without a ship or crew.

Cattle are prized possessions, and if cattle rustling is a dirty, mean and thieving thing to do, as it was in the early American West, even more despicable is it to try and steal the cattle of a powerful god like Apollo. This lesson would not have been lost on the readers of Homer's *Odyssey* either.

Book Three of *The Odyssey* contains a lengthy description of a bull sacrifice, of how a smithy places a foil of gold around each horn of the bull while the executioner with a sharp two-bladed axe cleaves the stroke of sacrifice while another catches the blood in a bowl and a third attendant sprinkles the barley grains and perfumed water on the bull in honor of the gods. The women spectators wail for joy while others cut the throat and dismember the carcass. The offerings are moistened with red wine and sizzle in the split-wood fire. (17)

Greek literature, its legends, dramas and plays, oratory, poetry, histories and Homeric epics became the core of secondary and collegiate education in the early Renaissance about 1500 until the latter part of the 20th century. The fables, like those of Aesop, became bedtime stories for children. Hence, we know more about Greek civilization than almost any other ancient culture. Thus, it is not surprising that the bull stories, the labors of Hercules, the rape of Europa, the encounters of Odysseus, the trials of Theseus, and the fable of the Minotaur, are so conveniently found as allegories, poems, critiques and stories in literature and in paintings. (18) Each succeeding civilization captures the spirit of the culture it absorbs, modifies it, and passes it along to the generations that follow. The Romans, whom we visit next, totally fell in love with Greek civilization and borrowed its culture intact, including its bull cult.

CHAPTER FIVE
ROMAN AND MEDIEVAL BULL CULTS

Cincinnatus, after whom the Ohio city is named, left his plow to fight for the Roman nation. Cato in *De Agricultura* encouraged the use of fertilizer to provide forage during the dry seasons. Virgil eulogized farming, the farmer and cattle in his long poem *The Georgics.* Cattle breeders established a great breed of oxen, the silver-gray cattle of the Campagna north of Rome. Even among later Roman writers there are prescriptions about the length of a furrow so as to give the ox time to catch its breath and occasionally to shift the yoke so it doesn't gall its back. The barbarian invasions in the 5th century CE slaughtered and scattered these magnificent herds and the centuries-old cattle breeding was set back a thousand years.

Like all ancient peoples the Romans relied on cattle as a food source among the tribes the Roman army fought, as Julius Caesar describes in his *Gallic Wars* in 65 BCE. Having defeated a people called the Menapii, Caesar carried off both men and cattle as the spoils of war. Caesar called the animal the Urus, the Latin word for wild cattle or aurochs, deriving from the Sanskrit word UR, meaning a forest where the animal was usually found, and described it as "approaching the size of an elephant, although in appearance, color and form they are bulls. Their strength and their speed are great. They spare neither men nor beast when they see them. In the expanse of their horns, as well as in form and appearance, they differ much from our oxen." (1) As indeed they did differ from oxen, because what Caesar was describing was not the bull we know today but its predecessor, aurochs, which became extinct in 1627 when the last one died in Poland. They are the ruminants depicted in vivid colors on the walls of the caves of Cro-Magnon Man in the southern France.

But the Romans also used literature to express how cattle influenced their lives. Virgil and Horace serve as examples.

Virgil and Horace: Poetic Celebrations of Cattle

Virgil (70–19 BCE), the most loveable and celebrated of Rome's glorious literary figures during the Augustan Age, whose *Aeneid* is one of literature's enduring epic poems and whom Dante chose as his guide in his own epic *The Divine Comedy,* spent a decade of his boyhood education on a farm where he wrote in leisure moments developing the sublime and engaging expression in his poetry. His popular and melodious *Eclogues* and

Georgics eulogized country life, earned him the adulation of Rome and more importantly the personal notice of Octavian Augustus. Virgil restored nobility to farming, helped settle returning soldiers on the land, and enhanced rural poetry as the Greek Hesiod had done centuries before. *The Georgics* written over a period of seven years in Naples consists of two thousand lines of poetry on the subject of agriculture and the cultivation of the earth, with multiple references to cattle and oxen and their usefulness to the farmer. Here are a few examples about cows from the *Georgics*.

> *Strait stamp their lineage with the branding fire;*
> *Mark which you'll rear to raise another breed,*
> *Which consecrate to altars, which to earth*
> *To turn its rugged soil and break the clods.*

> *Up then! If fat the soil, let sturdy bulls*
> *Upturn it from the year's first opening months.*

> *Be first*
> *To speed thy herds of cattle to their loves,*
> *Breed stock with stock, and keep the race supplied.* (2)

Virgil's poetic details are priceless and although he never belittles the hardships of farm labor, he is touchingly sympathetic to farm animals, even birds, marveling at the power of their instincts and raw passion. Unable to resist some gentle didactics, he also uses agriculture to promote moral character through hard work, showing how the example of sowing seeds, weeding and harvesting have their counterparts in developing character. He bemoans the loss of *pietas*, the old Roman virtues of patriotism and family and poetically links the work of the farmer to the cattle.

> *And the bull,*
> *Of victims mightiest, which full oft have led,*
> *Bathed in thy sacred stream, the triumph-pomp*
> *Of Romans to the temples of the gods."*

> *The husbandman*
> *With hooked ploughshare turns the soil; from hence*
> *Springs his year's labor; hence, too, he sustains*
> *Country and cottage homestead, and from hence*
> *His herds of cattle and deserving steers.* (3)

The Emperor Augustus gave him the commission of writing the mythical founding of Rome, resulting in *The Aeneid*, a ten-year long task in verse he did not fully enjoy. When Virgil died with what he believed was an unfinished assignment, schoolboys for the next 1900 years had the questionable privilege of memorizing its sonorous opening lines.

But if Virgil gave birth to bucolic and rural life, the poet Horace (65–8 BCE), Virgil's contemporary, painted an equally pleasant portrait of country life, complete with poetic oxen, from the farm where he lived about 45 miles from Rome.

Around thy home a hundred flocks are bleating,
Low the Sicilian heifers...

Blessed is he—remote, as were the mortals
Of the first age, from business and its cares—
Who ploughs eternal fields with his own oxen
Free from the bonds of credit or debt. (4)

Horace's mansion given him by Maecenas, trusted literary advisor to Augustus, consisted of 24 rooms, three bathing pools, mosaic floors, gardens and enclosed porticos. His farm employed eight slaves and five leasehold families, so although he probably never worked on the farm he could certainly rhapsodize about rural life and his cattle. (5) If his simple farm life appears ostentatious, his actual life was part skeptic, part epicurean, but mostly stoic as he shunned in his private life what he extolled in his chiseled and refined poetry. He was too cautious to marry and too urbane for the farm, yet wrote about it in exquisitely lyrical stanzas...and included cattle.

Loose strays the herd on grassy meads disporting,
What time December's Nones bring back the feast-day;
Blithe, o'er the fields, streams bring forth the idling hamlet,
    Freed—with its oxen. (6)

O that the bull were to my wrath delivered!
O for a sword to hack his horns, and mangle
The monster now so hated, though so lately--
    Woe is me—worshipped. (7)

Clearly, in this stanza above he shows his farm familiarity with the bull and its deliverance and his skepticism in the obvious reference to the Egyptian's devotion to Apis.

We may think poetry and the raising of cattle odd combinations. But Virgil and Horace were true geniuses of poetry even with royal patrons who had more imitators than Rome had soldiers. In them we have the example of two farmers who experienced both the pleasant quietude along with the odors of farm life and knew firsthand "the lowering herds that wander leisurely down the calm secluded vale," (8) and the poetic lionizing of rural life and the cattle.

Besides poetry, there are two cultural beliefs among the Romans that illustrate the significance of bull cults. One is the astrological sign, the Taurus or bull sign still fancifully retained in astrological predictions, and the now extinct Mithras bull cult with doctrinal and devotional similarities to Christianity.

Taurus: The Charging of the Bull

Astrology originated in ancient Babylon nearly 4,000 years ago and spread across the known world, and its influence did not even superficially wane during the high noon of medieval Christianity. The heavenly signs and their presumed predictive powers have grown from cocktail conversational openers to a cult status. Taurus the bull was one of those sky signs.

Zodiac comes from the Greek meaning life forms. There are twelve constellations in the zodiac roughly divided in sections of the sky about 30 degrees of arc each. Each interval, or constellation, was given a sign for that area of the sky it covered. The twelve signs of the zodiac, their order and animal representation, are: Aries (ram), Taurus (bull), Gemini (twins), Cancer (crab), Leo (lion), Virgo (virgin), Libra (balance), Scorpio (scorpion), Sagittarius (archer), Capricorn (goat), and Aquarius (water-bearer). This zodiacal arrangement is only a convenient means of showing the positions of the heavenly bodies, which originally included the sun, moon, and five known planets, although their procession in relation to each other has changed over the centuries. The positions of these constellations are calculated from the vernal or spring equinox when Aries begins.

Taurus the bull rises high in the east a couple of hours after sunset, usually in December. Within it reside two separate star clusters, the Hyades, near the bull's nose, and the Pleiades or the Seven Sisters near its flank. Taurus often contains two planets, Saturn and Jupiter. The star tips of Taurus's horns are known as Beta Tauri and Zeta Tauri respectively. Next to Zeta Tauri is the star cluster known as M1, the so-called Crab Nebula, the dying remnant of an ancient supernova, and at its center, a spinning neutron star, a dense mass of matter left over from a star explosion.

Taurus in Greek mythology represents Zeus in one of his many disguises. Aldebaran, the 14th brightest star in the sky and orange-red in color, also known as Alpha Tauri, is Taurus's left eye. The corresponding animal in the twelve-year cycle Chinese calendar is the buffalo or cow, and this astrological resemblance with the bull appears to be more than coincidental.

Taurus in astrology exerts an influence on human lives. A horoscope, from the Greek word meaning to look at the hour (or to spy), is a diagram of star and planetary alignments giving special readers the ability to forecast personality and character traits and their future day by day. The calendar time for Taurus, based on a person's birthday, is April 20 to May 20. Astrologers presumably predict what kind of day a person will have, usually on a scale of one to five, from dynamic to difficult. A special prediction occurs on a person's birthday. Horace was skeptical of divining the future and counseled his friend from placing any credence in astrological results. "Nay, Leuconoe, seek not to fathom what death unto me—unto thee, the goods may assign...Seize the present; as little as may, confide in a morrow beyond." (10)

The Chinese zodiac, visible on a paper place mat in many Chinese restaurants, has a 12-year cycle, each year named after an animal: rat, ox, tiger, rabbit, snake, horse, sheep, monkey, cock, dog, boar and dragon, the only mythical animal, all with distinct characteristics from which personality traits are deduced. For those born in 1937, 1949, 1961, 1973, 1985, or 1997, the ox means that you are bright, patient and inspiring to others. If you marry a snake or cock and avoid a sheep for a mate, you make a wonderful parent. It is simplistic to say that those who studiously follow the predictions of their stars are just superstitious. But the popularity of astrology, which has no scientific validation, does signal a deep and lasting cultural influence that, even though people may not know the origin, continues to exert an ingrained power over lives.

The Ptolemaic conception of the heavens and therefore the astrological views of the bull and other sky animals was razed by Copernicus who offered a more scientific explanation of astral movements, and in the following century by Galileo who, aided by the telescope, provided visible evidence of circulating orbs.

Nicholas Copernicus (1473-1543) forever revolutionized science, and in the process vaporized astrological myth, because he looked at old facts in a new way. He alone, of several who proposed that the earth rotated on its axis and that the sun stood at the center of the universe, proposed a system that attempted imperfectly to explain it. His famous book, *De Revolutionibus Orbium* (On the Revolution of the Orbs) went unread and un-noticed for over a hundred years. Our modern use of the word "revolution," which has come to mean a radical change in the course of events, but which literally means revolving in an orbit, comes from the title of this book.

Copernicus was a respected and internationally known mathematician. He had been asked by the Pope to reform the calendar as early as 1514, nearly 30 years before his stunning theory was published. He initially declined the offer arguing that the confused state of the astronomical sciences upon which the calendar was based prohibited him from any action. But spurred by an intellectual conviction that simplicity should prevail in heavenly bodies in nature just as it does in mathematics, and urged by the Pope's insistent request, he began a lengthy study of all existing books on the subject. By 1534, Copernicus's ideas about the heliocentric motion of earth and the planets around the sun, and not the motion of all heavenly bodies around the earth as the Ptolemaic system had predicted, was circulating throughout Europe. The Pope approved the theory when he heard it because Copernicus' responses conformed to the mathematical problems more precisely. At the time no one perceived how such a hypothesis that contradicted all known knowledge, and scripture, could actually be true.

And he has made the world firm, not to be moved. (9)

Copernicus contradicted the accepted thinking that the sun went around the earth, a belief postulated by Ptolemy in the 2nd century CE and universally believed for thousands of years because it easily conformed to sense perceptions. Copernicus suggested instead that the sun not the earth was the center of the universe and that the earth went around the sun. Throughout the ancient world it was obvious to all that the heavens rotated around the fixed planet earth. Only in the past five hundred years has humanity believed otherwise. Copernicus destroyed astrology but not the legend of Taurus.

A less well-known cult was extremely popular among Romans two thousand years ago. It centered exclusively on the slaying of a bull by a mythical hero, Mithras, a cult that today is similar to what the bullfight represents in life-threatening entertainment.

Mithras, The Bull Slayer

The Aventine hill, one of Rome's famous seven hills surrounding the eternal city, has a sweeping view of the west along the Tiber River towards Ostia on the Mediterranean coast. During the Roman imperial period this was the luxury quarter of Rome filled with the villas of the renowned and wealthy. The Roman emperor Trajan (53–117 CE) built himself a private villa here. On the site today stands the church of St. Prisca, named after a 2nd century Christian woman who used her home to entertain Christians. Archaeologists uncovered in the 1950s a *mithraeum*, or sanctuary devoted to the bull-slayer deity, Mithras, directly under the present church. According to inscriptions found on bricks from the site, the sanctuary had been dedicated on November 20, 202 CE. The cult of Mithras had been a favorite among Roman soldiers and practiced everywhere throughout the Roman empire until suppressed after Christianity was declared the state religion by the emperor Constantine in 325 CE. Another *mithraeum* lies under the church of San Clemente in Rome. (11)

But who was Mithras, what was the cult and secret mysteries surrounding him, and why is he represented as killing a bull?

The religious cult of Mithras, the oldest known god, was born in ancient Persia, today's Iran, and quickly became associated with the domination and worship of the bull. The first reference to Mithras occurs in a treaty in the 14th century BCE between the Mitanni, an Aryan peoples living in upper Mesopotamia, and the Hittites, the dominant people in what is today south central Turkey (12). The chief divinities in this Persian period were Mithras, the god of the sun, and Haoma, the Bull God who, when he was dying, rose again and gave his followers a command that they should drink his blood. The great King of Pontus, Mithridates (132–163 BC) took his name from this cult figure.

According to the legend, Mithras had a miraculous birth, appearing from the face of a rock on December 25th and nearby shepherds heralded the birth. Immediately, he was hailed as a god who would bring about the people's salvation, a god of justice, truth and light. He performed many miracles while on earth and ate sacred meals with his followers, chiefly bread and wine, symbolizing flesh and blood. After his death he ascended into heaven and will return on the final day to judge the living and the dead. (13) Mithras is always shown as a beautiful youth with a divine countenance, often with a halo of light around his head as a symbol of his affinity with the sun and its divinity. In sculptures he wears a loose, knee-length tunic, girded at the waist, and his footwear is leather laced to the ankle. He is sometimes shown with a cape. His hat, purportedly Phrygian (northern Iraq and Iran) in origin, is about 6 inches tall, with a round-pointed top, usually protruding or flapped forwards. His hair is a series of luxuriantly curly locks.

Mithras one day saw an enormous white bull grazing. He wanted to capture the bull without killing it, and so mounted the bull and rode it until the bull collapsed. But on receiving a summons from the sun god Sol, who ordered him to kill it so it could release its powers to regenerate new life, Mithras dragged the bull into his cave where he slit its throat with his dagger. Mithras is shown in stucco and stone carvings kneeling with his left knee embedded in the upper torso of the bull's back and thrusting his dagger downwards into the bull's neck while he tugs on the bull's nose or mouth. The killing supposedly represents the act of creation and renewal but also the ultimate domination of the bull, a high point in human civilization regardless of whether or not it is associated with religious purposes. The blood of the bull is precious and has to be ritualistically drunk among the devotees.

The cult of Mithras was perpetuated in a secret society based on astrological beliefs and not simply animal worship. One hypothesis is that the bull-slaying scene of Mithras represented in the equatorial summer constellations is Orion. A similar theory proposes that Mithras' slaying of the bull represents the movement of Taurus, the bull symbol in the ancient night sky, to Aries the Ram, and is a shift in equinoxes, as viewed by the ancient world that saw the planets and stars revolving around the earth. (14) The slaying of the bull by Mithras thus symbolizes that his power coming from outside the visible sphere of the heavens is so strong that it can cause shifts in the visible sky. Slaying the bull means that Mithras has caused the heavens to shift to the Ram and hence the bull is no longer in the ascendant. Other scholars opt for the philosophical adoption of Mithras into the pantheon of Greek divines equating him with Apollo. (15)

The Mithras cult spread quickly throughout the ancient world, to India, Turkey and then to Rome where it was enthusiastically adopted by Roman soldiers. Religious shrines of Mithras have been found wherever the Roman legions were, in Romania, Serbia, Italy, Germany, France and England. But prior to its suppression, the cult of Mithras, whose devotees were enlisted as soldiers in the fight against good and evil, light and darkness, spread rapidly among the virile Roman soldiers serving on the borders of the empire, including outposts next to Hadrian's wall on the border with Scotland where shrines have been found. Soldiers easily and quickly identified with the cult's manly virtues. The emperor Commodus (180–192 CE), son of Marcus Aurelius, portrayed in the film *Gladiator* as his father's murderer, was initiated into Mithraism and took part in the bloody ceremonies of its liturgy. (16).

Although several legends associated with Mithras, his birth from a rock, his command of water flowing from a rock, (17) and his ability to perform miraculous deeds, are extraordinary, it is his slaying of the bull that constitutes the essence of the mystery cult and its veneration. This act of slaying a white bull was repeated at least annually in special ceremonies by followers in *mithraeums* that resemble the design of early basilicas, including a narthex where the shrine of Mithras was venerated. "On the rear wall of every Mithraeum there was a representation of the bull-slaying, which was regarded within the Mithraic cult as the most outstanding of Mithras's deeds, performed as it were for the benefit of mankind." (18) The bull's death in this theology reputedly gave birth to the earth, the sky, the planets, all animals and plants.

The sanctuary had a vaulted, rounded ceiling and an exterior brick covering. The interior hall measured about 25 yards long by 10 yards wide. From a large anteroom, the central hall led to the altar and shrine at the rear of the hall. Two parallel, elevated stone benches on either side flanked the main hall. Along either side of the central aisle were arched openings at the entrance often decorated with statues on pedestals. Climbing three or four stairs, members could walk or sit along twin, low but elevated walkways or platforms running the length of the building, both facing the central aisle. The building resembled a grotto to simulate the cave in which Mithras killed the bull. A flame was kept burning before the crypt where Mithras was portrayed killing the bull. The sanctuary was decorated with inscriptions like hymns, and paintings, inlaid mosaics and marble sculptures and statues.

These sanctuaries, true to their purpose of representing a cave, never developed into great ornate temples large and luxurious in splendor but remained simple locations where the faithful gathered to perform their ceremonies as befits a religious cult with austerity and simplicity as its chief characteristics. The form of early Christian churches adopted the same elongated building design throughout the Roman empire until supplanted by the elegance of Gothic churches in the Middle Ages. But in a similar fashion, Christian churches kept an altar at the far end of the shrine or sanctuary where the celebration of the ceremonies occurred and with a flame burning.

The cult involved a pact, probably with blood, between the sun god Sol and Mithras showing how they were linked in the battle to do good and conquer evil. Mithras ascends in a chariot into the heavens at the end of his worldly deeds. Likewise, Elijah ascends into heaven in a "chariot of fire and horses of fire." (19) In other religions, Jesus ascends alone and Mohammed with his horse.

But before his ascension, Mithras finishes his sojourn on earth with a banquet in which he and Sol partake of the flesh of the bull. This bloodless sacrifice and ritual was accompanied by offerings of incense, prayers and hymns. Archaeological artifacts portray the divine meal almost as often as the bull slaying. Initiates into the sacred mysteries clearly partook of a similar meal, sitting along the raised platforms in the sanctuary. In one painting from 220 CE from the *Aventine Mithraeum*, Sol and Mithras recline on a couch joining their earthly followers in a meal. The divine presence is manifested while the followers and initiates into the mysteries perform the ceremonies in commemoration of the god's command and eat bull's flesh and drink its blood. A Mithraic saying on a wall under the Church of St. Prisca in Rome reads: "Us too you have saved by shedding blood which grants eternity." When bull's blood was unavailable, bread and water and wine were used. Initiates believed that by eating and drinking the bull's flesh and blood they would be born again. (20)

We may never know for certain all rites and ceremonies because votaries were all sworn to silence about the mysteries and most died with the silences of the congregation intact. Over 500 representations of the bull-slaying scene exist, but no literature. A part of the initiation includes a blindfolding and a member walking behind and pushing the novice, or, in one scene, standing on his calves while the novice is kneeling. The ceremony of induction was performed by two dignitaries, one known as Father and the other as Herald. Upon completion of the ceremonies, the initiate was known as Brother and the son of the Father. What is clear is that there are a number of trials, including branding or tattooing for which the initiate has to demonstrate his courage, after which he becomes a regular member of the fraternity.

There were seven grades of initiation: raven, bride, soldier, lion, Persian, Herald or sun courier, Father, or high priest. Coincidentally, these conform to the then known seven planets, and in several similarities with the seven gradations of a person intent on assuming clerical office in the Roman Catholic church, from initiate to the mysteries of faith, to baptism, to clerical novice, to lector, to sub-deacon, to deacon, to priest (father), at which stage one can perform the mysteries of bread and wine which turn into Christ's flesh and blood. There are also seven sacraments paralleling these seven steps in the Mithraic mysteries. The high priest, father, of the Mithras cult had as his symbols the Phrygian high hat, a dagger, a ring, and a staff, all closely resembling that of a Catholic bishop or abbot with his miter or liturgical headdress, crosier or staff, and bishop's ring.

The London Museum contains a marble carving of Mithras and a bull-slaying scene from the late 2nd to early 3rd century CE, surrounded by the 12 signs of the zodiac. An inscription in Latin reads: "Ulpius Silvanius initiated into a Mithraic grade at Orange, France, paid his vow to Mithras." Nearby, another delicate carved head of Mithras rests. It dates from about 1,800 years ago and reveals the head and torso of a handsome youth, adorned with a Phrygian cap, averting his eyes from the bull slaying as, reputedly, eternal life flows from the bull. Raphael's painting of *The Vision of Ezekiel*, believed to have been painted in 1518 and now in the Pitti Palace in Florence, shows a an elderly, bearded grandfather figure sitting on the back of a black eagle and three, cherub-like, angelic children swirling around him defying gravity, while a winged lion and a brown-winged bull look at him beseechingly. The bull has the same posture as a Mithraic bull. Raphael probably based it upon an ancient marble sculpted group of the Mithraic sacrifice.

The cult of Mithras as embodying the spirit of god rose and fell with the fortunes and whims of Roman emperors, though at the zenith of the Roman Empire it flourished in Asia and Europe for about 2,000 years. The Roman emperor Diocletian in 307 or 308 CE dedicated an altar to Mithras as "benefactor of the empire" at a shrine east of Vienna. A few years later, in 312, Constantine, at the battle of the Milvian Bridge on the Tiber into Rome, saw a sign in the heavens, a cross according to Eusebius, which indicated to him that he would be victorious in battle that day. (21) He captured Rome thereby insuring imperial control of the whole realm. He moved his capitol east to Byzantium which he renamed Constantinople.

The emperor Julian (331–363 CE) chaffed against his Christian upbringing and returned to the Mithraic code of truth and justice, dying a Roman pagan. Flavius Claudius Julianus (337-363), better known in the west as Julian the Apostate, was Roman Emperor in the east for less than two years. When Constanius, one of Constantine's sons suddenly commanded him to come and help defend the empire against the Persians, Julian, at the time in Paris defending the empire against European marauding hordes, and undecided about whether he should go to Constantinople and declare himself emperor, did what many other undecided men of eminence at the time did—he petitioned to the gods for help, sacrificing a bull to Bellona, the goddess of war. When news arrived while he was journeying back to the capitol because his adversary had died suddenly, he took this as a sign of the god's munificence and good fortune, especially when his troops unanimously proclaimed him emperor.

Julian had spent all of his young life in philosophical studies throughout the eastern empire, and though he had studied Christianity he decided it was not his preferred belief. He reverted to paganism and a form of stoicism. As emperor, he had laws enacted that made it difficult for his subjects to remain Christians. (22) Julian felt that Christian controversies were detrimental to the strength of the empire and that people had actually lost the older virtues of reason and duty, integrity, patriotism and honor, and in their place proclaimed meekness.

About 18 months later, when he and his army were on the march to battle the Persians who were in revolt against the empire in the east, Julian stopped in the city of Beroea (today's Aleppo in northern Syria). There he sacrificed a white bull on the acropolis as a tribute to the chief Greek god Zeus. Within a few days' march, he and his army found themselves across the Tigris River facing a division of the Persian army. There ensued a battle in which Julian's troops triumphed, killing over 2,500 Persian regulars. Later, when the army met the main force of the Persians, Julian himself, refusing to don his breastplate and charging into the thick of the battle, was hit with a flying spear in his side, piercing his liver. He died later that night, only 31 years old.

One interesting analogy Julian used in his extensive writings concerns a bull: "I am told that bulls which are weaker than the rest separate themselves from the herd and pasture alone while they store up their strength in every part of the bodies by degrees until they rejoin the herd in good condition, and then they challenge its leaders to contend with them, in confidence that they are more fit to take the lead." (23)

Less than two decades after Julian's death, the emperor Gratian in 382 issued an edict removing all state support for the religion of Mithras and thereafter the cult of the bull-slayer and bull-worshipper fell in dispute. A similar edict prohibited animal sacrifice. (24) Another edict under Theodosius in 391 forbade all pagan religions and formerly state-approved Roman ceremonies, severely punishing private non-Christian devotions. Christianity, with help from imperial power, thereafter ruled without dissent or religious plurality until the Protestant Reformation in the 15th century.

The triumph of Christianity by officially approved Roman imperial degrees by Constantine, Gratian and Theodosius eradicated the vestiges of a religion which celebrated the domination of the bull in civilization, except in maverick Spain which continued to celebrate bullfights. Christian iconoclasts then began systematically to destroy all remnants of the sanctuaries, images, statues and paintings, and to build churches over destroyed or abandoned Mithraic shrines. According to the devotees of Mithras, the slaying of the bull was an act of salvation. But the other idea that did not die with the destruction of the Mithraic cult was that the bull, apart from its religious connotations, had to be dominated and killed in order for the cow to remain domesticated for its incomparable benefits. Bullfighting is the partial remnant of the Mithraic cult of the bull, its symbolism and ceremonies embedded in Christian beliefs and practices.

The collapse of the Roman Empire literally dwarfed cattle. Although barbarous raids, warring and hungry tribes, bandits and thieves, and possibly climatic changes may have contributed to the diminution of the actual size of cattle—from an average 125 withers centimeters in Roman times to about 103 from the 4th through the 13th centuries—the loss of Roman animal husbandry practices certainly helped to slim sizes. Medieval cattle were actually smaller than cattle in prehistoric times and productive Roman breeds disappeared completely. (25)

When Theodoric, King of the Ostrogoths (454–526 CE), captured the military remnants of the fallen Roman Empire, he began an energetic re-building program and, to stimulate commerce with neighboring countries, revived the institution of fairs. The word fair has a dual origin coming from the Latin *forum*, or market place and *feria*, or holiday. Several of these fairs were held in northern Italy in 493 and among the items marketed, whose sales receipts usually exceeded that of luxury goods, were cows and work oxen. In the following centuries, France became the most productive country in the improvement and dissemination of cattle. French fairs, regulated by the crown, became permanent social institutions.

Romans had improved on the breeding of cattle but scientific methods were not established until the 18th century by Robert Bakewell (1725–95), English livestock breeder and agriculturist who bred cattle from his

Leicestershire farm for meat, developed new breeds, including longhorn cattle. He introduced the progeny test for selective breeding and also improved methods of housing stock, cultivation of suitable grasses, proper irrigation of the grassland gave him four cuts a year, and manure procedures. His improved methods of raising beef is said to have actually contributed to a rise in human population by a decline in mortality rates.

Medieval Bull Realities and Myths

Just when you think you have heard everything, you encounter a more fantastic story that makes you realize that the full scope of the human imagination has not been exhausted. Medieval fables of bulls and oxen that give a new meaning to fiction fall into that category.

The centuries of the Middle Ages were a time of faith and superstition. Some of the writing of the age is not historically believable but written for amusement, much like our cinema, comics, video games, or science fiction. For example, Giraldus de Barri, known as Gerald of Wales (1146-1223) who wrote *The History and Topography of Ireland*, the earliest account of Ireland used reliably as an historical, if fanciful, source of life forms in medieval Ireland, was one such chronicler. This fanciful account of Ireland and the Irish people, surely to help glorify the virtues of the English at the expense of the Irish, was read at Oxford on or about 1188. The book is remembered less for its historical accuracy than for its fantastic storytelling. Barri's accounts today are as laughable as they must have been for the readers of 900 years ago. Especially in light of the advances in genetic engineering and cloning, his descriptions of the crossing of certain animals, including humans, with a cow are literary hallucinations. Here are three examples.

A man that was half an ox and an ox that was half a man.
In the neighborhood of Wicklow...an extraordinary man was seen—if indeed it be right to call him a man. He had all the parts of the human body except the extremities that were those of an ox. From the joining of the hands with the arms and the feet with the legs, he had hooves the same as an ox. He had no hair on his head, but was disfigured with baldness both in front and behind. Here and there he had a little down instead of hair. His eyes were huge and were like those of an ox both in color and in being round. His face was flat as far as his mouth. Instead of a nose he had two holes to act as nostrils, but no protuberance. He could no speak at all; he could only low...He came to dinner every day and, using his cleft hooves as hands, placed in his mouth whatever was given to him to eat. The Irish natives of the place, because the youths of the castle often taunted him with begetting such things as cows, secretly killed him in the end in envy and malice--a fate he did not deserve. (26)

> *Shortly before the coming of the English into the island a cow from a man's intercourse with her, a particular vice of that people, gave birth to a man-calf in the mountains around Glendalough. From this you may believe that once again a man was half an ox, and an ox that was half a man was produced. It spent nearly a year with the other calves following its mother and feeding on her milk, and then, because it had more of the man than the beast was transferred to the society of men. (27)*

> *A cow that was partly a stag. Near Chester in Britain a cow that was partly a stag was born in our time from the intercourse of a stag with a cow. All the fore parts as far as the groin were bovine, but the thighs and the tail, the hind legs and the feet, were clearly those of a stag, especially in quality and color of hair. But since it was more of a cow than a wild animal it stayed with the herd. (28)*

If it wasn't Irish whiskey, it must have been the impertinences of the wet wind and long winter nights that allowed the imagination of the author of these stories to roam so freely without critical hindrance. Of course, the occasional perversity of the English, always ready to make the Irish look foolish, could have prompted these tales. What became of the creatures described here and their descendants is the untold story.

Erik the Leif: Bringing Cattle to the Western Hemisphere

Erik Thorvaldson (a.k.a. Eric the Red, 950–1004?), Viking explorer, was the first European in Greenland and the Western Hemisphere. Geographers may debate whether or not Greenland is actually a part of the western hemisphere, but if it is then Eric brought the first domesticated cattle to it far in advance of the cattle the Spanish introduced. His son Leif was to extend the range of cattle even further westward. (29)

Eric was born in southern Norway in the early part of the 900s, leaving with his exiled father to journey to Iceland. A feud in which he killed two men caused him to flee or be banished and he left with a party of family and friends sailing westward. After a journey of about 800 sea miles avoiding countless icebergs, he colonized Greenland in 981, the world's largest island about three times the size of Texas. The improbable name Eric gave the land was clearly deceptive advertising for attracting settlers as it was largely ice, and Iceland as a name was already taken. The settlers were not deluded when once they arrived and may not have been inclined to laugh about the slick advertising gimmick.

Eric returned to Iceland to encourage new settlers and returned to Greenland in 986 with 14 ships with about 500 settlers and provisions. Exploring the numerous fjords he established a farmstead in southern Greenland and cattle were among the basic livestock imported. The island at the time was warmer than it is now, with spring grasses, and could support a medieval livestock husbandry of dairy farming, leather working, spinning and weaving and fishing. Excavations reveal stables that contained as many of 104 cattle stalls and nearby grasses and herbage provided animal feed. The Greenland colony, scattered among several farming sites along the fjords and reaching a population of 9,000, lasted for about 500 years eventually disappearing without a known cause in 1480, but presumed to have become extinct through Inuit Indian raids or an epidemic.

By the year 1004 the region was explored by Leif Ericsson, son of Eric the Red, and Thorfinn Karlsefni who roamed the warmest climes they ever traveled, Labrador, Newfoundland and Nova Scotia, where 160 men and several women landed with their livestock, the first cattle to set cloven foot on the North American continent. When Quebec was founded in 1608, French colonists imported cattle of Normandy-Brittany stock, closely allied with the Alderney cows of the Channel Islands, and these provided the foundations for later Canadian herds.

Wherever colonists dropped anchor and built homesteads in the New World over the past thousand years they brought cattle with them. In the long journey of mankind through the vicissitudes of war, famine, disease, religious persecutions, the search for food and land it is comforting to know that cattle were always accompaniments in that journey. Cattle had their own discomforts in exploitation, slaughter and enslavement. But in some cultures they were idolized as gods and goddesses, worshipped like rock stars and entertainment celebrities, besieged with prayers and besotted with the tears of petitioners. But as with all humankind, they have been returned to a degree of normalcy and are once again a global commodity.

Collapsing centuries into a few paragraphs is not so much an injustice to history as it is a synthesis of a cultural impression about how central to human life cattle have been. The bull and the cow have become not just a source of protein, land and agriculture lauded in bucolic poetry like Virgil and Horace, but symbolic of a very old story of powers and mysteries, of heroic deeds, of divinities, of astral personalities.

The cow is integral to our history. In the following chapters I will show its contemporary relevance to our lives.

Figure 2: Zeus Camouflaged as a Bull Carrying Europa From Pompei in the Naples Museum

CHAPTER SIX
AMERICAN BISON

A cold wind blows across the rolling hills of native grasses that cover the prairie near the Smokey Hills in Kansas. Nearby, a herd of about 250 American bison, or *Bison Bison* a genus different from the buffalo, loiter munching the grasses. Another 60 elk warily eye visitors. This is the Maxwell Wildlife Refuge in south central Kansas, a 2,250 acre bison and elk refuge, one of the last such remaining wild herds in America, with a bison band that came from three bulls and seven cows Kansas obtained in 1951 from Oklahoma. The refuge is named after a generous businessman, Henry Irving Maxwell, whose estate donated the land to create the refuge in 1944. Over 600 volunteers donate their time to provide tours. So many bison are now born that the state has been auctioning them off for decades, controlling the herd by eliminating ageing animals and maintaining the number of bulls to reduce fighting among them, some sold to ranchers who start their own herds for commercial purposes.

The bison herd closely together while grazing and can act very aggressively towards humans who do not carefully read their body language for clues of what distance and actions they will tolerate, revealing a primeval yet fearsome presence and deadly power that has fascinated and endangered men since Paleolithic times. Bulls can also act aggressively towards each other. Alexander Majors, of the firm Russells, Majors and Woddell that held the contract for the pony express, close friend of Buffalo Bill Cody who gave him his first job as messenger, in his 1893 book *Seventy Years on the Frontier*, described witnessing a head-butting fight among bulls in which the combatants, struggling too close to the precipice of a cliff, fell into a river below and were swept downstream. (1)

The bison, the only bovine with fur, is an exceptional forager and can survive and adapt heartily in severe cold and in semi-arid climates. Bison which live to be about 45 years old have been unsuccessfully crossed with cattle, where both males and females are fertile as the hybrids are often sterile. The few resulting beefalo have the genetic traits of bison. They make use of roughage for food better than cattle and produce leaner meat with higher protein content. American bison now live in confined domains on public and private reserves and discriminating consumers prefer their choice meat.

Ted Turner since 1987 has purchased 1.7 million acres of land, in 14 different ranches, in several states (Kansas, Nebraska, South Dakota,

Montana and New Mexico). One of the largest private landowners in the United States, in New Mexico alone he owns (at this writing) 1.1 million acres, or 1.5 percent of the whole state, and New Mexico is the fifth largest of all the states. His largest ranch is Vermejo Park Ranch in northern New Mexico and encompasses 580,000 acres, or three times the size of New York City. With over 20,000 buffalo scattered throughout all his ranches, the biggest commercial herd in the U.S., Turner banished the cattle and put buffalo to roam instead. (2)

The American bison, commonly referred to as the buffalo, the hunchbacked figure on a nickel and a city in New York, is integral to the story of the bovine species in America. When Cabeza de Vaca, a recorder traveling with the Spanish Coronado expedition, first saw them in 1533 in the southern U.S. he saw a "brown sea," a moving mass that covered the earth as far as the eye could see. From 1824 to 1836 a traveler could start from any point in the Rocky Mountains heading east and during the entire distance never be out of sight of buffalo until he arrived at the Mississippi.

The American bison has an enormous head covered with clotted hair, with huge, pyramid shoulders, a hump of muscle. It has short horns and small hindquarters with a high waist compared to its frontal appearance. This is a front-end loaded animal, easily recognizable in the American West. Bison mature in about eight years. Males will stand six feet at the shoulders and weigh 1,800 pounds; females about 1,000 pounds. They shed their thick winter coats as summer approaches and grow it again as fall nears. They have a longer breeding life than cattle and live between 20-40 years. The European bison, *Bison Bonasus*, disappeared from France and Britain by the 11th century and Germany by the 16th, hunted to extinction as were the American cousins.

About 90 percent of the world bison population is found on private lands in Canada and the U.S. Russia plans to introduce bison to native steppes where woolly mammoths once roamed. Most national parks, forest and grasslands are unable or unwilling to introduce or commit the use of public lands to bison because of poor rates of economic return.

The buffalo was not domesticated by native peoples nor worshipped as a god, though rituals did celebrate their creation. The Sun Dance, an occasion of gratitude for the Creator's gift of the buffalo, was most common among the Plains Indians and the Blackfeet in Montana. In a trial of strength and endurance, the skin of the breasts of a young warrior were slit, attached to bone skewers, and then attached by rawhide rope to a central pole. In the warrior's back similar slits were made and a buffalo skull hung from bone skewers. Thus, pinned, the skewered warrior danced until exhaustion and then the buffalo skull and skewers were ripped away freeing the dancer but forever scarring him. (3)

The Lewis and Clark exploratory party, the Corps of Discovery, saw the impressive herds grazing in the summer of 1804, a scene no other easterner had ever witnessed. The immense sky, the gentle rolling hills, the endless horizon and grasses, the buffalo—it was a paradise of land wealth. The land of the Great Plains that Lewis and Clark saw was as monotonous as the Indians tribes were diverse, from the peaceful and friendly Mandans to the vicious and aggressive Lakota Sioux who then controlled trade on the Missouri.

As the largest and most numerous animal in North America since the extinction of the woolly mammoth, the buffalo was hunted solely as a major food source by the American natives and by the gray wolf which preyed on the weary, young and disabled in the herd. It's doubtful a matador would dare challenge the ferocity of an angry buffalo. Daniel Boone's Wilderness Trail followed a buffalo path from the Cumberland Gap to Kentucky. (4) After millions were slaughtered in the latter half of the 19th century, by 1900 only two wild herds remained, one of wood bison in Canada and one in Yellowstone National Park, both now protected.

But the only free-ranging herd made a remarkable recovery in the 20th century in Yellowstone National Park, beginning from 23 animals in 1902 to over 4,000 at the turn of the 21st century. The bison have no known predators except for savage winters and predatory men. Their fierce independence makes them disrespectful of humans who attempt to corral, restrict their movements or domesticate them. But lately their numbers have increased beyond the park's management capacity and, though some are regrettably shot if they stray beyond the confines of the park, they pose other uncertainties. When they roam from the national park, as hundreds of them often do into neighboring Montana and Idaho, the cattle industry and park and agriculture officials worry because they may carry the infectious brucellosis disease beyond the perimeter of the park where it can quickly spread to the cattle, elk or even canine populations. (5) Cattle with brucellosis can transmit the disease through their milk, unless pasteurized, and through cheese. Even though American Indian tribes have offered to take some surplus buffalo, park officials still fear the disease can infect other animals and even humans.

Living in surrounding states, the livestock industry stands tall in the political saddle and believes that culling bison is a management issue. Arrayed against the shoot-first ranchers are the environmental and animal activists who prefer to use the term massacre to describe the killing of the buffalo and want to let the bison roam free without restriction. (6) The story of the American bison is an example of the venality, callousness and unscrupulousness of those who ventured into the West and saw its plenitude as a way of satisfying greed.

European immigrants inherited the bull and beef cult and brought this cultural imperative with them to America but in the process eliminated the native protein source. The buffalo, so far as we know, had no such history, except as providers of meat and household amenities. Yet the profusion of their presence in the Great Plains (in 1851 there were more buffalo in the U.S. than people) is a testimony to their endurance in an environment friendly for sustenance even as they were hunted by humans. How long does it take to bring a large, mammalian animal that has existed for millennia to extinction? Since there is no literary history to follow regarding these prodigious beasts, we have to rely on the interpretation of the last hundred years to piece together their near extinction.

Imagine the bitter consequences. For about 15,000 years your native people have lived on the protein of this animal, using its hide as clothing and tepee tarpaulin, its bone as sewing awl and cutting tool, buffalo chips as fuel. After a successful buffalo hunt, your people made pemmican, lean strips of buffalo meat dried on a rack of wickerwork and pounded into a flaky mass and mixed with an equal amount of melted buffalo fat. Packed into bags of buffalo skin, this compact food, sometimes mixed with nuts and dried berries, was nutritious and sustaining during winter months and between encampments. (7)

At the height of the reign of American bison, before the railroad brought visitors who killed them for sport, there were an estimated 60–75 million buffalo roaming the central plains. On Tuesday, October 27, 1878, a round trip excursion train left Leavenworth, Kansas for Sheridan at the Kansas border near Colorado, then the terminus of the Kansas Pacific railroad, to return the following Friday. Passengers paid $10, had refreshments, and shot buffalo from the train. One can picture these enormous animals as they sped across the landscape full tilt avoiding capture or death, often outrunning the horses of pursuing hunters, looking both fearful and comical, their huge shoulders heaving, their tails pointing directly to the sky, their eyes black from the depths of fear.

A part of the accelerated decline was the increase in demand for buffalo furs. An estimated accounting taken from the records of the American Fur Company reveals that this one company traded approximately 70,000 buffalo robes annually. (8) New Orleans received 199,870 buffalo robes in 1828. When unemployed young men discovered they could make $3 for a buffalo robe, they quickly entered the killing spree. Absent animal protection and endangered species laws, and with buffalo skins selling smartly in the eastern U.S. and Europe, teams of men, aided by the army which wished to eradicate the food of the Indians, became buffalo slayers in unprecedented proportions. Brick Bond, one such buffalo hide hunter, killed

5,855 from mid-October to mid-December, 1876 or about 100 a day, the carcasses left to the wolves and other scavengers.

Entrepreneurs saw the skeletal remains and found a market for the whitening bones littering the land. Stacking these bones at railroad stop sites, and shipping them east to fertilizer and potash plants, they make money from buffalo demolition. Even the Indians sold bones to army posts to survive.

Lewis and Clark and the Buffalo Dance

Vast herds of unmolested buffalo, elk, antelope and prairie dog loped and rambled over the prairies and provided exclamatory expressions in the Lewis and Clark journals as sheer numbers of wild animals inspired their amazement. After the beaver, buffalo steaks, tongue, and hump were the meat of choice of the Lewis and Clark expedition in 1804, though the explorers ate more elk than any other food. (9)

The buffalo dance was a ceremony among the northern tribes for invoking kindly spirits for buffalo hunting success. Lewis and Clark, leaving Ft. Mandan in west central North Dakota for the Yellowstone, were witnesses at a buffalo dance on January 5, 1805 in a village that describes how the erotic blends with the reverence for the elders' prayers for divine intercession in buffalo hunts. The description with all the characteristic diary entries of grammatical and spelling errors makes for an unusual and contrasting anthropological study regarding buffalo rites.

A Buffalow Dance for 3 nights passed in the first village, a curious custom   the old men arranged themselves in a circle & after Smoke a pipe which is handed to them by a young woman Dress up for the purpose, the young men who have their wives back of the Circle go to one of the old men with a whining tone and request the old man to take his wife necked except a robe and—(or Sleep with her) the Girl then takes the Old Man (who very often can scarcely walk) and leads him to a convenient place for the business, after which they return to the lodge; if the old man (or a white man) returns to the lodge without gratifying the Man & his wife, he offers her again and again; it is often the case that after the 2nd time without Kissing the husband throws a new robe over the old man &c. And begs him not to despise him & his wife (We sent a man to this Median Dance last night, they gave him 4 Girls) All this is to cause the buffalow to Come near So that they may Kill them. (10)

On Saturday, April 13, 1805 Lewis and Clark while traveling up the Missouri river in North Dakota saw buffalo carcasses lying along the shore. They had been drowned by falling through the ice in winter and had lodged on the shoreline when the river broke in early April. On May 29th that year they were frightened when a bull buffalo charged the camp at full speed coming within inches of men sleeping and trampling a rifle useless before the sentinel could sound an alarm.

In May, in one of the more vivid narratives of the journals, Lewis describes how a great number of buffalos were killed by running them over a cliff. One young, athletic warrior is selected who is clothed in the skin robe of a buffalo, complete with buffalo head and horns. He placed himself between a herd of buffalo and the designated cliff. Other Indians surrounded the herd from the rear and flanks and at a signal all moved towards the herd. The disguised decoy so placed himself as to be observed by the herd to be able to run before them as if leading them, and dangerously not to be able to be trampled by them. He must be agile and fleet of foot. He raced at full speed toward the precipice as the herd gallops after him and dove into a crevice he prepared in advance while the herd flew over the cliff to sudden death about 120 feet below. (11)

Lewis and Clark, and soon thereafter English and French trappers, were the first white men to encounter these buffalo herds and discover how useful are hearty was their meat. Yet unlike Neolithic counterparts none of these explorers thought of domesticating them. Within years, and in a positive and swift culture adaptation, Indians did not have to wait in hunger for the buffalo to come to them but acquired horses which gave them mobility to follow the buffalo herds.

Washington Irving's Buffalo Encounters

Washington Irving (1783–1859) was a diplomat who served in the American embassies in Madrid and London and writer of easy but sophisticated informality and graceful style, the author of *Rip Van Winkle, The Legend of Sleepy Hollow*, a five–volume biography of George Washington. However, another popular book was *Tour of the Prairies*, an account of his 1832 journey through Kansas, Oklahoma and Texas in a region then inhabited with Osages, Pawnees and herds of buffalo, deer, wild horses, elk, and moose. Irving, lately returned from 17 years in Europe, was invited to attend a government inspection team commanded by Henry L. Ellsworth (after whom Ellsworth, Kansas is named) who had been appointed Commissioner for western Indians. Unlike the Lewis and Clark expedition, this was not a scientific excursion. The party saw no hostile Indians and failed even to survey the land.

Irving describes several encounters with buffalo in the plains and grassy meadows, hunted, as were deer, bear and wild turkeys, as daily food and provisions for the traveling party. Indeed, on some occasions no game were killed and the amateur but enthusiastic company went hungry until there was a successful hunt. Three descriptions are illustrations of Irving's writing style and experiences with this immense and lordly animal.

Of all animals, a buffalo, when close pressed by the hunter, has an aspect the most diabolical. His two short black horns curve out of a huge frontlet of shaggy hair; his eyes glow like coals, his mouth is open, his tongue parched and drawn up into a half crescent; his tail is erect, and tufted and whisking about in the air, he is a perfect picture of mingled rage and terror. (12)

"We heard a trampling among the brushwood. My horse looked toward the place, snorted and pricked up his ears, when presently a couple of huge buffalo bulls, who had been alarmed...came crashing thorough the brake and making directly toward us. At sight of us they wheeled around, and scuttled along a narrow defile of the hill. In an instant half a score of rifles cracked off; there was a universal whoop and halloo, and away went half the troop, helter-skelter in pursuit, and myself among the number. The most of us soon pulled up, and gave over a chase which led through birch and brier, and break-neck ravines...One of them returned on foot; he had been thrown while in full chase; his rifle had been broken in the fall, and his horse, retaining the spirit of the rider, had kept on after the buffalo. It was a melancholy predicament to be reduced to; to be without horse or weapon in the midst of the Pawnee hunting grounds." (13)

The repast of eating the juicy rump of a buffalo in the evening camp, or the choicest and most prized tongue, was the culmination of a successful hunt. However, there were gainsayers. "I do not like the flesh of this wild ox," wrote Horace Greeley, "it is tough and not juicy...I would rather see an immense herd of buffalo on the prairie than eat the best of them." (14).

Carrying the carcasses of buffalo into camp for the night brought additional discomforts: the all-night howling of ravenous wolves that smelled the uneaten buffalo morsels. Losing a horse or mule that decides to run excitedly with the stampeding buffalos is another disadvantage of buffalo hunting. Fremont notes: "One of our mules got a sudden freak into his head today and joined a neighboring band (of buffalos) today." (15) Such escapees were almost impossible to rescue. Fremont's memoirs of his first expedition from Missouri to Fort Laramie (July 1,1842) has the most enthralling description of a buffalo hunt.

As a genre writer of clarity and vivid description, Irving describes a frontier of wild abundance and his powers of observation and recording are unparalleled. He is an American writer imbued with the sensibilities of European romanticism and a return to nature blossoming at the time in the writings of Schiller and Goethe, and William Blake, Coleridge, Wordsworth, Byron, Keats and Shelley. Though traveling as part of a government mission to determine the condition for relocating thousands of

Indians, he describes picturesque groves of cottonwood trees, the eccentricities of his companions, the horses, the buffalo, and the physical conditions of the Indians he encounters, but nothing about their status, sociological conditions or speculations about their future. The odious legislation known as the *Indian Removal Act* of 1830, enforced by President Jackson, was intended to reallocate all Indians along the eastern seaboard to Oklahoma and the purpose of the expedition Irving accompanied in 1832 was to survey this territory. (16)

Irving is a storyteller and an amateur hunter not a social commentator. Perhaps he did not want to offend his generous hosts by offering analysis of what such government actions would ultimately entail. Forcibly relocating the Cherokee, Choctow and other eastern Indians to Oklahoma to face the more aggressive Kiowa, Pawnees, Comanches and other Plains Indians was tantamount to guaranteeing internecine conflict. What was to be expected when thousands of Indians from the Carolinas move into the territory of the Pawnees and Comanches and compete with them for the available buffalo? What would all Indians in the territory live on when their principal food sources become threatened?

Unlike Irving's descriptions, the argument in this book is to reveal the deep cultural long-term relationship between bovine animals like the American bison, cattle and humans. The elimination of the bison from the landscape in the expansion of the West was meant to reduce warfare by destroying the principal food source of Indians. Neither the government nor any serious settlers ever considered raising bison for domestic consumption. Instead, domesticated cattle replaced the native bison and in effect devastated the symbiosis of people and a centuries-old plentiful native food. The introduction of European cattle and sheep was simply another manifestation of the importation of protein sources with the culture of western civilization and the failure to perceive an indigenous American meat source that might be more readily adopted. Commentators have written frequently about the conflicts between the cow and agricultural cultures during the time of western expansion. (17) But no author to my knowledge has focused on a colonizing culture bringing its own preferred meat.

The Indians in southern New Mexico were the first peoples to raise corn in 4,000 BCE, a food that became the staple product of the early colonists and saved them from starving, while imported potatoes from the Americas kept Europeans from starving. American Indians had no domesticated animals except dogs. Ducks and deer were added to the American protein intake in the first years of settlement and wild American turkeys became a seasonal holiday food. But American buffalo meat was never cultivated. Cattle were everywhere substituted instead and have had such a long and intense relationship as a source of so many foods and domestic benefits that humankind cannot accept replacements. Humans may occasionally use sheep's milk to make Roquefort cheese but never dream of any permanent substitute for cow cheese, milk or meat on a regular basis.

From 1867 to 1883 the number one food source Indians relied for as long as there is memory was totally decimated by colonizers eager to eliminate them and their commissary so they could introduce their own chosen animal, the cow, to graze on the now devoid prairies. An English visitor writing in 1879 decried the cruel excess. "Such has been the miserable and wanton destruction of this fine beast during the past fifteen years, and the apparent apathy of the Government in checking it, that but a short time must elapse before it will be difficult to obtain a buffalo by any method." (18)

The preferred weapon of mass destruction was the Sharps Rifle, a .44 caliber rifle with a 29.5 inch barrel known as 'Old Reliable' for its big, long-range cartridges and accuracy, used extensively by buffalo hunters who could bring down a bison at about a 1,000 yards. A second favorite was the Winchester Rifle which had several caliber models and barrel lengths and an ease of loading via the port in the side of the breech that held multiple rounds in the magazine below the barrel. It was a popular gun with Indians and lawmen alike.

The breech-loaded 50-calibre Springfield rifle was the favorite of William F. Cody, "Buffalo Bill," who earned his nickname by shooting about 12 buffalo a day as food for Kansas Pacific railroad workers in 1867, an occupation which paid him $500 a month salary. "During my engagement as hunter," Cody writes in his autobiography, "for the company (the Kansas Pacific railroad), a period of less than eighteen months, I killed 4,280 buffaloes." (19) Indians were still troublesome, food was scarce but bison were plentiful and free. Cody had to go from five to ten miles from the railroad line each day accompanied by a man with a light wagon for transporting the meat. (20)

The killing resulted in dwindling supplies for native peoples. The Sun Dance of the Kiowa had to be postponed in 1881 so the tribe could eventually find a solitary bison after a two-month search. The following year none were found. Natives slowly starved, some surviving by selling bison bones. Today bison conservation initiatives help preserve a destroyed way of life and the once nearly extinct but revived American bison.

Buffalo Roundups

The annual bison roundup occurs in mid-October on Antelope Island in the Great Salt Lake where bison, deer and pronghorn antelope roam freely since first discovered on the island by Mormon pioneers. When the lake levels drop during very dry seasons it's possible to walk gingerly from the mainland to the island, as these animals must have done centuries ago. The 200+ head each fall is herded with horses and helicopters, corralled, weighed, examined for diseases and pregnancies and vaccinated against various infectious diseases. The herd is protected from hunters and poachers and managed by the Utah Division of Wildlife Resources.

I came to observe this unusual western wildlife phenomenon one October afternoon. Courteous park rangers allowed me to approach close to pens and corrals. There I saw the wild-eyed, snorting, wheezing bison paw the ground as they waited inspection, had a blood sample extracted, and, like children in line for their smallpox shot, were vaccinated against the brucellosis bacteria, a disease ironically brought to America by European cattle. Bison are also susceptible to tuberculosis. Once vaccinated, they are then tagged and released back into the spare landscape of the island.

According to U.S. Dept. of Agriculture scientists a new vaccine against brucellosis in cattle may also protect bison. Bison and elk are the last major sources of cattle brucellosis in the United States. Brucellosis is caused by the bacterium *Brucella Abortus* and costs U.S. beef and dairy farmers about $30 million annually. The response of vaccinated bison female calves was comparable to that of cattle vaccinated. (21)

While some states preserve buffalo, others hunt them for wildlife management and revenue. Once-in-a-lifetime hunting permits are drawn randomly for public hunts of bison to control the population of managed herds. Hunters can also shoot bison on private ranches for about $2,000 for a two-day hunt. (22) These bison, of the over 200,000 in most western states and Canada's western provinces on about a thousand state public reserves and private ranches, are all that remain of the approximately 50 million that once inhabited the Great Plains. There were less than 2,000 at the turn of the 20th century, but today one of the largest herds is in Yellowstone National Park where approximately 2,500 graze in a protected environment. If it were not for private efforts and public statutes prohibiting and limiting hunting and protection of this species on natural barriers like Antelope Island, American bison would exist merely as faded photographs of an extinct western animal, the largest in North America since the woolly mammoth.

Occasionally, smaller roundups are necessary. In upstate New York in 2000, a few American bison were roaming the woods near Malone, a small town near the Canadian border. For a few days the loose buffalo had stamped through fields of corn, potato and diary farms, often getting the cows to follow them like sheep. Sixteen had escaped from a nearby farm where they were kept as pets. Buffalo can't be herded any more than cats can. Seven were shot and killed by U.S. Border Patrol agents as the buffalo attempted to cross a busy highway. A neighboring farmer had killed two bulls figuring it would be easier to herd the cows but several took off into the woods lead by a female. At 1,500 pounds each, buffalo cannot hide for long. Several cows trotted after the buffalo cows as they traversed a dairy farm. Another farmer tried to herd them with his truck but instead scared them back into the woods. Meantime, someone had called the police who eventually shot all but two of the loose cows whose carcasses were taken to a meat rendering plant to be cycled as dog food. (23)

Native Land and Cattle Loss

By 1842 the Sioux were beginning to feel the depletion of their principal subsistence, leading to increased rivalry among tribes for buffalo products and raids on settlements for food and supplies and resulting in the establishments of more military outposts farther west. The army wanted to crush Indian resistance and one fast way was to eradicate buffalo herds.

By the end of the Civil War, when the U.S. Cavalry and troops became available for curbing Indian rights and prerogatives and when tribes were herded onto reservations, the cattle that Indians owned, like the bison, had become nearly extinct, as had many Indians. The most notable massacres of Indians, a genocide condoned by most Americans, were at Camp Grant in Arizona Territory, Washita in Oklahoma by Gen. Custer, and Sand Creek in Colorado. (24)

Native Americans had commercial ties with neighboring tribes throughout the West. Archaeological evidence exists in the Salinas ruins area of New Mexico, an hour's drive south of Albuquerque, of shells, flints and dried buffalo meat that were traded for salt and maize. (25) Buffalo were plentiful and their use as a food source widespread among early natives. Besides meat, the buffalo provided Indians with robes, teepee covers, blankets, skin for moccasins, sinews for cordage, and bone tools and utensils. (26) But when the Spanish first introduced cattle, natives also adopted the practice of domesticating them.

With the disappearance of the buffalo, Indians sued for peace and cattle replaced the buffalo herds. After the completion of the wars with southern Indians and the attendant treaties in the middle of the 19th century, most Indians were settled on what became known as the Cherokee Strip in today's Oklahoma. While thus restricted to reservations they could no longer support themselves on the sparse game and needed to rely on government rations to survive. (27) The fact that the government had to contribute rations to the Indians meant that a whole new market had been created for beef.

In 1890 the federal government condemned the practice of unscrupulous businessmen, under the guise of legitimate state corporations, leasing Indian land for grazing for egregiously large prices. The practice was declared illegal and void. A proclamation by the Dept. of the Interior in 1890 directed that no more cattle be brought upon the Cherokee Outlet, then in Oklahoma. Cattlemen were avoiding taxes on cattle and on land by bringing such cattle grazing on Indian land into competition with legitimate ranchers who paid taxes on land and cattle. (28)

Canyon de Chelly (a Spanish corruption of the Navajo word "Tseyi" which means "in the rock") in northeast Arizona on the ancestral land of the Navajo people is a national monument, a place of uncommon natural beauty and intensely angular red rock formations and brazing colors. The monument is actually two adjoining canyons, Canyon de Chelly and Canyon del Muerto, an aptly named canyon of death, a forbidding reminder of the raid by the Spanish in 1805 which massacred 105 women and children hiding in caves and along a ledge while the native men were on a hunting expedition. The canyons today are the repositories of cliff dwellings of the ancient Anasazi peoples related to today's Hopi Indians who inhabited this region from 350–1275 AD. Navajos, related to indigenous peoples in Canada, arrived later, in the 16th century.

Navajos have always respected the former residence of these ancient ones as they call them and have not built over or near them in successive generations of occupation. But Standing Cow ruin in Canyon de Chelly, not far from the rock known as Navajo Fortress below the Antelope House overlook, is an exception. In the 1700s, Navajos built a hogan, a ceremonial house on existing pueblo grounds that still stands. On the red rock wall behind the hogan, about 15 feet above ground level, is a white pictograph, a wall painting about a hundred years old, of a life-sized cow. Standing Cow represents a fairly modern artistic representation of the usefulness of the cow, introduced by the Spanish, as were horses and sheep, to the lifestyle of the Navajos. The relatively recent introduction in visual form of a domesticated cow alongside antelope, bison and other animals significant to the history of native peoples drawn (pictographs) and carved into rock walls (petrographs) illustrates how central this animal was to the life of the Navajos. The domesticated cow, before the Spanish and Americans killed them to reduce natives' fighting spirit, allowed the Navajo to rely less on hunting deer and rabbits as a protein source.

Ruthless and regrettable as was the slaughter of millions of bison and thousands of Native Americans and their cattle, a sense of the landscape as a religious element was also lost. American Indians have an intense, personal and religious relationship with the land. Clearly, none of their principles were incorporated into federal land use policy or legislation until very recently. To avoid the despoliation, destruction and vandalism of select locations in the West, portions had to be placed in perpetuity as reservations, national parks and monuments to preserve scenic and historical features.

In September, 2002, 40 heavily armed agents from the Bureau of Land Management, fortified with helicopters, descended on two elderly Shoshone sisters in rural Nevada and confiscated 232 cattle, later sold to pay for grazing fees. Lost land and lost cattle among native peoples is not just a tidbit of history but is played out today in parts of the west, like Crescent Valley Nevada southwest of Elko. (29). The Shoshone sisters, Carrie and Mary Dann, living without electricity, hot water or furnaces, graze their cattle on what the U.S. Government considers federal land and what they consider Indian land. They want only to graze their cattle on land they believe is theirs and have refused for 30 years, in language so acrid it would curdle mouthwash, to pay grazing fees on what they think is traditional Shoshone land. The Ruby Valley Treaty concluded between the Shoshone nation and the U.S. Government in 1863 granted access for settlement, but not title, to Shoshone lands. In the 1970s the Indian Claims Commission decided that through gradual encroachment the Shoshones had indeed lost the land and agreed to settle for $26 million in compensation. The U.S. Supreme Court ruled in 1985 that the tribe in effect lost title when money was deposited as payment, though the Shoshones had not accepted it. Because of confiscations, half of Carrie and Mary's herd is gone and the government says they owe over $3 million in back taxes. The government has agreed to settle the land dispute with the Shoshones at 15 cents an acre, a price first set in 1862. Many Shoshones are exhausted from the lingering claims and counter-claims and in June, 2002 tribal members voted 1,647 to 156 to accept the payment of about $20,000 each.

TC and the Buffalo Harvey Wallbanger

TC knows all about buffalo. He is a world and national riding and shooting champion, and state champion in Arizona and California in 2001, shooting with his single action 45 pistol at balloon targets held by young attendants. When I called to set up an appointment to interview him at his buffalo farm, the Living Legend Ranch in Scottsdale, Arizona, and asked what time would be convenient to see him, he replied "dusk." Men close to the land and nature think of time in terms of the sun not the clock.

TC is friendly and generous, as big as the West, a rare man whose life has been engineered around his love of buffalo and other animals. He has raised buffalo, ridden them in races against horses, in rodeos, and entertained thousands at wild west shows, at half-time at football games, and appeared in movies and TV ads. He has earned his living from entertaining with buffalos and in the process has demonstrated that a man can have a special relationship, as few ever have or will, with this awesome animal. TC, like the buffalo itself, is a symbol of the pride of western independence and its living vitality.

Colin "TC" Thorstenson was born in 1956 in South Dakota of a Norwegian-American ranching family with 24,000 acres and 700 head of cattle. Together with his older brother he learned to break Shetland ponies purchased from a local auction barn. When he was nine his father bought four buffalo, and at 15 he and his brother bought the small herd from their father. When he was 18 there were 20 head of buffalo, and he bought out his brother's portion and added 10 more buffalo to the herd. As a young man, he married, moved out of the homestead, sold the herd to a veterinarian, and branched out into several incoming-producing enterprises, riding broncos in rodeos, trucking, fencing, and working for four years in the coalmines of Wyoming. Soon, his occupations were to change completely.

In 1978, poachers attacked a small herd he was again cultivating, leaving a three-day bull calf orphaned. TC bottle-fed the calf that imprinted itself upon him as if he was its mother. A unique and special bond was created, the key to the buffalo's training, and for the next thirteen years TC and the buffalo he named Harvey Wallbanger were inseparable. Responding to an inquiry from Gillette, Wyoming about racing a buffalo as an added attraction with horses, TC began a career racing Harvey throughout the West, Canada and Mexico, winning 76 of 92 races against quarter horses and thoroughbreds.

While on a spring trip to Tucson in 1991, Harvey happened to eat oleander, a toxic substance when ingested. It was unintentionally mixed in with his hay and he died within days. TC was devastated at the loss of his special companion and livelihood. TC today keeps the complete stuffed remains in a loft visible from the living room in his expansive home. He quickly recruited Harvey Jr. from his herd as a replacement for competing in the 35 rodeos he competes in annually, in wild west shows and private corporate performances, and his occasional appearances in feature movies.

TC himself had once been gored by the horn of one of his buffalo and suffers the scars still. But in 1992 a buffalo gored his father to death, a tragedy that still elicits strong emotion from the man who left his own imprint of determination and championship courage on his son. You would expect no less from someone whose family name refers to the ancient Norse god of thunder, Thor, after which Thursday is named, is the patron and valiant protector of peasants and warriors.

Norse Cow Myths and Babe the Blue Ox

The old Norse world of a thousand years ago contains creation stories and stark images of ice and fire, the birth of gods and, not surprisingly as we've seen throughout this narrative, the wonders of a cow. In the beginning, so the Nordic legend goes, there was nothing. North of the void was ice and mist and to the south fire. When fire melted some of the ice, the Frost Giant Ymir appeared. As more fire spread, Ymir melted more that spawned a race of giants and the enormous cow, Audumla whose milk nourished them. While the cow Audumla licked a salty stone, on the first day hair sprouted from its top, on the second day a head, and on the third an entire being arose named Buri who fathered three sons, Ve, Vili and Odin, all gods. The sons killed Ymir, plundered his corpse and with his organs fashioned the world.

The Prose Edda by the Icelandic poet Snorri Sturlasson written about 1220 contains a variety of such Scandinavian lore and narratives of Norse mythology. For example, in a slightly different narrative explaining how the different races arose one story begins: "Now it is said that when (Vafthrudnir, the Giant) slept, a sweat came upon him, and there grew under his left hand a woman, and one of his feet begat a son with the other; and thus the races are come; these are the Rime-Giants. The Old Rime Giant, him we call Ymir." When Gangleri asks, "Where dwelt Ymir, and wherein did he find sustenance? Harr (the narrator) answered: Straightway after the rime dropped, there sprang from it the cow Audumla, for streams of milk ran from her udders, and she nourished Ymir." (30)

There are variations on this Norse creation myth. In the middle of Ginnungagap the air was mild and quiet, and when the rime met the warmth and melted and a creature like a human being was created drip by drip, a giant called Ymir. While he was asleep he started sweating and from his armpit a man and a woman grew up, and one leg begets a son with the other leg and thus the family started.

Where the rime in Ginnungagap melted a cow called Audhumla were shaped, and Ymir (the giant) absorbed nourishment from the four rivers of milk, which sprang from the udder. The cow was licking the salt and hoar-frosted stones, and the first day it was licking, hair from a human being came out of the stone; the next day a head appeared, and the third day the whole body appeared. His name was Buri and he was tall and handsome. Their three sons are the gods: Odin,Vile (Vili) and Ve. These three killed Ymir. Odin is the head of all Norse gods and sits in Valhalla in the Hall of the Slain where he awaits with his hand-picked warriors the Twilight of the Gods, the end of the world.

This story has deep roots that undoubtedly go back to in Indo-European history, where there is a similar story about the primordial giant and the primordial cow. Parallels to Ymir, the progenitor, and Audhumla, the fertility cow, can be found in Indian and Iranian as well as Scandinavian myths. Audhumla is the primordial cow whose milk sustained Ymir, created at the beginning of time, from melting ice. She took her nourishment from the salt and ice that made up hoarfrost. From her udder sprang four rivers of milk. Her name means "Old Honey Land" or "Old Little Fish Eggs." She was created after Ymir, yet she nursed him. She is the Primal Goddess, a Goddess of fertility, far older than the first God.

The big American Midwest has its own Herculean tall stories involving a cow legend undoubtedly brought to America by Scandinavian immigrants and the symbolism includes some of the same cold, ice and snow imagery. (31)

Paul Bunyan was born big. Nobody knew exactly how big, well perhaps his tailor, but the stories that grew up around him were as big as he was, big as all outdoors. His parents weren't that big so nobody knew where he got his growth hormones, or even how his mother delivered him. But boy, could he swing a mean axe and cut down trees, a whole forest before lunch and another one afterwards. He cut down all the trees in North Dakota. But he was lonely.

One especially cold winter Paul was walking in the blue snow, blue because of the exceptional cold, and saw a small, unfamiliar mountain start to move. When he got close he saw two huge horns sticking out of the blue snow. Struggling to get free, a huge ox, whose eyes were 27 axe handles across (plus one plug of tobacco) came forth. Paul called the blue ox Babe and it stayed with him the rest of his life as his best friend. Babe could carry on its back all the lumber Paul could cut so houses could be built.

Paul and Babe carried water to the logging camps from the Great Lakes. Wherever Babe walked she left huge footprints in the ground and that is why Minnesota has so many lakes. The water Babe spilled from the huge tank on her back once when she tripped started the Mississippi river. When Paul went to California to look for more trees to cut, he got careless and dragged his axe on the ground forming the Grand Canyon. And Babe faithfully followed him wherever he went. Some claim that Paul and Babe wound up in Alaska. But wherever they are, they would be together. Paul and Babe were best friends all their lives.

So the American imagination via Scandinavia is as fertile as the Babylonian, Egyptian or Greek and equally playful. Not a collection of men, but a true American hero, bigger than life, cut down the forests for people to farm the land. And the companion is not a goddess, a woman guide, a oracle, an Ariadne, but humanity's long-time faithful animal friend, the cow, with its distinctive blue hide employed, as it was from the beginning of domestication, as a draft animal. Children's literature is enriched and unusual formations in the earth explained without geological complications. Northern European immigrants brought this Old Norse legend to New Amsterdam, Old New York. Although they changed her name to "Babe the blue Ox," and some minor details, the essential legend did not change.

A clash of civilizations is a confrontation of other peoples' history and ideas. Europeans, Danish, Spanish, Dutch, French and English, who disembarked on America's shores since the late 16th century brought with them their cultural baggage, including domesticated cattle. This is not unusual. What is surprising is that in the interim they destroyed the native herds of environmentally adapted native buffalo, a potentially lavish food source, and failed to accommodate to this new and plentiful protein provision.

As this study of domestication and deification shows, once a bond has been identified and coalesced as strong as that between humankind and cattle, it will not be torn asunder even in the presence of innumerable, similar animals. Suppose for a moment that the animal roles are reversed, and the bison has the long history of domestication in the Middle East and Europe that cattle have, and that explorers, adventurers, hunters and settlers found millions of native cattle instead of bison roaming the Great Plains. Despite the presence of a new and renewable food source, the cultural bond would be so potent and the taste of the known meat source so ingrained that the similar meat on the hoof would be slaughtered for occasional food or sport. Even today, when it is commonly known that buffalo is less fatty, has less cholesterol, four times as much protein as beef, no one can find buffalo in the meat section of a supermarket. Buffalo is better as a protein food but Americans and the world are accustomed to beef. (32)

Figure 3: Mithras Slaying the Bull In the London Museum

CHAPTER SEVEN
CATTLE DRIVES
AND GREAT PLAINS GRASSES

Land is as meaningful as the values and beliefs we give it. We can wax rhapsodically about land as Tolstoy when he described its meaning through a Russian peasant in *Anna Karenina* who dug his hands into the rich, raw soil extolling its virtues. Or we can view land as we do about the West as romantic nostalgia for a bygone past shaped by sentiment, heroic figures and endless, desolate vistas, awesome scenic beauty, and celebrated by unique voices in American literature, in western movies and pulp fiction.

Land is what we own, a property right derived from English royal aristocratic traditions, to be cherished or bartered like any commodity. Land is what we use for seeding and harvesting of grains, crops and fruits for consumption. Land is for investment, to hold indefinitely, to pay the requisite property taxes, and then to sell when the conditions or price is right, or to bequeath to posterity. Land is for preserving, holding in public trust for future generations.

But land is also where ancestors lived and are buried and therefore it possesses a sacred quality. Land is Mother Earth, containing the spirits of wind and water and earth, like the roots of a corn stalk bringing its regenerative life force over and over again for human survival and benefit. "What is this you call property," questioned Massasoit in mid-17th century Massachusetts. "It cannot be the earth, for the land is our mother...How can one may say it belongs to him?" (1) Native Americans could not believe how anyone could own land, ideal pasture for the millions of buffalo grazing on millions of acres of wild grasses that provided support for peoples in the Great Plains. But with the destruction of the buffalo herds and the concentration of Indians on reservations, the land became occupied with residents who introduced cattle where buffalo proliferated. The study of American land and the environment is intimately associated from the history of bison and cattle, symbols of the American West.

Newcomer Americans walked through the West and called it exploration. Succeeding generations called it discovery. Natives called it robbery. Lewis and Clark and the 1804 Corps of Discovery precipitated by Jefferson's deft Louisiana Purchase, brought preconceived perceptions of what the land should look like: gentle hills like the Appalachians in Virginia, productive soil for growing tobacco and cereals, a predictable climate. What they found was endless prairies of wild grasses, a sky that could swallow you, a constant and unavoidable wind, no accustomed trees, and more buffalo that the east had people. The party's preconceptions literally fell apart when they reached the Rocky Mountains. The promise of land did not dissuade those who followed who brought the same preconceptions of tillable land with them and in addition had to confront people already living there who did understand the land and its rhythms and did not welcome those who broke promises and knew nothing about the land, its people or animals.

Colonizers traded rum and the Bible with natives and then gradually usurped the continent. Intruders believed that failing conversion to Christianity Indians would soon become extinct anyway. Hardscrabble settlers squatted on the land without analyzing its unique properties. Consequently, few made a decent living in the West because beyond the 100th meridian, about central Kansas, the land gets less than 20 inches of annual rainfall, only enough to support unaided agriculture. A few monopolists, individuals pursuing only self-aggrandizement, expropriated and then hoarded the valuable resources, wasted the land's subtle and distinct ecological balances, and vehemently resisted government intrusions that attempted to impose the valued idea of public domain and common interest on them. (2)

Farther west, in the Four Corners region, the land rises to awesome heights and plunges to spectacular depths. This is the region of the Grand Canyon, Zion National Park and Monument, Bryce Canyon, Capitol Reef, Cedar Breaks, Arches National Monument, Natural Bridges and Rainbow Natural Bridge, an unparalleled sand and rock wilderness where raw material has been exported by the gods of geology, extravagant in natural beauty displaying the colorful aesthetics of stone, sedimentary strata and lava coulees, a place of natural beauty unlike anywhere else in the world. The emerging science of geology in the late 19th century would discover in this region a whole new playground and experimental laboratory while poets and travelers would laud its primitive and sublime extravagance.

Only an impatient itch by a Fremont or a Powell could possibly justify the adventure into such high plateau country, a forbidding, arid landscape where settlement meant hardship, prospecting promised lingering poverty, and exploration meant finding the quickest way out. Only the Mormons made settlement possible as everyone else hustled to California.

Reasonably accurate descriptions of the arid West existed from the days of Lewis and Clark and Zebulon Pike (1779–1813), but the facts got mixed with fiction, congealed, and then the West assumed an idealistic glow it did not possess, and the glaring hardships of moving through it or living on it toughened travelers and settlers encouraging them to move elsewhere, anywhere horse, train or wagon would carry them. Fantasy and legend, the romance of the West, were mistaken for the reality of permanent aridity. It was as if subsequent travelers were seeking the seven cities of gold that Coronado pursued. The West is a desert where settlers made cattle king despite the land's resistance to European cattle and European squatters.

This is the partial story of the land that held millions of American bison with which the Native Americans had a cultivated and intimate relationship and the European cattle that replaced them. It is the story of how two civilizations clashed and how the invading culture brought its own cultivated protein food source, cattle, and ignored the resident and abundant meat the continent provided, the American bison or buffalo.

Primogeniture and Land Laws

Quadrupeds, whether domesticated or wild, need grazing land to survive. The story of the supplanting of buffalo with cattle in the West begins with the canonical, European idea of private property and the ownership of land.

In England, a unique policy existed known as primogeniture, whereby the eldest son inherited all the property of the landed aristocracy leaving nothing to siblings. When new lands were "discovered" in the Americas, the king was able to give land to aristocratic sons who were unable to inherit land in England. English kings did this generously in the early days of American colonization. Lord Fairfax, for example, was given 5 million acres by Charles II, today Fairfax County in northern Virginia. Emissaries and explorers of the French and Spanish did the same for their kings claiming, as La Salle did in 1680 for Louis XIV, the Mississippi river drainage and all its tributaries.

When the Europeans arrived in the New World, finding no property titles or deeds, they claimed the land on behalf of the King or Queen who supposedly conferred it to them, a dubious power of attorney and legal assertion. Native Americans could not understand this because even their chiefs could not own land. They believed that the land belonged communally to everyone and was not the personal property of any individual. So how could a foreign king who has no right to the land, and who was not even present on the land, own it? (3) Except for some vagabonds in New Orleans and a few anti-social beaver hunters in the mountain west, few French had ever seen the land Napoleon sold to Jefferson.

The idea of private property, a uniquely western idea became the major sticking point between the newcomers to North America and the Native Americans, a classic example of a cultural clash and as deleterious to the natives as a toxic virus. Private property, enshrined in the principles of the

Declaration of Independence and U.S. Constitution, and based on European political ideas from David Hume, Adam Smith and John Locke, soon became the American Natives' undoing as more land was staked out by more immigrant claimants settling on land, owning it and taking from it what they wanted, its grasses, minerals, timber, water and buffalo, returning mostly ignorance and greed as weak vices to to replenish the land's munificence.

Alexis de Tocqueville published his seminal, definitive and classic treatise *Democracy in America* on American life in 1835. He noted then, describing the character of men who would enter the western plains, that: "It is difficult to describe the rapacity with which the American rushes forward to secure the immense booty which fortune offers him. In the pursuit he fearlessly braves the arrow of the Indian and the distempers of the forest; he is unimpressed by the silence of the woods; the approach of beasts of prey does not disturb him; for he is goaded onward by a passion more intense than the love of life. Before him lies a boundless continent, and he urges onward as if time pressed, and he was afraid of finding no room for his exertions." (4) But a half-century prior to de Tocqueville, the states were really not that united. After the Declaration of Independence in 1776 delegates established the *Articles of Confederation*, binding the states together contractually in a partnership. There was no executive and no judiciary, only a non-binding legislative body known as the Continental Congress.

The Ordinance of Congress on Public Lands (1785) and the *Northwest Ordinance* (1787) both passed by the Continental Congress provided the legislative basis for land expansion westward by prescribing how land would be parceled (a rectangular land survey system) and for what purposes, schooling for example. (5) The Continental Congress viewed land known as the "Public Domain" as a bounty to be sold to raise revenue and to reward soldiers who had served in the Continental Army, a legislative attitude of the U.S. Congress too until the late 19th century. The sale of public lands and its illegal appropriation, through theft and legal loopholes, became so blatant that in 1885 the Commissioner of the General Land Office, William Sparks, reported gravely on the situation. Note that he makes special notation about the cattle interests.

At the onset of my administration I was confronted with overwhelming evidences that the public domain was being made the prey of unscrupulous speculation and the worst form of land monopoly through systematic frauds carried on and consummated under the public land laws. In many sections of the country, notably throughout regions dominated by cattle raising interest entries were chiefly fictitious and fraudulent and made in bulk through concerted methods adopted by organizations that had parceled out the country among themselves and enclosures defended by armed riders and protected against immigration and settlement by systems of espionage and intimidation. (6)

Texas is central to the story of cattle and how settlers brought their food as investments with them to settle the immensity of the West. The story begins in the 16th century when cattle were introduced from Mexico and Spain then accelerated in the 19th century with a newspaper story.

The concept of Manifest Destiny was a political slogan, first coined by newspaper editor as an argument for annexing Texas, a territory that would be crucial to the expansion of the cow culture in the West. Here is what John O'Sullivan wrote in *The Democratic Leader* in 1845: "The far-reaching, the boundless future will be the era of American greatness. In its magnificent domain of space and time, the nation of many nations is destined to manifest to mankind the excellence of divine principles." (7) The slogan was catchy journalism, part idealistic and part religious in origin and tone, but wicked as policy initiative, extolling an assumed superior civilization's presumed right to territory it neither owned nor had legal claim to. The slogan became a social and political movement emphasizing that if the blessings of democracy are good, they should be extended to everyone who doesn't yet have them. "Manifest Destiny" did not long remain simply political but became a government cry for pre-emptive expansionism and land acquisition, exploited by President James K. Polk militarily when the U.S. in 1848 acquired Texas from Mexico in the Mexican American War, a vast territory which became grazing pastures for millions of cattle.

With territorial expansion blessed by military victory over Mexico in 1848, sanctioned by invocations of Divine Providence and the advantages of democracy, the U.S. became one of the largest landowners in history. Its terrain in the 1850s exceeded even the most bountiful of expectations. The congressional legislation that followed aimed at giving western land away, for a price, and only much later at regulating its use. Without cognizance of their contributions, and with its meat as sustenance for humans, cattle's symbiosis with nature amid a unique environment for rapid growth empowered a government and emboldened the raw and unrestrained elements in American society now permanently ingrained and popularized in the western, America's contribution to a cinematic art form. The presence of cattle on the North American continent convinced Congress to pass laws on grazing and created federal and state government agencies to protect and regulate the industry supporting them.

Here is a brief summary of land acquisitions that created the pasturage for cattle through the West. Prices paid then for entire states would today equal the prices spent on large single homes in certain upscale neighborhoods.

- *The Louisiana Purchase* includes most of the western drainage basin of the Mississippi River; price paid, $27,267,621.98.

- Florida from with Spain; price paid Spain, $6,489,768.

- Oregon; title established on basis of exploration and occupation, thereafter in dispute with Great Britain; embraces the present States of Oregon, Washington, and Idaho.

- This cession gave the United States the States of California, Nevada, and part of Colorado, Utah and part of Arizona and New Mexico; acquired from Mexico by *Treaty of Guadeloupe Hidalgo* at the close of the war with Mexico. The price paid Mexico: $15,000,000.

- The United States purchased from Texas the claim to certain lands now included in Kansas, Colorado, New Mexico, and also the "public land strip," for which $16,000,000 was paid.

- From Mexico, by purchase, known as the "Gadsden purchase." Territory in New Mexico and Arizona; Purchased to rectify the southern boundary of the United States; purchase price, $10,000,000.

- Alaska, the State of Alaska, purchased from the Empire of Russia; purchase price, $7,200,000.

The laws Congress passed in the 19th century favored settlers, ranchers, homesteaders and railroads, not necessarily in that order. In 1875, cash sales from public lands totaled 745,061 acres; to homesteaders, 2,357,057 acres; but to railroads, 3,107,643 acres. Common schools received 142,388 acres and land grant colleges 22,321 acres. (8) Because of the availability of land, the cattle trade expanded because money was to be made from them. Railroads grew because of the cattle. Ranchers and settlers moved wherever the railroads went.

The Preemption Act (1841) was a statute passed by the U.S. Congress in response to the demands of the Western states that squatters and pioneers, who often settled on public lands before they could be surveyed and auctioned by the government, be allowed to purchase land. The act permitted settlers to stake a claim of 160 acres and after 14 months of residence to purchase it from the government for as little as $1.25 an acre before it was offered for public sale.

The Homestead Act of 1862 has been called one the most important pieces of legislation ever enacted. Signed into law in 1862 by Abraham

Lincoln, it sold off 270 millions acres, or 10% of the area of the United States. A homesteader had only to be the head of a household and at least 21 years of age to claim a 160-acre parcel of land. Each homesteader had to live on the land, build a home, make improvements and farm for 5 years before he was eligible to "prove up." The filing fee was $18. Homesteads were few and usually far apart, often just a sod house crouching in a hollow. The land overwhelmed these pathetic beginnings in the vast silent wastes, the land that just wanted to be left alone, to preserve its own fierce strength, its savage beauty, its uninterrupted pathos. *The Homestead Act* succeeded in transferring land through deception, thievery and fraud into the hands and deep pockets of monopolizing interests, the very private interests government had sought to prevent from acquiring land.

If you wanted to avoid summers while homesteading your land in the late 19th century, Red Lake Falls, Minnesota might just be the place. Albert Blondin (1854–1919), my great, grandfather, originally a leather-maker in Montreal, immigrated to Minnesota in 1881 to homestead one of those 160 acres. He operated a farm with cattle, unsuccessfully, then a grocery store in Crookston in northwestern Minnesota before moving on to Yakima, Washington. My grandmother, Emma, was born in 1890 on that homestead. She told stories about how the snow has so deep in the winter that she and her siblings (8 of 11 who survived childhood) had to dig a tunnel through snow to the barn to feed and milk the cows.

The Desert Land Act of 1877 was a signal victory for the cattle barons who used it to acquire over 3 million acres of valuable rangeland. The word "desert" was intended to deceive potential purchasers of the true nature of the awarded land though not its actual properties. The law provided that anyone over 21 years of age had a right to 640 acres at 25 cents an acre provided he irrigated the land within three years. For an additional dollar he would then receive a deed to the land. The law contained no provision for actually occupying the land. Through dummy entries, cattlemen, who had early purchased all the prime land nearest water sources, acquired title to millions of acres.

The Taylor Grazing Act in 1934 provided a way to regulate the occupancy and use of the public land, preserve the land from destruction or unnecessary injury, and provide for orderly use, improvement and development. When overgrazing, drought, unsuitable soil and poor farming methods reduced Western rangelands to a dust bowl, Congress approved the *Taylor Grazing Act* which for the first time regulated grazing on the public lands through the use of permits, favoring cattle owners by excluding sheep ranchers who did not have a private land holding. *The Taylor Grazing Act* ultimately closed public lands to homestead settlement.

The course of empire through land acquisition and misdirected congressional regulation shaped the history of western America. The migratory explosion into a few fecund landscapes and large tracts of desert shaped the singular, frontier western character. That cattle played such a complementary role in this movement should not be surprising since the history of cattle runs parallel with human history in the West.

California, spurred by the 1849 gold rush and the incorporation of the Mexican territory as a state, quickly needed meat to feed its rising population influx. (9) The Franciscan missions and the once thriving agricultural ranchos they operated, including large cattle and diary herds, had been abandoned by 1850. Unattended stock died from lack of suitable forage and available water from 1855 to 1859 in Los Angeles County, where many of the missions were, though cattle population doubled in California. (10) Texas drovers soon assessed the conditions and realized they could make a hefty profit leading herds to California where prices were rising.

The Plains Truth

A book published in 1848 noted that the Western Plain, the frontier, was *east* of the Mississippi covering the western states of Kentucky, Ohio, Indian, Illinois, Michigan, and the western parts of Pennsylvania, Tennessee and Virginia. (11) It would be another two decades before the concept of the West actually referred to the Great Plains where vast territories beckoned the restless and reckless alike, a vast area stretching westward from the Mississippi river to the Rocky Mountains and from Texas north to the Dakotas comprising 20% of the land area of the United States. The Great Plains became in the 19th century the point of origin for further movement west, a gathering place for people and animals, and the site of the greatest manmade animal slaughter since the extinction of the dinosaurs, which occurred through a natural disaster.

It is difficult to imagine in today's visual overload of shopping malls, fast food restaurants, asphalt parking lots, neon signs and vehicle-bloated highways that the Great Plains were, less that one hundred and fifty years ago, an enormous landscape of wildlife abundance, rich in resources and natural beauty. The eastern seaboard in 1670 had been similarly spoken of in such glowing terms, as a land filled with green grasses, abundant timber and wild game. (12)

The Great Plains became a magnet for its share of vagabonds and the venturesome, wayfarers and wanderers, pioneers and the pious, brigands and bandits, homesteaders, cowboys and cattlemen. Finally, Americans and new European immigrants found a terrain to match their individual expansiveness and freedom from restraint and migrated in droves despite realistic accounts of hardships and deprivations.

The western population movement in the 19th century was a mobile generation seeking opportunities for land, a bonanza the government suddenly had in excess. Wherever they went, these immigrants took cattle with them but did not readily adjust to the climatic conditions, the long, cold winters, and provide for their cattle accordingly. Barely half of the settlers in the 1860s made provisions for wintering their cattle by laying up prairie hay when it was abundant. As a result, the calves were weak and unhealthy, the cows unfit to make butter, and the oxen half starved from the winter just when draught work was needed (13). Nor did most settlers usually make shelters for their cattle. The majority of small farmers relocating to the midwest were not richly endowed with the necessary physical prowess for the desolate conditions nor the intellectual nimbleness and so rarely adjusted to the hardships, hoping someone would buy up their unprofitable claim so they could push farther west in desperate land speculation.

Major John Sedgewick's 1857 supply expedition to Cheyenne left Fort Leavenworth, Kansas with 300 head of cattle, some of which were killed for food, and acquired another 150 head at Fort Laramie in Wyoming. A large herd of cattle were driven along with Pacific Telegraph workers laying the first transcontinental line in 1861. (14) A typical westward-bound wagon train or outfit consisted of about 25 wagons each with a six-yoke team of oxen as engines of movement and one mess wagon carrying the spare parts. Oxen were the preferred bovines, not mules or horses, as they ate grass along the route and did not need hay.

The rolling Plains of the Midwest were laden with wild grasses on which bison grazed for millennia and which today yield the nation's supply of corn, wheat and grains, America's breadbasket and cornucopia. In the Southwest, the land is textured with a mixture of chaparral, impenetrable thickets of shrub pines, dwarf evergreen, oak and mesquite trees in the highlands, and saguaro, prickly pear and barrel cactus in the Sonoran desert. In each of these ecological environments, beef cattle can be left alone for long stretches without management, and are an advantageous investment because they are more economically valuable than sheep.

Today's Plains are undergoing another transformation of migratory movement. The same general area stretching from central Texas to the Canadian border and from about 750 miles at its widest, an area five times the size of California, containing all or most of ten states and one-fifth of the land of the continental U.S., has only twelve million people, or four percent of the population. (15) Metropolitan Los Angeles now has more people than the Great Plains. Over half the U.S. population lives along the coasts where immigrants, constituting 70% of population growth, tend to congregate. This demographic dispersal means that the mid-west is again facing the prospect of more ghost towns as older residents succumb and younger ones seek jobs elsewhere. In just over 150 years, the Great Plains, as it rapidly depopulates, is well underway to becoming another frontier for discovery.

The Longhorn Long March

Cattle are not indigenous to America. Longhorn cattle, initial descendants of the Iberian aurochs and the open-range crillo Spanish cattle imported from the Antilles islands by the Spanish were eventually interbred with Shorthorns, Herefords, Polled Angus and Galloways introduced by the British and Scottish. Herefords were winning popularity in the 1880s from their fine, uniform appearance and fatty taste. Longhorns were long-legged and long-horned, lean, wild and mean looking, vicious when challenged. Longhorns weighed about three hundred pounds less than a shorthorn at two years of age. Jaramillo, a captain in Coronado's expedition in 1541, reported: "Indeed there is profit in the cattle ready to the hand, from the quantity of them, which is as great as one could imagine." (16)

Cattle were settled by the Spanish and Mexican missionaries wherever missions were established. Spanish and Mexicans built herds along the Nueches River in southern Texas where fertile valleys produced a hardy, wild breed. When Spanish missions in Texas were abandoned during and after the Mexican American War in 1848, the cattle became feral and descendants roamed for hundreds of miles across the Texan and mid-western prairies. Cattle were so plentiful in the mid-19th century in Texas that beef was hardly worth salvaging. Instead, whole industries grew up around hides, tallow for candles, hoof and horns for making glue. (17). In 1851 Gail Borden began condensing milk and selling it in tin cans.

From 1846 to 1865 only sporadic and irregular attempts occurred to bring cattle to far-removed markets. Some cattle were driven to New Orleans in 1842, to Ohio in 1846, to California in 1850 to accommodate the miners flocking there, and to Chicago in 1856. (18) The Civil War that began in earnest in 1861 stopped all cattle traffic as confederates would not have southern cattle driven north to feed Union armies, and Union soldiers halted the delivery of beef to the Confederates by blockading the Mississippi river and the Gulf coast. (19)

When Civil War veterans returned in 1865 they found the Texas plains alive with cows, proliferating during the war years, most unmarked and unclaimed. Unofficial estimates had 100,000 cattle in Texas in 1830 but 3.5 million by 1860. Cattle had little value unless a way could be found to transport them to where eager palates awaited. Soon, cattle were driven from southern Texas north to St. Louis and then Kansas where they fed on the abundant grasses and corn to increase their weight and transported by rail to Chicago where they were killed and their meat shipped to eastern cities and England. "The receipts of cattle at Chicago, Kansas City, St. Louis and Omaha," crowed an agricultural report in 1890, "are more than three times as large as fifteen years ago, and have increased more than 70 percent in five years." (20) Cattle drives headed north to Kansas along the Chisholm Trail, named after Jesse Chisholm, part Cherokee and part Scot, a trader who inaugurated the route in 1866 between the Trinity and Brazos rivers north of Houston, Texas to Wichita.

In 1870 in Texas there were 3 million beef cattle and 600,000 cows, or four cattle for every person; in other states the proportion was reversed. Texas ranchers were averaging 30-40 percent profit on the sale of beef cattle. But also by 1870, The Society for the Prevention of Cruelty to Animals had been formed and argued vigorously and fearlessly for the abominations occurring in abattoirs, to keep pens ventilated and not to perpetrate frauds on the public.

The longhorn was a clever and resilient animal, perfectly suited to the demanding Texan summers and winters, able to withstand the climate and topography of the landscape. But its meat was unappealing. In one of the more creative ventures in the cattle culture, a marriage was made on the grasslands of the prairies in which abundant corn was fed to cattle to fatten their loins, then transported by rail to the stockyards in Chicago, thereby altering the dietary habits of Europeans and Americans forever. From selective inbreeding with British and Scottish brands, the true longhorn now has become as scarce in the West as the bison once was. What was a small business of eastern farmers became a major industry unlike any other from 1865 to 1890. The cow industry, not automobiles, became the forerunner of contemporary automated business standards.

England's beefeaters had grown accustomed to the succulence of beef and longed for the untapped reserves of Texas cattle for their dinner tables a hundred years earlier. Live cattle, and salt and pickled beef were exported from the American colonies during the 18th century. Charleston, South Carolina exported nearly 2,000 barrels of pickled beef in 1747; Philadelphia shipped 3,500 barrels annually beginning in 1753. (21)

After John Bate of New York in 1875 perfected the experimentation of ice-cooled circulating fans in refrigerator rooms, keeping the temperature for the beef carcasses at 38 degrees Fahrenheit, Bate Shipping cooled beef carcasses to Britain. Proving that refrigerator methods were successful, he sold the patent and the process to Timothy Eastman of New York who, by December of 1876, was shipping 3 million pounds of beef to England annually. (22). Soon every shipping line leaving Philadelphia or New York had refrigerated beef in its cargo.

British and Scottish cattle farmers were so alarmed by this influx of American beef, selling in Britain for less than British beef, that they commissioned James MacDonald to journey to America and study the American cattle industry and to gauge the extent of the competition. (23) MacDonald concluded that the best quality beef did not come from Texas but from Illinois, followed by Kentucky, Ohio and Indiana. He had to conclude that the best of British beef still was preferable to American beef. "The class of cattle which produce the best quality of American beef are decidedly inferior in almost every point to the best beef cattle of Britain." (24) He decried Americans for emphasizing the weight of the cattle and not its quality, but could not explain how so much of it was transported to England and why English beefeaters could not seem to get enough of it. But he knew a desert when he saw one. "The soil on the prairie seemed light...it is unsuitable for cultivation and that it is best adapted for what it is now occupied as—an immense cattle range." (25)

British funding of ranches and colonizing of the western range lands for cattle once the American Indians had been subdued and the bison eliminated is a story of greed, complicity, illegal land acquisition and environmental exploitation. The destructiveness, cruelty, criminal negligence and avarice of the fur traders and miners followed. Lured by the great rates of interest to be earned, estimated at about 33 percent, from moving existing cattle north for sale, British investors sent teams of financial analysts like James MacDonald to scout out the investment possibilities. These advance teams wrote such glowing reports like *The Read and Pell Report* that aristocrats and financiers alike ventured like immigrants, either bodily or with their money or both, into the virgin territories and established cattle companies.

Within two decades after the end of the Civil War a marriage of convenience and capital would change forever the shape of the vast prairies Thomas Jefferson had purchased from Napoleon in 1803 and the grasslands and deserts won from Mexico. That marriage would bring Texas cattlemen together with American, Scottish and British financiers to consolidate the cattle industry and confiscate and condition the land for cattle. Corporate ranching and the beef industry were born from the union of American and British money, cheap land and available cattle. (26) According to the Secretary of the Interior, in 1884 there were 29 foreign companies owning over 20 million acres of land with millions of dollars of investments in land and cattle in most of the mid-western and western states. In 1883 there were 20 cattle companies with $12 million in assets operating just in Wyoming (27). Baron Walter von Richthofen (1848-1898), the German aristocrat rancher who settled in Denver, used as many financial spread-sheets as he did narratives in gushing over the cattle business.

Foreign investment fuelled the enterprise. The Scottish-American Investment Company was founded in 1872, financing cattle companies throughout the West, followed by the Scottish-American Mortgage Company with ranges in Texas, Kansas, Colorado and New Mexico, with over 140,000 head and $4 million in assets. The Matador Land and Cattle Company was incorporated in 1881 by English financiers. (28) Within a year it had $11 million from overseas investors, purportedly over 60,000 head of cattle on 300,000 acres and access to 1.8 million ranging acres. In 1880 the British owned Prairie Land and Cattle Co. Ltd. with 124,212 head of cattle and the Swan Land and Cattle Co. Ltd. had 123,460 head. By 1882 there were ten major British and American joint ventures in cattle and western land, with Americans as vendors and the British as venture capitalists.

Not all British came as financial speculators. Reginald Aldridge from Bristol, England, down on his luck in 1877 and restless for adventure but clueless about his future occupation or activity, came to Denver in 1877, found a compatriot who interested him in becoming cowboy, and within a year found himself raising cattle in Texas. (29) He had an unobservant eye and his narratives of typical western experiences are droll and uncreative. Written for an English audience, his *Life on a Ranch* has useful information about the manner of investing in cattle, guns, traveling conditions and the search for business partners.

Some spirited British aristocrats actually came to live not just find excitement and built expensive ranch houses and imported furnishings and servants to tend them. Lord Dunraven owned a large part of what is today Estes Park, Colorado. The Marquis of Tweesdale held title to 1,750,000 acres. Lord Dunmore had 100,000 acres. Walter Baron von Richthofen built a ranch in Colorado and wrote a small book about cattle ranching. (30) Antoine de Vallambrosa, a French nobleman, went to the Dakota Territory in 1883 to ranch on the Little Missouri, near the ranches of Theodore Roosevelt.

Into this mix of adrenalin stimulation, lawless adventure and excitement over a raw, seemingly boundless land, Midwestern towns suddenly appeared and some just as quickly became deserted, depending on how the town originally formed. Some settlers just wanted to till the soil. Cattle ranchers wanted to drive their cattle through to market to reach a transporting railroad and towns sprang up where the cattle assembled. Rail lines were built and towns grew along the tracks to accommodate the cattle trade.

Missouri and Kansas ranchers tried to stop Texas cattle from coming into their territories in the late 1850s and early 1860s because of transmission of a fever they believed were killing off local cows. Resident citizen farmers who were filling up the formerly empty pasturage through which the cattle had to traverse to reach the cattle towns, and farmers who were also raising cattle to meet the demand for beef, organized into associations and rallied in Abilene in 1871 to enact county herd laws prohibiting large cattle drives from Texas. Many farmers who had planted corn or hay saw their crops trampled by invading herds. Farmer settlers around cattle towns like Abilene reminded people that the profits from the Texas cattle trade went to a minority of non-residents. Moral reformers noted that the creation of social but demoralizing influences, mostly gambling and prostitution, would also be diminished. (31)

But because many towns were built on cattle trade, prospering as the railroads followed when the single industry moved elsewhere, the towns quickly decayed. Abilene's last year as a cattle town was 1871; Wichita's in 1876. Two-thirds of businesses left town or closed. A growing political groundswell against large cattle drives throughout the West crested by 1885 when Kansas, Colorado, and Nebraska enacted tough laws against driving "foreign" cattle across state borders. Soon, the railroad lines were stretching into Texas itself thereby eliminating the need for cattle drives north altogether. Each town's love affair with the cattle trade lasted but a few short years occasioned by settlement on available farmland thus reducing cattle pasturage and ending rapid and expansive town development built on cattle as a single industry.

When the long drives ended the cattle free grass industry also collapsed. (32) The revolutionary technology, as dramatic as the rifle, six-shooter, telegraph, windmill, and locomotive, was simple and ingenious—the barbed wire fence. Wood for fencing was scarce and costly, and rocks for walls such as are found in England, Scotland and Ireland are infrequent in the Great Plains. Wire was relatively inexpensive and quicker to install. According to the National Archives, nine patents for improvements to wire fencing were granted by the U.S. Patent Office to American inventors, beginning with Michael Kelly in November 1868 and ending with Joseph Glidden in November 1874. (33)

The fence, which did not always make for good neighbors, was a way of saying No Trespassing, of emphasizing property boundaries and of keeping cattle from wandering where they might be stolen. Use of barbed wire as barriers or enclosure pens for cattle did not diminish the landscape of the West but did restrict the openness of space, land and water, thus antagonizing cattlemen, homesteaders and nomadic Native Americans who were cut off from access to streams prompting the range wars between various stake holders. Barbed wire is still used to control access and as a prison and military weapon.

The West is as much a state of mind as a landscape. It is filled with its myths, legends and stories, the real and imagined gunslingers and sheriffs, the hardy squatters and sod-busters, the lone rangers and lonely farmers who scratched a living from the resistant soil. But the most heroic figure from the West is the image of the cowboy. Howard Hawks' 1949 classic, academy award winning movie with John Wayne and Montgomery Cliff, *Red River*, is an authentic re-creation of the cattle drive from Texas. Texans on horseback learned to rope, brand and herd cattle and eventually drive them to northern sites for sale. The cowboy was the rough-rider of the plains, the hero with a rope and a revolver, just as the frontier backwoodsman from the Allegheny had been with his axe and rifle, and perhaps fiddle. His was a storybook, nostalgic romance immortalized by Fredrick Remington in drawings and sculptures. E. C. Abbott as a young man was such a cowboy on one of the long cattle drives from Texas. His harrowing account from 1883 narrates the tribulations of the cowboy life on the move, devoid of the sentimentality found in fiction and cinema.

> *One night at sundown, after we had been working cattle in the brush all day, we came to a little open prairie just about big enough to bed down the herd. I tied my night horse to the wagon, took off my chaps and laid down on them, pulled my slicker over me and went to sleep. About nine o'clock a clap of thunder woke me up, and somebody hollered: 'They're running.' I grabbed my hat and jumped for my horse, forgetting to put on my chaps, and I spent half the night chasing the cattle through that thorny brush. When daylight come (sic) and we got them together, we hadn't lost a head. But I was a bloody sight. I had a big hole in my forehead, and my face was all over blood, my hands was cut into pieces—because I'd left my gloves in my chaps pocket—and my knees was the worst of all. I was picking thorns out of them all the way to Kansas...I believe the worst hardship we had on the trail was loss of sleep. (34)*

Others journeyed West seeking freedom from religious persecution. Fringe religious groups, unwelcome in most settled communities, drifted westward in search of their own private religious preserves like the Mormons, or to establish sites of fervid missionary activity, like the missionary Whitmans who settled in Walla Walla and then were murdered by natives for reputedly transmitting diseases to the tribe. The burgeoning cattle industry saw the limitless opportunities as well, flooding the sweeping grass vistas with millions of grazing cattle for settlers of all religious persuasions.

But some cowboys did not drive cattle but herded them alongside wagon trains because settlers journeying to California and elsewhere were also seeking grazing land for their cattle. On September 11, 1857, a party of prosperous pioneers from Arkansas heading to California and the open ranges for their cattle camped in southern Utah after being denied supplies from the Mormons. According to children eyewitnesses and forensic evidence, all 120 men, women and children were massacred, stabbed, clubbed and shot at point-blank range by Mormon militia. (35). All except 17 children, all under seven years of age taken in by local families, were spared. Known as the Mountain Meadows Massacre, the Mormon Church, The Church of Jesus Christ of Latter Day Saints, has adamantly refused hierarchical complicity in the killings, especially any connivance on the part of the church leader, Brigham Young, claiming instead that the murders were the work of local Indians and leaders only. Money, arms and cattle, over a 1,000 head of longhorns, "rich, very rich in cattle," according to Mark Twain, were taken as booty by the local Mormon settlers. (36)

Horace Greeley, who interviewed Brigham Young and spent time in Utah on a journey to San Francisco in 1859, remarks: "That a large party of immigrants...were foully massacred at Mountain Meadows in September, 1857 under the direct inspiration and direction of the Mormon settlers in that vicinity...is established by evidence that cannot be invalidated." (37) John D. Lee, local Mormon militiaman at the time, adopted son of Brigham Young, was tried and executed by a firing squad for orchestrating the Mountain Meadows massacre. (38) But no church authority was questioned, indicted or convicted as Mormon juries would not bring charges.

The organized migrating parties of settlers leaving Independence, Missouri and heading to California and Oregon, traveling in company with about a thousand people and a hundred and twenty wagons, would not have succeeded into their journey had it not been for oxen and cattle. A few individuals had only their teams of oxen to lead them. But others had large cattle herds that had to be driven separately and allowed to graze near the camps each night. Parties this large had to break into smaller groups and those who had more than four or five cows formed a cow column. Jesse Applegate, a commander of an Oregon-bound party, described one such cow column as it left Missouri in 1843.

> It is four o'clock; the sentinels on duty have discharged their rifles—the signal that the hours of sleep are over; and every wagon and tent is pouring forth its night tenants...Sixty men start from the corral, spreading as they make through the vast herd of cattle and horses that form a semicircle around the encampment, the most distant perhaps two miles away.
>
> The herders pass to the extreme verge and carefully examine for trails beyond, to see that none of the animals have strayed or been stolen in the night...and by five o'clock the herders begin to contract the great moving circle, and the well-trained animals move slowly towards the camp...In about an hour, five thousand animals are close up to the encampment, and the teamsters are busy selecting their teams and driving them inside the "corral" to be yoked. The corral is a circle one hundred yards deep, formed with wagons connected strongly with each other; the wagon in the rear being connected with the wagon in front by its

tongue and ox chains. It is a strong barrier that the most vicious ox cannot break.
(39)

The worse dread for a cowboy was a stampede occurring when cattle became spooked by unaccustomed noises, like howling wolves, thunder and lightning or by unusual happenings. One such strange event was St. Elmo's fire, a flame-like phenomenon like tongues of fire occurring during stormy conditions when discharges of electricity could be seen between cattle horns, scary enough to send the cattle thundering away in the night and the cowboys scurrying to contain their runaway fears.

Financing a Boom

Two major technological innovations changed the beef business in the late 19th century: the completion of the trans-continental railroad in 1869 at Promontory Point in Utah, and the development of refrigerator railroad cars in 1875. The steam engine and the telegraph had already been established as improvements in transportation and communication. Prior to the railroad spanning the continent, cattle had to be driven north from Texas directly to the stockyards in Chicago or the mid-west where they were fattened with corn then shipped from Chicago via the Great Lakes waterways to the east coast and England. By boxing the beef contents into saleable parts, distributors could take up less space than by shipping the whole cow by railroad thus reducing costs. But with the introduction of refrigerator cars on railroads, beef could be slaughtered wherever they were fattened and kept from spoilage and decomposition until their destination and served fresh after arrival.

The railroads helped destroy the habitats of the buffalo ruminants, brought the cow industry into existence, and, within two decades as prices fell, helped destroy the cattle industry and the towns created for them: Abilene, Wichita, Caldwell, Ellsworth, Dodge City, Ogallala. In a public relations struggle for the cattle business, a cattle town's preoccupation was to preserve and expand the cattle trade as other towns offered monetary allurements, favorable prices and entertainment.

One form of entertainment was a bullfight. Alonzo B. Webster, a saloonkeeper and mayor of Dodge City, proposed that to celebrate the Fourth of July celebrations in 1884 the town sponsor a bullfight. He found no statutory prohibitions against it. Respected clergy, a few editors and outraged citizens objected but were outvoted. The business community quickly and enthusiastically raised $13,000 to advance the scheme. A Texas drover selected the bulls and five part-time bullfighters were lured from Chihuahua, Mexico and an amphitheater hastily constructed. Big city newspapers sent reporters to cover the unusual story, the first bullfight on U.S. soil. (40)

As the railroads expanded westward, pork, wild game and whiskey were more plentiful and cheaper than beef and the cattle towns lost business and population. Speculation in railroad stocks and bonds made millionaires of British investors.

By the late 1870s cattlemen's associations dominated the plains and prairies ruthlessly through bribes, payoffs, political corruption, intimidation and with organized militia murdering settlers who attempted to settle on or near their bosses' property. In 1892 cattle barons in Wyoming organized a party of 24 six-shooter vigilantes together with 30 other ranch guns, calling themselves regulators, to raid on nesters and settlers in an around Buffalo, Wyoming, accusing them of rustling their cattle. Justice was to be administered in lead, as law was often unavailable and order was scarce. This Johnson County war in 1892 involved several hundred men, and though few were killed, the ferocity of the battle is legendary. The venomous power of the cattle ranchers was, however, eventually broken and law and order restored to the territory by the turn of the century. (41)

Over-grazing on the central plains in the late 19th and early 20th century caused severe ecological damage before wheat and corn became the principal crops. The result was the calamity in the spring of 1934, the greatest dust storm in history when wind kicked up the dust left from over-grazing and agricultural gang-plowing of sod holding only a fine web of grass roots. The "Dust Bowl" had arrived in the middle of the Great Depression. Men and cattle had obliterated the prairie grasses that had nourished millions of bison for thousands of years. The Dust Bowl, the century's worst farming, ecological, economic and social disaster, was named for the billions of tons of dry soil blown away during years of drought in the over-cropped Plains. After the Dust Bowl disaster, much of the Plains returned to soil-protecting plants and grazing. (42)

Agricultural scientists have concluded recently that a moderate level of cattle grazing makes for a more diverse ecosystem—at least on the Great Plains. The Central Plains Experimental Range (CPER) near Nunn, Colorado is managed by the U.S. Dept. of Agriculture. The High Plains station was established in 1937 on ploughed or overgrazed lands abandoned by farmers and ranchers during the Dust Bowl years. In a study comparing rangeland grazed by low to high numbers of cattle, researchers found that plant biodiversity, and ranch profitability, is highest on land grazed moderately, defined as one yearling heifer for every 16 acres for 5 to 6 months annually. Plant biodiversity is highest when high numbers of plant species are combined with a more even distribution of production among species. Moderate grazing offers the best compromise when balancing numbers of plant species and their beef dominance and forage production. (43)

Mavericks and Managers

Some of the more courageous, industrious and public-spirited individuals comprise the history of the West, as describers of the decimation of the buffalo herds, the rise of the cattle industry and the reporting of the ravishing of land. Horace Greeley, Sam Maverick, Gus Swift, Richard King, John Wesley Powell, Zebulon Pike, John Charles Fremont, William Cody, Theodore Roosevelt are a few of the more colorful. Here are their short biographies.

The year 1859 was auspicious: Charles Darwin published *The Origin of the Species*, John Dewey was born in Vermont and the most renown newspaperman in America, **Horace Greeley**, who urged all young men to go west to seek their fortunes, made a trek from New York to San Francisco. Greeley founded *The New Yorker*, a weekly magazine devoted to literature, the arts and sciences, which proved unsuccessful financially, and *The New York Tribune*, a newspaper he edited for 30 years and whose major foreign contributor in London was Karl Marx. Greeley believed in the organization of labor unions, profit sharing for employees, ideals he put into practice with his own business, and was virulently anti-slavery. He served one term as a congressman but was defeated in a senate race and for a nomination to the U.S. Presidency. He formally nominated Abraham Lincoln for President in 1860.

On his western journey across the plains Greeley marveled at the magnitude of the herds. "What strikes the stranger with most amazement is their immense numbers," he wrote of the buffalo herds. "I know a million is a great many, but I am confident we saw that number yesterday...I doubt whether the domesticated horned cattle of the United States equal the numbers." (44) Greeley foresaw the eventual decimation of the vast buffalo herds. "Take away the buffalo and the Plains will be desolate far beyond their present desolation; and I cannot but regard with sadness the inevitable and not distant fate of these noble and harmless brutes, already crowded into a country too narrow for them, and continually hunted, slaughtered, decimated, by the wolf, the Indian, the white man." (45).

About 1870 **Samuel Augustus Maverick**, a San Antonio lawyer who preferred his legal practice to running a cattle business, realized his stock was running wild and decided to sell. The buyer then placed his brand on every unmarked cow he found in the region believing them to be "mavericks," a term which originally meant an unbranded cow now refers to a non-conformist. The fastest branders became the largest stockholders and in the next 15 years over 5 million head were driven north, 700,000 Texas cattle passing through Abilene, Kansas in 1871 alone. (46) A hundred years later Wichita commissioners had a contentious time deciding whether to promote tourism in the town as a cow town or a 21st century city. The grain trade was not an option. A pioneer village was hastily reconstructed for the bicentennial. Old Cowtown Museum, a living history village providing an authentic western experience, revived the city's reputation. (47)

Gustavus Franklin Swift was a Yankee from Cape Cod, apprenticed in the butcher business and learned early in his young life that waste was bad but could be profitable. He observed that people in Massachusetts who lived near slaughterhouses regularly made a living from the refuse of the butchers, their offal, and processed it into soaps, tripe and oils. When he came to Chicago to organize his stockyards and meatpacking plants he would not let one iota of the cow go to waste. To satisfy the hunger of easterners for beef in 1866, he would slaughter the cows in Chicago and ship them dressed to eastern outlets. The Union Stockyards opened in Chicago on Christmas Day, 1865 by P. D. Armour and that day can stand as the beginning of the American cattle empire. Swift and Co. soon followed and established the business end of the beef industry in Chicago just as the long cattle drives, about a quarter million cows in 1866, began arriving from Texas.

Richard King (1824–1885), born of poor Irish immigrants in New York City, made his first fortune as a steamboat entrepreneur, graduating to livestock capitalist by founding the King Ranch in Texas. As a lad he stowed away on the *Desdemona* for Mobile and was taken on as a cabin cub and schooled in the art of navigation, becoming a pilot by age sixteen. In 1842 he enlisted for service in the Seminole War in Florida and again in the army for Mexican War service in 1846.

King stayed on the Mexican border after the war and became a principal partner in a steamboat company, dominating the Rio Grande river trade for about two decades. Soon he began speculating in lots around Brownsville, choosing carefully because of the disputed legality of land transactions, and buying land in the Nueces Strip, 15,500-acre Rincón de Santa Gertrudis grant, in 1853, and in 1854 the 53,000-acre Santa Gertrudis de la Garza grant held title under an 1808 grant from the crown of Spain. These two irregularly shaped pieces of wilderness became the nucleus around which the King Ranch grew. By the time of his death in 1885, King had made over sixty major purchases of land and amassed some 614,000 acres.

During the Civil War, placing steamboat companies under Mexican registry and moving operations to Matamoros, he and his partners contracted with the Confederates to supply European buyers with cotton, and, in exchange, they supplied the Confederates with beef, horses, imported munitions, medical supplies, clothing, and shoes. Union forces captured Brownsville in late 1863 and raided the King Ranch, looting and destroying it. King escaped to Mexico and returned after a pardon by President Andrew Johnson in late 1865.

With their war profits, King and his partners built the most famous cattle industry in the American West, driving more than 100,000 head of livestock along trails to mid-western transportation towns and then to eastern markets. King's capital assets fuelled his ambitious expansion in land and livestock, allowing him to out-compete Anglo and Mexican ranchers. By 1870 he had 84,132 acres of land stocked with 65,000 cattle, 10,000 horses, 7,000 sheep, and 8,000 goats. To manage this industry, he employed 300 Mexican herdsmen and kept 1,000 saddle horses in constant readiness. Yearly he branded 12,000 calves and sold 10,000 cattle investing the proceeds in more stock cattle. (48)

As a strong believer in private property, an idea incompatible with the open range, he mobilized armed forces when necessary to protect his vast interests. He wisely integrated his livestock business with his other businesses, railroads, storage facilities, shipping, and championship show horses, while preserving his sizeable fortune in land and cattle. Before it was broken up in 1935, the King Ranch holdings consisted of over one million acres, larger than the state of Rhode Island. (49)

John Wesley Powell (1834–1902), a scholar, military officer, explorer and government administrator whose travels down the Green and Colorado rivers and through the Grand Canyon was one of the most extraordinary feats of the 19th century, was an astute political figure and visionary scientist of the West. Major Powell lost his right arm in the Battle of Shiloh in 1862, was a Professor at Illinois Wesleyan (1865) and Illinois State (1867) and published an early definitive study: *An Introduction to the Study of Indian Languages* in 1877. Powell did more than anyone, including Jefferson, to bring vision and prophetic understanding to western lands. (50)

After extensive explorations in the 1870s throughout the West he organized a report for the Department of the Interior that became a publishing sensation, uniting both advocates to his cause of reasonable land use, and his multiple enemies composed of special interests in land, timber, cattle, mining, speculation, government bureaucracies and the politicians who represented them. Powell's *Report on the Lands of the Arid Regions of the United States* published in 1878 argued for proper surveying of land prior to settlement, the disuse of rectangular plots of land allotments, and the use of water sources for legal boundaries. Powell wanted the government to withhold giving unsuitable agricultural, desert land away to settlers and to arrange the land differently for farming than for ranching, and not simply by acreage allocation. (51)

Powell was a realist un-persuaded by blandishments of popular journalism and promotional brochures that raved about the availability of land and its richness to easterners unfamiliar with arid regions and accustomed to perpetual wet seasons without droughts. Powell saw that much of the West was desert and would yield unproductive fruits and products unless plentiful water was nearby. He wrote in the *Report*: "Compared with the whole extend of these lands, but a small fraction is immediately available for agriculture; in general, they require drainage or irrigation for their redemption." (52)

Powell was the first to place the West as a whole in an ecological context. In the end, he was politically defeated by feudal cattlemen, vindictive and powerful land speculators, men with money but without his prescience and foresight. He foretold the calamities that would befall settlers who believed in eternal water for farming but ignorant of the unforgiving land they were entering. Settlers who came west often had their deluded optimism punctured with reality. He recommended that cattle be grazed in common and not fenced in around private lots. The following quote, sounding eerily contemporary, is a tribute to his powers of observation and the consequences of over-grazing.

> *The area affected by grazing is far greater than that affected by farming. Cattle...have ranged through all the valleys and upon all the mountains. Over large areas they have destroyed the native grasses...Where once the water from rain was entangled in a mesh of vegetation and restrained from gathering into rills, there is now only an open growth of bushes that offer no obstruction...The treading of many feet at the boggy springs compacts the spongy mould and renders it impervious. The water is no longer able to percolate and runs away in streams. The porous beds of brooklets are in the same way trampled and puddled by the feet of cattle and much water that formerly sank by the way is now carried forward.* (53)

Powell had recommended to the Montana State Constitutional Convention in 1889 a plan so reasonable and intellectually persuasive it was rejected in favor of special interests. He recommended that the new state of Montana be organized into counties divided by hydrographic basins and not just political boundaries. Surveys had by then already plotted water drainages systems and county lines could simply follow them he argued. Counties would then have timber, drainage, agriculture and grazing regions organized around water, a natural self-governing system. Instead political persuasion prevailed. (54)

Powell's founding governmental legacies are the Bureau of Ethnology, for the study of American Indians located in the Smithsonian, and the U.S.

Geological Survey, both surviving agencies of authoritative scientific investigation.

Zebulon Pike (1779–1813) was a career military man of limited education whose early assessment of western desert lands as a hindrance to settlement and reckless expansion was based on his exploratory trip through Colorado and New Mexico. Nevertheless, Congress fifty years later wanted settlement plots for small farmers but decided in favor of the unfounded romantic belief that western lands had a similar topography and climate as eastern lands. Even faced with Powell's compelling manifesto in 1879, Congress still adamantly refused to reverse its misguided legislation and continued to hand over 160 acres to any prospective farmer willingly to cultivate resistant and unyielding soil. By 1900, two-thirds of those who tried farming under *The Homestead Act*, including my great, grandfather, had failed.

Until the latter half of the 20th century when it entered scientific research on its own and through funding agencies, the federal government had little interest at all in scientific research. There is still wrangling by those who advocate limited government. Yet today the complexity and variety of governmental agencies associated with the West reads like a Who's Who of government itself: the U.S. Geological Survey, Coast and Geodetic Survey, Bureau of Land Management, U.S. Army Corps of Engineers, Bureau of Reclamation, Bureau of Indian Affairs, and Soil Conservation Service among others.

John Charles Fremont (1813–1890), endearingly known as The Pathfinder late in his career, after graduation from Charleston College, became an assistant on a survey mission between the upper Mississippi and Missouri rivers and later commanded an expedition on the Des Moines River. With Kit Carson (1809-68) as guide, he explored the Rocky Mountains, Oregon and Nevada Territories and entered California, at the time a part of Mexico. In 1850 he was elected California's first Senator and Republicans nominated him as a presidential candidate in 1856. Fremont is the first scientific explorer in the history of the West, carrying sextants, barometers and thermometers and using Latin terms for botanical specimens he found. Accounts of his journeys furnish rare insights into the disappearance of the buffalo. A remarkable entry from August 30, 1842 reveals his keen powers of perception.

The extraordinary rapidity with which the buffalo is disappearing from our territories will not appear surprising when we remember the great scale on which their destruction is yearly carried on. With inconsiderable exceptions, the business of the American trading posts is carried on in their skins...the Indians derive their entire support from them, and slaughter them with a thoughtless and abominable extravagance...It will be interesting to throw a glance backward through the last twenty years and give some account of their former distribution through the country and the limit of their western range. (55)

A biography of Fremont in 1850 describes how he found the Indians in northern California dying from lack of food as the miners had driven them from their traditional lands in the San Joaquin valley during the gold rush. Commissioners appointed by the President came to California to survey the condition and recommended they be given beef so they could transfer to new lands not yielding any gold. Col. Fremont won such a government contract and, as he was then in the cattle business, personally herded cattle from southern California, hired drovers and superintended the drive. The cattle were herded about 300 miles in a dry season and about 400 died, but he delivered 1.2 million pounds of beef on the hoof in 1854. It was widely recognized that he helped prevent Indian wars. (56) He was appointed Governor of Arizona Territory from 1878 to 1883, and in 1890 belatedly received a government pension but did not live out the year to enjoy his retirement.

About the same time Powell was writing his report, **William Frederick "Buffalo Bill" Cody** (1845–1917) was going into partnership with one of his old friends from the Army, Major Frank North in 1877. The two bought land on the Dismal River about 65 miles north of North Platte, Nebraska, his home until he died, and purchased cattle in Ogalalla, then the most western and northern headquarters of the Texas cattle drovers. After a long summer season of roundup with his companions and hires, Cody purchased another herd of cattle. (57)

Cody killed more buffalo, many for food for railroad workers and others for sport, in his lifetime than anybody in his generation or since. His exploits retold in his autobiography are tales of daredevil adventures, near-death experiences, comical incidents and bold escapes. His name was popularly associated with frontier heroism, trick horsemanship, and buffalo and Indian killing. But he remains the epitome of the outdoorsman and life on the frontier.

Theodore Roosevelt (1858–1919), the 26th president of the U.S., had two cattle ranches in 1883 in Dakota territory, Chimney Butte and Elkhorn, where he lived in his mid-twenties. There he was a cowboy and rancher before returning to New York and politics in 1886. His ranches are today three sections of national memorial parks in west North Dakota near the Montana border and the Badlands. (58)

His earthy descriptions of western experiences in his autobiography in the chapter *In Cowboy Land* read like a picaresque novel filled with adventures with characters who later served his administration as marshals, postmasters, forest rangers and soldiers in the Spanish American war. "No guests were ever more welcome at the White House than these old friends of the cattle ranches and the cow camps—the men with whom I had ridden the long circle and eaten at the tail-board of a chuck-wagon—whenever they turned up at Washington during my Presidency." (59) He relates shooting distraught bears, buffalo hunts, fist fights with drunken cowboys, nights spent with wanted outlaws and train robbers, encounters with hostile Indians, and the dangers of the cattle roundup. Other presidents have been farmers and ranchers but none except Teddy Roosevelt was a true cowboy and full-time cattle rancher.

His legislative proposals and proclamations saved large swaths of the West from despoliation, enlarged the parks and monuments, salvaged much of the land from expropriation, and directed his energies to defending public lands against fraud and theft. *The Reclamation Act of 1902* authorized the Secretary of the Interior to develop irrigation and hydropower projects in 17 Western States, reclaiming arid wastelands for farmers to raise crops and graze cattle. (60) The year 2002 marked the centennial of the Bureau of Reclamation. Here are a few of some of his more notable presidential accomplishments in favor of public land, individual farmers and of the preservation of buffalo herds.

1907   Roosevelt created Yellowstone National Park and a federally protected buffalo herd in it.

1908   Upton Sinclair's book *The Jungle* published; Congress passes *The Federal Meat Inspection Act* and *The Pure Foods and Drug Act.*

1909   Roosevelt gets the Forest Service to collect leasing fees for sheep and cattlemen for grazing livestock on federal land.

1910   He increased the national forests by 43 million acres.

1911   He created the National Bison Range in Montana with 18,000 acres from the Flathead Indian reservation.

Western land has vastly changed since the first settlers, adventurers, speculators and ranchers altered its appearance, importing foreign grasses, diverting creeks and lakes into irrigation canals, pumping water from artesian wells and oil from derricks, extracting the natural wealth and resources irrespective of ecological constraints. Re-engineering the landscape to maintain its productivity without disastrous social consequences will demand the finest political skills and determination in the future.

The cattle barons, adventurers, speculators and buffalo hunters in the 19th century did not arrive as caregivers of the land but as usurpers. They came as prospectors for gold, miners of minerals, exploiters of land for livestock and real estate speculators. And the land was abundant: 2.3 billion acres or 3.6 million square miles. (61) The speculators and settlers who came were often ignorant of land use and would plant where they should have grazed, dry farmed where irrigation was necessary, or sod-busted where grazing made sense.

Wealthy cattle ranchers, early purveyors of animal trade and plunderers of the plains, did not have a wholesome respect for the forces of nature and its influence on the life of the plain. One who marveled at the excess of cattle and expressed outrage at the damage done to bison was paradoxically an English hunter named Vivian, a Member of Parliament, who journeyed west to shoot wild game and write about his experiences. He wrote in 1879: "The head of cattle possessed by individuals is enormous, sometimes over 10,000 bear the brand of the same stock-owner! As may be supposed, this cattle trade is a great source of traffic to the railroad company who, I am told, convey over this line more than 400,000 head of cattle a year! These vast, monotonous, bare-looking prairies are already the great beef-producing regions of the United States." (62)

The American West was not really settled as much as it was raided. When the resources dried up, the buffalo, the gold, silver, the grasses, the water, the raiders moved on leaving the evidence of their destruction on a fragile frontier like a crumpled calling card.

CHAPTER EIGHT
COLONIZERS AND CATTLE

I'm going out to fetch the little calf
That's standing by the mother. It's so young
It totters when she licks it with her tongue.
*I shan't be gone long. You come too. (*The Pasture, *by Robert Frost)*

The small but maneuverable caravel with its broad bows and high narrow poop deck, the kind appearing in most pirate movies, was the common ship of Spain as it maneuvered in Caribbean and Gulf waters to stretch its foothold in New Spain. In 1521, two years after Hernando Cortes conquered the Aztec capitol near the eastern coast of Mexico, its bulwarks gleaming with polished brass, such a caravel landed with its usual occupants of soldiers, merchants, priests and adventurers. But rocking in their stanchions and bracing against the deck roll, was a small group of calves from Santo Domingo, a Spanish breed, the first to establish a cloven toehold in the New World, the forbearers of the Texas longhorns. Gregorio de Vallalobos, the industrious shipper of this small band of calves, foresaw a cattle industry where his fellow Spanish saw only the glitter of Aztec gold. (1)

Cattle grazed in the New World wherever immigrants and colonizers landed and settled. The first breeding stock of cattle loaded from Cuba landed with the Spanish military expedition, which routed the French from Florida's east coast in 1565 at St. Augustine, the first permanent settlement of Europeans on U.S. Territory. A second Spanish settlement at Tallahassee was more successful but eventually, because of Indian attacks and British expeditionary incursions, and until Florida became a British territory in 1763, the cattle disappeared as suddenly as the Spanish.

Across the continent, Juan de Onate brought Spanish cattle with him to northern New Mexico in 1598 and the Native Americans learned cattle husbandry with their catechism lessons from the missionaries in Santa Fe. Spanish criollo cattle became well-stocked in south western Louisiana after New Orleans was settled in 1718 and the city and area depended on the locally grown beef from southern rangelands. (2) Spanish cattle followed settlers to San Diego in 1769, the first cattle to reach California, where missions with large cattle herds and haciendas expanded. Besides the Spanish place names, Colorado, Arizona, Los Angeles, San Francisco, Las Vegas, and words for the flora of the southwest, mesquite, chaparral, pinon, cattle or ranching words with Spanish origin define the vocabulary of the West as much as do corral, bronco, frijoles, lasso, rodeo, pinto, lariat, ranch, rangers.

The first Spanish to cross into Texas was Cabeza de Vaca, literally "Head of Cow." Cortes himself, before he retired to his hacienda in Cuba, named it Cuernavaca, or Cow's Horn, a city just south of Mexico City. By about 1583 herds of Spanish cattle were grazing north of the Rio Grande, then a part of New Spain. Shakespeare had not yet written his greatest dramas. It would be a quarter of a century before the English would establish the first colony in Jamestown in 1607. Cattle roamed all along the southern plains range undeterred from theft by nearby Indians who did not yet possess the horses of the Spanish. Evolved into a fierce animal, resistant to the heat and dry plains, later bred with English shorthorns, these early Spanish longhorns came to overrun the American West, but not before the Native Americans were reduced to starvation when rugged American bison were wantonly slaughtered.

Together with horses and merino sheep, the Spaniards introduced longhorns from Andalusia in the south of Spain into the West Indies and the coast of Mexico. (3) The Portuguese brought cattle to Nova Scotia about the same time, long before the Puritans landed at Plymouth Rock. Andalusian stock, accustomed to the dry climate and parched land of southern Spain, easily adapted to the dry environment of Texas and the American southwest and flourished where grasses prevailed.

One of those grasses they found was astragalus, from the Greek word for ankle-bone, the shape of its seed, also known as locoweed, characterized by delicate pink, purple and yellow flowers on stout stems proliferating in arid and semi-arid climes and woodlands. There are about 60 species and dozens of sub-species of this threatened plant stretching from Tennessee to California. Astragalus is beneficial to cattle because it absorbs toxic substances like selenium from the soil which can give cows ailments. Chinese herbalists use it to tone blood as root extracts help increase T-lymphocyte white blood cells for immunities. (4)

Kino and Kine

Father Eusebio Francisco Kino, S. J. (1645–1711), missionary priest, explorer, colonizer, historian and first rancher of the American Southwest, was born in Segno, in the Tyrolean Alps of northern Italy. (5) He was admitted to the Jesuit order in 1645, and after the extended study of a Jesuit education in German universities (Freiburg and Ingolstadt) in which he specialized in mathematics, was ordained a priest in 1677. After a short stint as mathematics instructor and the publication of a book on comets in Spain, he came to Mexico in 1681 as a missionary in New Spain and was appointed royal cosmographer to accompany the expedition to colonize Lower California.

He journeyed to Northern Sonora and Southern Arizona, then known as Pimería Alta, in 1687 bringing with him his unique skills as map-maker, mathematician, and astronomer and the specialized tools of astrolabe, telescope and sextant. From 1687 until his death in 1711, he made more than 40 expeditions inland, exploring southern Arizona, traveling from his first mission in Sonora along the Rio Grande, Colorado and Gila Rivers eventually establishing 27 missions more than half a century prior to the first California missions. (6)

From his base in Sonora he established agriculture at the missions he founded and brought in cattle, horses, and sheep and gave cattle and seed grain to the Native Americans, stocking all *rancherias* with hundreds of cattle. The cattle and wheat, together with the novelty of a new religion, had a strong appeal among the natives. Kino's stock raising industry existed at each of 19 documented missions and the economic independence and prosperity of these sites supplied food for the indigenous natives. By contrast, Father Junipero Serra founded nine missions in California bringing 400 head of cattle to San Diego on his first visit.

Since Kino was the first person to chart the American Southwest in 1702, his detailed map based on direct observations not errant speculation became the only authentic reliable map of the region for the next two hundred years. He claimed that as evangelist and missionary he instructed 30,000 Yuma and Pima Indians and solemnized over 4,000 baptisms. He experimented with science but not on human activity and did not catalogue anything about the life and customs, the meaning of native dances and songs, adornments, weapon-making skills, object venerations, or medicine men of these native peoples referring to them in his correspondence only as "souls." (7)

He wrote to his Father General Superior on October 18, 1701: "I have provided for this purpose a new ranch with more than two thousand head of cattle."(8) Though he did not personally own any of the cattle, he was America's first large cattle rancher, not only introducing cattle and animal husbandry to the Indians, but demonstrating feats of horsemanship to admirers. He trained Indians in life-sustaining skills like carpentry, blacksmithing and baking, and introduced the native farmers to European fruits: apricots, citrus, figs, peaches, pears, quinces, pomegranates, and, most importantly, wheat to supplement the local beans, corn and squash. His missions were self-sustaining enterprises.

Kino, a contemporary of Marquette and La Salle, does not fit the typical profile of an early American explorer, colonizer and stock raiser. He was Italian, not Spanish; a black-robed Jesuit, not a brown-mantled Franciscan; a scholar not a soldier; a settled agricultural livestock man and not an exploiter of the land or its resources. He came to the region to bring faith and practical living skills to the natives. By stocking the missions with cattle and other livestock he made them self-sufficient environments, a significant accomplishment compared to the history of later American settlements. He died poor as he had lived. His deathbed was as his bed had been: two calfskins for a mattress, two Indian blankets for covers and his saddle for a pillow. On February 14th 1965, in the "National Hall of Statuary" a statue of Father Kino, was dedicated in the Capitol Building in Washington DC.

The Spanish brought their cattle to the southeast and the southwest via Mexico. The English brought their breeds on ships with them in 1620 and thereafter launched a cottage industry of dairy farms and meat processing to feed resident Puritans and immigrant arrivals.

Colonial Cattle

Arriving with the Puritans was an imported educational device made from cattle residuals, an instructional aid in colonial education was known as a Hornbook made of wood about the size of a paddle and inscribed on a sheet attached with letters of the alphabet, some numbers, syllables and passages from scripture. This sheet was covered with a thin, transparent plastic-like substance made from pounded cattle horns, hence the name. Hornbooks were common as instructional aids in the Middle Ages in Europe and followed schoolmasters to America. (9) Cows provided the means for children to become educated. Though no longer made of cattle horns, hornbooks kept the familiarity of the name as a textbook and were in use into the 20th century. As late as 1949 there existed *A Hornbook of Virginia History* issued by the state enumerating all the important facts of Virginia history.

But like the Spanish, the English also imported breeds of cattle, over time interbred from among Danish, Dutch and other English stock. By 1620 the Jamestown Virginia colony boasted of 500 head of cattle, and 2,000–5,000 by 1627. Shortly after the establishment of the charter for the Massachusetts Bay Colony in 1620, subsequent emigration from England to the new colony in Massachusetts brought six small sailing ships with 406 people, 300 men, 80 women and 26 children. As a necessary accompaniment to developing a productive new community, 400 head of cattle were also shipped to the New World. Coming from Devonshire and adjacent counties in England, cattle were introduced in 1624 with the Puritans after the first settlers had established a working community. (10) According to one historical record, 100 head of oxen were driven into Boston in 1628 and sold for 25 pounds per head. By 1635 Trelawney's Plantation near present day Portland, Maine listed 58 cattle with other assorted livestock. (11) By 1634 there were 20 villages in Massachusetts

Bay Colony with about 4,000 inhabitants and 1,500 head of neat cattle. The Dutch brought Holland cattle to the New World and to New Amsterdam. By 1694, over 4,000 cattle were being slaughtered annually below what is now Wall Street on the Brooklyn shore.

As immigrants like Thomas Hooker moved into the interior they drove cattle with them. Thomas Hooker, born in Leicester, England, in 1586, became a fellow of Emmanuel College, Cambridge and for three years worked as a Puritan lecturer in Chelmsford in East London. Hooker immigrated in 1633 to Massachusetts Bay and became a minister in Newtown, later Cambridge. In 1636 he moved with his congregation to Connecticut, leading 106 cattle before him and founded the town of Hartford and the new state of Connecticut.

By the middle of the 17th century men were hired to graze and superintend a common herd held by townsfolk. According to a 1661 record for Ipswich, Massachusetts, "Daniel Bosworth is to keep the herd of cows on the north side of the river from the 1st of May until the 20th of October. He is to go out with them half an hour after sunrise and bring them home a little before sunset." (12). The colonists were following an ancient English tradition from the 5th century of pasturing village cattle under one herd supervised by one person. (13)

Not content with simply introducing cattle as food for the immigrants, thus carrying on a cultural tradition, the colonists also had to import the best food for their cattle. Because the colonies possessed no good indigenous forage for beef cattle, English bluegrass and white clover were imported by 1663.

By the early 18th century, farmers in western sections of the colonial states had an abundance of cattle but slim local markets so they walked their cattle to more populated areas along the coastal towns where meat was scarce. As farmers spread into western Massachusetts and beyond, and with the distances increasing to get cattle to abattoirs, farmers "drove" cattle to market, and hence the new term arose for getting cows to towns for slaughtering, the designation extended later for western cowboys who became "drovers."

Getting cattle to distant markets meant inaugurating trails to drive them there. By the early 1700s the Bay State Cow Path had been established running from the Connecticut river valley, and even down from Vermont and New Hampshire, through Worcester to Boston. A similar cow path known as Three Mountain Trail ran cattle from the Ohio River valley from around Pittsburgh to Harrisburg in eastern Pennsylvania. A third cow path called the Wilderness Road led from Louisville in Kentucky through the Cumberland Gap across Virginia to Richmond. (14) By 1860 when the railroad pushed through to Pittsburgh, the cattle trails in the east became history.

The single most important livestock throughout the colonies were neat cattle, including beef cattle, working oxen and dairy cows. From 1820 to 1840 British breeds were shown at annual animal exhibitions, state fairs and cattle markets, like the Brighton Cattle Market outside of Boston. They included North Devons, Alderneys, Guernseys, Durhams and Shorthorns. Auctions of stocks were social events and the British breeds then would decades later figure in the crossbreeding with longhorns from Texas to create a more appetizing marbled beef. From 1840 to 1860 cattle in the Scioto Valley in Ohio, cattle were fattened on cheap corn and Kentucky bluegrass and driven to eastern markets. (15)

Less affluent settlers in the early 18th century, like those unable to afford the plantation land laced with tobacco in Tidewater Virginia, migrated westward to the Piedmont and in the valleys west of the Blue Ridge mountains driving cattle with them. Many of these poor pilgrims drifted into the Carolinas where rice and indigo planters had already established plantations along the coast and found large bands of stray Spanish cattle roaming the interior woodlands. Through judicious interbreeding with domestic herds, they became cattle-rich in lands ideally suited for pasturage and soon acquired enough herds to supply the meat cravings of Charleston, Norfolk, Baltimore and Philadelphia. (16) Cattle survived and increased because these cattlemen, who averaged about 100 acres of land, let half the cattle wander in the nearby woods while the other half the herd were kept in cow pens, corrals, stock pens or enclosures. By spring when all cows had calved, farmers rounded up the wild, foraging cattle and placed them in enclosures and branded the calves. These cattle owners and managers, even then known as cowboys, lived in cabins adjoining the cow pens, sometimes tending patches of corn, but subsisting on butchered older cows and lean calves likely to die. In the fall, they drove the fat steers to markets in the large coastal cities.

The upper coastal plains of the Piedmont region of Virginia and North and South Carolina provided suitable grasses, climate and range forage for cattle and their numbers quickly achieved impressive proportions. In the northeast, cattle were supplemental food and income to subsistence farming. But in the inland region of the mid-Atlantic coast raising cattle became, next to tobacco farming, the largest industry in the 18th century.

Meanwhile the trails were expanding in the West. The transportation firm of Russell, Majors & Waddell operating out of Kansas and westward to Salt Lake City and California under government contract in the 1850s, and for whom William "Buffalo Bill" Cody worked when a young teenager, operated 6,250 wagons using 75,000 oxen and employed 8,000 men, with a capital investment of over $2 million. (17) The trail followed the rivers of the Big Blue in Kansas, the South Platte, the North Platte in Nebraska, and the Laramie, Sweet Water and Green rivers in Wyoming. On his first trip in 1856, Cody recounts in his autobiography how a buffalo stampede crashed through the wagon train, got entangled in the gearing, frightened the drivers, and scared off some oxen that broke and ran dragging the loaded wagons behind them.

Men like Kino and Cody were adventurers even if they had other occupations. But whether the American arrivals were priests (Kino) or postal riders (Cody), Governors (John Winthrop) or tenders, cattle became the indispensable commodities that nourished communities.

Having soaked up some of the early history of the continent about cattle, I next wanted to see what other individuals had felt and believed about cattle in contemporary times. For this study I interviewed selected representatives of the cattle business to collect a personal perspective of the cattle business.

Cattle People

We don't raise cattle for their looks, their speed or agility. They're big, often ungainly, and can't run to escape predators. We breed them for food and drink and eradicated those traits that, in earlier generations, would have allowed them to survive on their own like elk cows. Cows yield the leather belts that keep a man's pants from falling down.

Here are the stories of men I interviewed: Earl Petznick, feedlot owner, Ron Hunsaker, small cattle rancher, Gene Courtney, small slaughterhouse owner, and Bill O'Brien, cattle rancher extraordinaire. Long conversations with these cattlemen who had spent their careers in the cattle business, all now in their late 60s or older, gave me new insights into raising cattle and meat processing.

Earl Petznick is about 6' 3" with a shock of wavy white hair and has been in the cattle business all his life. Born the son of a preacher in 1939 south of Chicago, he moved with the family to Phoenix in 1948 and immediately went to work for Olen Dryer, his future father-in-law. He has a no-nonsense business manner and is forthright in his conservative political views not shared by other family members. I liked his easy manner and direct responses to questions when I interviewed him one warm spring day at the Biltmore Hotel in Phoenix near where he lives. The cattle business was his only job.

Olen Dryer, born in 1915, was a fourth grade dropout who owned Pinal Feeding, a cattle feedlot. Like so many others of his generation in the fast food and franchising business who never finished, or in some cases never began, high school, men like Ray Kroc of McDonalds, Dave Thomas of Wendys, J. R. Simplot, the Spud King, Frederick DeLuca of Subway Sandwiches, or Harlan Sanders of Kentucky Fried Chicken, Dryer substituted hard work and perseverance to achieve financial success in the feedlot cattle business. (19) "He was the best cattle trader in the business," Earl recalled. But as a strapping young man, Earl put up fences, branded Holstein cows, loaded feed, and helped in the office. He learned and relished the business top to bottom. When Dryer died, Earl, with his wife and sister-in-law assumed control of Pinal Feeding in 1978. In 1978 there was 35 to 40 feedlots in Arizona. By 2001, there were only seven. By 2001, Pinal Feeding had a capacity of 120,000 cattle. "Why are Holsteins prized more than others," I asked. "Holsteins are available," Earl replied, "predictable in behavior, have little outside fat, are well-marbled and taste good. This combination is not easily duplicated by other breeds."

The cattle industry, Earl noted, is not vertically integrated. In the course of its short life of about a year, a calf born to a dairy farmer will travel to a calf raiser, then to a feedlot rancher, and finally to a meat packing house. Dairy farmers sell calves from the herd to Calf Raisers who keep them for two to three months, or until each calf weighs about 300 pounds. The Calf Raisers sell yearlings in the fall to feedlots that feeds them for about a year, each calf gaining nearly three pounds daily. Feed costs about $140 a ton and is a mixture of alfalfa, hay, corn, tallow and nutritional supplements. Packinghouses slaughter the cow, cut it into sub-primal sections, box and freeze the contents for shipping to fast food franchises and supermarkets.

Earl lived and breathed cows. I ordered a chicken salad for this luncheon interview. Earl had a hamburger.

Roy Hunsaker, 70, by his own admission is a hobby cattle rancher, a man who raises 12 head of cattle on 30 acres in Moxee, Washington. Like small farmers everywhere, he is a dying breed apart, according to him yielding profits to conglomerate interests. "I never had to make a living from raising cattle, just lucky to make expenses," he told me one summer afternoon when I interviewed him on his farm.

Roy's' grandfather had shorthorn dairy cows on a farm in Ava, Missouri during the height of the economic depression in the 1930s. Roy inherited his grandfather's love of the farm and soil and always wanted to spend a part of his life around cattle. Roy came west to Wenatchee, Washington for a short time, returned to Missouri to finish high school and then moved west permanently. At first he worked odd jobs in the fruit orchards, as a building contractor and painter. He didn't buy his first cow until 1970. Unfortunately his purchase corresponded with the beginning of an economic recession. He bought cows for 55 cents a pound and had to sell at 25 cents a pound, had to refinance his house and property, and climb out of a $10,000 debt before returning to parity.

Roy sells calves at the Toppenish Livestock Commission auctions and receives on average 75-95 cents a pound on the hoof. His heifers, which will sell about 10 cents a pound less than steers, weigh in at about 600 pounds or better; his best weighed 905 pounds. His cows weigh 1400–1500 pounds so the calves are larger and more succulent for veal.

Roy's water expenses average $3,600 a year for the farm and that adds an additional cost to his feed. Without his other business interests and income he could never earn enough to survive. When he does decide any time now to retire permanently he will sell the property and live off the proceeds. His three children who enjoy farming will have to have the kind of passion for cattle that Roy has if small ranchers maintain any presence in the future.

Roy is an example of the fast-disappearing cattle raiser who loves farm work, and some working cattle, but doesn't realize a large enough profit margin to stay in business without other income.

Gene Courtney, 71, started summer work at H & H Meat Packing in Yakima, Washington in 1952, only retiring in 2002, a half century in the meat packing industry. By the time he retired Gene knew meat packing like few ever had with experiences as janitor, delivery man, salesman to owner of cattle and meat packing plants.

He began as a part of a clean-up crew, sweeping floors and hosing down equipment after a day's slaughtering in a meat packing plant. He followed his grandfather and uncle in the livestock business. Both had two of the 20 sheep ranches in central Washington State about 1930. They shipped sheep by rail to St. Paul and Chicago. "Lamb production is almost dead in the US today," Gene noted.

When he started work full-time in a meat packing plant in 1957 after a stint at Washington State University he was delivering hind and front quarters of beef directly to butchers and retail outlets. Swift bought H & H Packing in 1960 among a half dozen other meat plants in Washington and Oregon, and immediately began upgrading the processing of cattle coming from feedlots, installing coolers, rubbing down the meat to soften it, preparing it for USDA grading and placing the meat directly in boxes for distribution. The most advanced technology, according to Gene, was the hide puller installed by Swift in 1960. Previously, and even later in smaller meat packing plants, the felled cow was cradled, lying flat while its hide was stripped. But the hide-puller could strip a cow's hide while it was hanging in about a minute and then quickly move on down the line saving valuable processing time.

Gene then became a salesman for Swift selling to supermarket chains like Thriftway, Associated Stores and Safeway just as food distribution sites were consolidating in the early 1960s and the corner grocery store was disappearing.

Some of Gene's more interesting episodes were with rabbis who had to insure that meat for Jewish customers was kosher. Rabbis individually killed the cow, bled it, and blessed the whole operation including the delivery trucks, for the sale of kosher meat. Rabbis brought their own sword to slay the cow because it has to thoroughly bleed prior to quartering, and blessed the meat after the USDA inspector finished his job.

By 1977 Swift abandoned its Yakima meat venture and Gene left Swift to become a partner in Grandview Meat Packing in the lower Yakima valley. His Grandview plant processed about 500 head a week, half what the Swift plant was producing, but assuming many of the former Swift clients.

One night a truck unloaded live cattle at the plant. Someone neglected to close a gate at the far end of the corral and 40 head of cattle took advantage of their freedom to roam Grandview city streets in the early morning hours. By dawn nearby ranchers on horseback, adventurers on motorcycles and drivers of recreational vehicles were chasing cattle. Soon, the cattle were bunched and corralled in their pens. Eight to ten bulls once similarly escaped on a hot mid-summer Sunday and rampaged through the city, not bunching together as cows do, and ran so hard and fast to avoid capture that two died of heat prostration. One cowboy also fell exhausted near a bull and as others approached and cried out, "I'm a born-again Christian," and invited those gathered to place their hands on the bull to revive it. The blessing was unsuccessful and the bull succumbed.

The Grandview plant was old, needed improvements which were not made, and Gene closed it in 1986. He became a salesman again for Washington Beef relocated in the old H & H plant from 1986-91 and bought a Toppenish feed lot buying and selling his own cattle. A Japanese beef craze began in the late 1980s and a Japanese firm bought Washington Beef in 1990, shipping beef to Japanese consumers, eventually selling the plant in 2002.

Like other livestock and most commodities, Gene observed, beef prices go up and down quickly and are not reflected in supermarket prices as stores hold prices stable to retain customers. And to keep their customers, meat packers often sell at a loss and on occasion less than what the meat is worth. Meat processing prices are affected by drought conditions, when water for feed becomes more costly, when shipping and transportation costs rise because of increased gas prices, or when there is a heightened social consciousness about high fat content in beef and other nutritional dangers such as depressed the meat industry in the mid-1980s. Gene thinks that from a high of about 135 pounds of annual per person meat consumption in 1980 that the average amount had dropped to about 100 pounds per person within a decade.

Beef regained some of its dramatic losses, increases partially offset by a rise in population and not just meat popularity. The increase in chicken and pork consumption challenged the supremacy of beef as a protein source just when consumers were becoming more aware of nutritional values and potential threats of high fat intake. "Unless you're big in the meat business these days," Gene concludes, "you won't survive."

William Howard "Bill" O'Brien was born Irish to the core in 1923 in Hollywood, California. Today, a young man of 82, small in size but huge in stature, he leases 420,000 acres, over 700 square miles, of cattle range land a few miles west of Phoenix. His journey has encompassed a lifetime with ruminants. I spent nearly five hours over lunch one afternoon (we had to leave the restaurant so the crew could set the place for supper) letting him instruct me in the details of his life spent in the animal husbandry business: trapping, horsemanship, wool buying and cattle ranching.

He left high school at the height of the Depression, hitched a train going east, and with his bedroll on his shoulder worked as a cowboy and rode bulls at a rodeo for a bunk berth and change. Chore boy work was tiresome, "the loss of my dreams," he said, and Bill quit to run trap lines for a living to catch coyote, fox, badger, and skunks for their hides. He boxed the skinned and dried hides and shipped them to a trapping firm that paid him $2 a hide, $4 for a bobcat.

With his buddy Chuck McKeen's encouragement, he entered the University of Arizona to study agriculture. Because of his horsemanship ability he was chosen to enter the U.S. Army's Horse Cavalry reserve while a student in Tucson and played polo to maintain his skills.

When war was declared on Japan in 1941, his friend Sam Sloan convinced him to join the Navy as there was better food and shore leave. But because he was already in the Army cavalry he had to seek an honorable discharge, which, with his theatrical skills he convinced an officer to do, and then joined the navy where his first job in Lafayette, Louisiana was cleaning and cutting beef, sheep and hogs for navy mess. He was elated because he didn't have to wear the uniform and could sport jeans and a tee shirt and nobody knew he was military. Within weeks he was shipped to Coronado near San Diego where he learned underwater demolition. He put to sea as an Ensign on the U.S. Meriwether, an attack transport carrying 3000 marines into battle.

At the end of the war he enlisted for another year and a half and was sent to Aruba, an island off the Venezuela coast as commander of a staff of twelve. The post-war army occupied the island because the Germans during the war had shelled the oil refineries there and the military needed to protect its oil supply.

Bill returned to Tucson on the GI Bill to complete his degree, writing his senior paper on the wool trade and graduating in agricultural economics. While visiting a sister in Boston he encountered Paul Draper who owned a wool market and offered Bill a job. He met his wife Sadie there and they were married in 1948.

He was sent West to buy wool and learn the business, like preferred shearing techniques and commodity prices, but then within months was sent to South Africa buying wool and shipping it to Boston, often hiring a tramp freighter like the *African Rainbow* to haul an entire shipment. Within a year he was assigned a similar purchasing position in South America living in Peru from 1948 to 1959 but traveling throughout the continent and selling llama, alpaca and sheep wool to firms in Italy, France, Germany and Canada as well as his home company in Boston. During the interim he bought a farm west of Boston in Millis, Massachusetts.

Tiring of the wool business, in 1960 he returned again to Arizona, bought and sold houses, trading land for land, ran for the state legislature but lost the election, and purchased 200 acres of his eventual enormous cattle ranch west of Phoenix.

Bill's knowledge of breeding practices is legendary and I could tell he was anxious to share his knowledge with me. Select breeding is critical because in the dry southwest, unlike the lush grass plains of Montana for example, water is scarcer and cows will not graze too far from a water source. Cattlemen in the southwest have to dig wells about every eight miles to satisfy herds. So if you can breed a cow that has the water endurance of a camel you enjoy a preferred cow. Bill fed his cows juices from the locally available jojoba bush with edible seeds. When crushed they yield a rich liquid containing about 26% protein.

The cross between a Brahma, disease resistant, and a Hereford, good marbled meat, is an F1 (first cross) and the calf is stronger, healthier and smarter than either parent. A cross between an F1 and either a Brahma or a Hereford is an F2. By the time you reach an F4 you have exhausted the good genes and the resulting progeny are dumber, wander off, trail at the end of the herd, and usually die if bitten by a rattlesnake, which is not true of F1s. A summer intern could not learn as much as the accumulated experience of a man like Bill who has spent his life with animals like horses, sheep and cattle.

We will never know the faceless men who first domesticated cattle but we can get close to the men who breed and nurture them, ensuring a continual supply of protein so humans can continue to have a reliable protein source. I learned from these interviews that the cattle business requires both entrepreneurial and managerial skills, but that the men who have spent their lives in the business are also tenders of livestock, like shepherds or animal lovers of a kind pet owners and vegetarians might not understand, yet tied to the ancient and honorable trade, like hunters, of raising food for human sustenance. Their personal stories would guide me towards a more humane view of the facts and figures of the meat processing industry, as we shall see in the following two chapters.

Figure 4: Heifer Fight In Santa Marie del le Mere, southern France

CHAPTER NINE
MEAT AND MILK

Restaurant patrons are never given an invited tour of the kitchen prior to dining. Some may find the experience unappetizing and forget their hunger regardless of the ambiance, floral arrangements, candles or quality of service. This chapter explores the equivalent of a symbolic tour of the sometimes odious side of the meal beginning with the slaughter of the animal. A few conglomerates dominated the beef industry in the 19th century, as they did in the 20^{th,} and recent revelations have uncovered disquieting flaws in the operations of the industry, some of which have been corrected but all requiring continuous and stern vigilance and not just voluntary compliance in food safety requirements.

Among the most consumed beef products is the hamburger. German immigrants introduced ground steak in the Hamburg style, thereafter known as hamburgers and were on the menu at Delmonico's in New York City as early as 1833 for 10 cents a plate. Today, hamburgers appear to be the favorite lunch and often dinner selection. (1) How that delicious and succulent patty gets concocted, however, is another story. First, you have to kill the cow, skin it, dismember it and send its parts everywhere without any micro-organic infections. Any part of the process can go wrong. Here's how it can go bad.

Examining the Carcass

Under pressure from McDonalds, packinghouses began changing their slaughtering practices, brought about by an E. Coli outbreak at a Jack-in-the Box franchise in Seattle in 1993 when seven hundred people became sick, enough to require medical assistance. McDonalds, sending a strong message to the industry, established animal-handling standards for its beef and chicken suppliers, auditing sites where meat is slaughtered, and suspending deliveries from plants it found inadequate. McDonalds was prompted to act by an 11-month protest by the People for the Ethical Treatment of Animals whose campaign included handing out boxes labeled as "Unhappy Meals" with vivid illustrations of slaughtering plants. Of course, U.S. Dept. of Agriculture inspectors are also on site at all meat packing plants to insure that contaminated meat is not supplied to an unsuspecting public, but regrettably, some contaminated meat is finding its way into the kitchens and stomachs of the American public. (2)

Meatpackers, realizing that food safety is a hot political issue, began retraining workers and installing new equipment. The federal government even helped the American Humane Association launch a program labeling food as "Free Farmed," a designation that rewarded food processors that adhered to strict animal welfare standards.

Animal rights' groups (American Society for the Prevention of Cruelty to Animals, The Humane Farming Association, Animal Welfare institute) would have us take the cow's perspective for a moment. Imagine being hit on the head and finding yourself dazed but conscious. You are suddenly lifted up by a chain around your leg and hoisted to a moving conveyor assembly line, your throat is slit and your blood drains into the lower floor. Your lower legs and arms are snipped off and the skin stripped up to your neck. But you are still conscious. Giving cows the equivalent of human consciousness, or even souls like the Hindus, would make vegetarians of anyone.

The first expose of the slaughterhouse practices in the U.S. came from Upton Sinclair. For a gut-wrenching literary experience, and an example of an expose of failed and unhealthy business practices, read *The Jungle*, a 1906 story of an immigrant meat packer, even if you read only the section entitled "the Nation's Food Supply" for a description of a meat-packing plant's swindles. (3) This book convinced President Theodore Roosevelt to order stricter regulations on the packinghouses and for Congress to pass the *Federal Meat Inspection Act* of 1906.

In the century since, because of the meat industry's control of the market and its collusion with the U.S. Department of Agriculture to reduce and minimize government regulation and oversight over several administrations, contaminated beef and beef byproducts have increasingly compromised food safety programs and jeopardized the health of American consumers. Recent books describe in detail the meat industry's callous disregard for safety and health regulations, for not protecting low-paid, usually immigrant labor employees, for the inhumane treatment of the animals, and the sloppy regulation and control over animal parts, including fecal material, incorporated into beef byproducts. (4) Each of these is contrary to provisions of *The Humane Slaughter Act* of 1958.

Vincent Erthal was a federal meat inspector with 18 years experience for a Pennsylvania poultry plant owned by Wampler Foods, a division of Pilgrim's Pride. He raised questions about sanitation problems in September 2001, months prior to an outbreak of bacteria that killed eight and sickened 54 people. (5) The sanitation conditions in a plant where he worked consisted of mould and algae on the walls and condensation dripping from the ducts over processing tables. Listeria monocytogenes, the deadliest type of Listeria, was fond in the plant's drains. Wampler Foods recalled 27 millions pounds of meat, one of the largest food recalls in American history.

In 2000, according to the Food Safety and Inspection Service of the U.S. Department of Agriculture, 3,746,214 pounds of beef products were recalled of which 968,409 was eventually recovered, a 25% success rate. (6) In other words, 75% of meat with E.coli, listeria, salmonella, and such indigestible materials as plastic (Owens Country Sausage) and metal (Berks Packing, Campbell Soup), were not recovered from distribution. The largest was Green Bay Dressed Beef which had 1.1 million pounds of beef recalled because of E. coli. Packing and food houses in 24 states had meat recalled in 2000. And this list includes only beef products, not chicken, pork or turkey, or the glass found in seafood (Allen's Family Foods, Delaware), the undeclared substances in chicken, or the oil found in sliced turkey (Farmland Foods, Minnesota). The names of companies with "grandma," "farm," "family" or "ranch" in them, meant to appeal favorably to the product, somehow ring hollow when contaminating substances or harmful bacteria are found in their foods.

On the one hand it is gratifying to know the inspection service is finding and identifying violators. However, it is discomforting to realize that the problem of lack of food safety is so pervasive, stubborn and threatening. There were 18 beef and beef products recalls in 1998, but 41 cases in 2000, a doubling of the problem, the majority of the active recall cases with ground beef. Yet as early as the turn of the 20th century this caution was administered: "In the general meat inspection much time has been devoted to improving the system and perfecting checks to guard against the use of unscrupulous dealers of meat which has been condemned as unwholesome." (7)

Meat Inspectors classify meat as: Healthy (no disease), Sound (clean, sanitary), Wholesome (not adulterated), Properly Labeled (it is what it says it is). Additionally, Individual states have statutes regulating licensing, inspection and production of meat and meat byproducts. For example, Mississippi's Meat Food and Poultry Regulation and Inspection Law deals with the licensing and regulating of those in the business of manufacturing, slaughtering and preparing meat and meat-food products. *The Meat Inspection Law* (1968) focuses on the inspection of animals, carcasses and products and upon the sanitary requirements for slaughtering establishments. (8) In addition, the Food and Agricultural Organization (FAO) of the United Nations devises laws and regulations to protect the health of consumers and insure fair trade practices worldwide. Such programs promote coordination of standards in food safety, pesticide residues, food additives, and hygienic practices for fresh meat. (9)

The Centers for Disease Control estimates that about 6.5 million cases of food poisoning occur annually in the U.S. Food-borne diseases cause an estimated 5,000 deaths, 76 million illnesses, and 325,000 hospitalizations annually. (10) In the decade from 1984 to 1994 deaths from food poisoning quadrupled. The Centers for Disease Control had earlier identified a Michigan plant as the source for an outbreak of listeriosis, harmful to pregnant women, the elderly and children. Prior to renovations at the adulterated plant, government inspectors had said that the plant was infested with roaches, old meat and debris.

Coincidentally, this was a period of deregulation of the meat industry. The National Academy of Sciences in 1995 warned the U.S. Dept. of Agriculture that meat contamination was imminent and recommended immediate action such as new science-based methods for meat inspection and not just sight and smell techniques. Current inspections methods cannot detect the presence of harmful pathogens. In 2002 Pilgrim's Pride had to recall 27.4 million pounds of ready-to-eat chicken and turkey products. (11)

In the United States in any given year, 101 million pigs, 37 million cows, 4 million horses and sheep and 8 billion chickens and turkeys are killed annually to feed meat consumers. In 1980 there were 103 meatpacking plants. Because of the increase in beef consumption and consolidation of meat processing plants into agribusiness conglomerates, by 1996 only 11 meatpacking plants were responsible for slaughtering one million animals each year, about 40 percent of the cattle killed each year. (12)

In June, 2001, Sara Lee, one of the country's largest producers of hot dogs and deli meats sold in supermarkets, pleaded guilty to a misdemeanor charge of making and distributing tainted meat from which 15 people died and scores were sickened. Sara Lee's tainted meat led to a nationwide outbreak of listeriosis in which 15 people died, 6 had miscarriages, and 80 people became seriously ill. The contamination came from meat produced in deli and hot dogs in 1998 in Borculo, Michigan. Managers knew that there was contamination eight months before the outbreak. The recall of the meat eventually totaled 35 million pounds. Sara Lee paid $4.4 million to settle damages awarded compared to its sales of $17.7 billion in 2001. The company had recalled 15 million pounds of meat in 1998 after the U.S. Department of Agriculture determined that a rare strain of listeria bacteria was in the meat. (13)

Informed citizens are aware of the dangers of food contamination and how necessary it is to rely upon the dissemination of scientific knowledge about the ingesting of potentially health damaging food products, and equally aware of the disregard for human health standards practiced for decades by the tobacco industry, the silicon breast implant industry, the drug industry, and government agencies concealing or minimizing adverse public press about food consumption, like the dangers of eating contaminated beef.

Let's take a case study example. ConAgra Foods is one of the world's largest food companies. In July, 2002, it was forced by the U.S. Dept. of Agriculture to recall 19 million pounds of beef from a plant in Greeley, Colorado because of an E.coli outbreak in which 19 people fell ill. The 2002 recall was the second largest ever for a meat product. From 2001 to 2002, the agency halted processing at two ConAgra plants unless safety regulations were more stringently enforced. One ConAgra plant had the highest rate of salmonella among all turkey meat processors. (14) The U.S. Dept. of Agriculture could not close Supreme Beef in Dallas in 2000 because, according to the court, it lacked the legal authority even though Supreme Beef had failed tests for salmonella three times that year. (15)

The failure of meat processing plants to satisfactorily monitor safety standards consistently is only matched by the laxity and tardiness of federal regulation and congressional failure to give adequate enforcement powers to the U.S. Dept. of Agriculture. For example, after a year of record recalls for contaminated meat, the Bush Administration sought legally to close Nebraska Beef, an Omaha slaughterhouse for continued violation of food safety rules. (16) Hamburger contaminated with E.coli was discovered in its beef and fecal material on carcasses.

Here is a scenario of the more serious mishaps at the opening of the 21st century:

- Jack-in-the-Box chains sold hamburger tainted with the E. Coli bacteria. Four children die and hundreds become ill

- Hudson Foods began a modest patty recall that ballooned into a 25 million pound recall causing it to lose its biggest customer, Burger King

- Tyson Foods purchases Hudson Foods

- Sara Lee pleads guilty to selling tainted meat that apparently caused the deaths of 15 people and 6 miscarriages. It paid a penalty fee equivalent to about two percent of its sales in 2000

- ConAgra begins a recall of 19 million pounds of beef. The U.S. Agriculture Department goes to court to close Nebraska Beef

Profitability is the meatpacking industry's primary mission. It constitutes half of all employment in agriculture and has annual revenues higher than any other agricultural commodity in the U.S. Health and safety appear to be relevant only when public consciousness and corporate image are damaged or endangered. The pattern has been established over the past century: when health issues are raised, there is predictable denial from the beef industry, the impugning of motives of the critics, the vigorous pursuit of federal deregulation of the industry, and the shifting of safety and health costs to the consumer. (17)

The consolidation of the meatpacking industry means that unskilled labor comes at a price for the worker. According to the Bureau of Labor Statistics, meatpacking is the most dangerous occupation in the U.S., estimated to be six times more dangerous than working in a coalmine. As line speeds have risen systematically to process more meat more quickly thereby increasing profits, workers are urged to work faster imperiling their safety.

Federal prosecutors took Tyson Foods, the nation's largest meat supplier, to court for conspiring to suppress workers' wages, for smuggling illegal immigrants into the U.S. from Mexico, and for obtaining false documents such as social security cards and driver's licenses for smuggled aliens. In 1997 Tyson Foods make illegal gifts to the Secretary of Agriculture, Mike Espy. In 1999 it was fined by the government for violating child labor laws (one 15 year-old immigrant boy died and a 14 year-old was seriously injured). Again in 1999, Tyson settled federal charges it had discriminated against blacks and women in Mississippi. In 2001 the company was fined for cheating workers out of wages. (18) In February, 2004 a federal jury in Alabama awarded a group of ranchers $1.28 billion against Tyson Foods for using illegal cattle contracts to hold down prices it paid them artificially. Tyson purchases a third of all cattle in the U.S.

The 2000 census detailed what many already knew about the expansion of the agribusiness meat industry: that rural communities were losing population, small cattle ranchers included. But driven by old-fashioned greed and excitement, cattle–rustling was on the rise. Susie McKay of Spring Hill Kansas, for example, was awakened at 4 AM one morning to the sound of trailers. Cattle thieves were hauling off 24 of her specialty bred cattle. Armed with her cell phone, she jumped into her vehicle and followed the rustlers and saw them captured by sheriff deputies. David Zortman of Jerome, Idaho lost 1,600 cattle to thieves posing as employees over a one-month period until he discovered what was happening. (19) Some ranchers use DNA fingerprinting to keep track of their cattle. Still, with all the precautions, it is often difficult to detect lost from stolen cattle over broad pastures and cattle grazing lands. In response, regulations, policies and laws are being upgraded to reflect the seriousness of the crime. Stealing a calf in

South Dakota had been a misdemeanor but the state legislature made cattle theft a felony in 2001 and appointed a special prosecutor for livestock crimes.

After 112 years of continuous operation, the Union Stock Yards in San Antonio, Texas, where President Lyndon Johnson's cattle were sold in 1973 after his death, closed permanently. More than a million head of cattle moved through the yard by 1950. By 1998 the hogs, sheep and goats were gone and within three years the cattle. The cattle market's ageing building, once surrounded by 35 acres had shrunk to 9 acres. Once, the stockyards had been outside San Antonio's 30,000 population and cattle grazed nearby. Today, with an urban population of over one million, San Antonio has overwhelmed the stockyard location. Urbanization drove away rural farmers who found getting their cattle through urban and suburban sprawl to the stockyards too difficult. The stockyards had been a cherished link to the city's past and the state's 20th century love affair with cattle. (20)

The Internet has also changed the way cattle are bought and sold, eliminating the need for a middleman or a cattle auction. Videos of cattle can be seen by potential buyers and the cattle shipped directly to the buyer avoiding a central location where cattle are congregated.

The meat industry in the U.S. is one of the most powerful lobbies among legislators and regulators, chiefly those in the Food Safety and Inspection Service in the U.S. Dept. of Agriculture. (21) Using the services of the American Meat Institute, the National Meat Association, and the National Cattlemen's Association, the industry has successfully lobbied to prevent new safety standards for meat in the last few years as it did in 1995 to delay the regulations for testing for salmonella safety and the E. Coli bacteria after the Jack-in-the-Box outbreak in 1993. The meat industry ultimately won when Tyson Foods, a poultry producing meat source, merged with IBP, a major meat producer. Suddenly, Senators from poultry producing states allied with representatives from meat producing states were persuaded to delay and eventually kill safety amendments regulating the level of salmonella allowed in food products.

Over the years the small cattle rancher has been supplanted by the large, industrial meat processing company that can more efficiently raise thousands of animals through automation. Through the apparent virtues of what is known as vertical integration, or birth through butchering and packaging, a few corporations control nearly all beef production in the U.S. In Iowa, for example, in 1980 there were 64,500 hog farms. Although the number of hogs in Iowa remains the same, at 15 million, by 2000 there were only 10,500 hog farms. (22) The residual of such enormous feedlots is manure in greater quantities than can be distributed as fertilizer to local farmers. Typically, when big business collides with the environment, clean air, clean water, uncontaminated ammonia gases and other toxic substances usually lose.

Beef remains relatively inexpensive but health risks are greater. In 1999, the U.S. Department of Agriculture allowed irradiated meat to be served to the public but prohibited it for use in school lunches until 2002 when it was approved in a farm bill passed by Congress. By using low levels of gamma rays, irradiation sterilizes bacteria such as E.coli and salmonella. Irradiated beef has been found suitable by the World Health Organization and the American Medical Association though some advocacy groups claim irradiation also destroys vitamins and nutrients. (23) Yet in spite of these large feedlots and the takeover of beef production by huge industries, small farmers are reviving the raising of grass so cattle (and sheep, chickens and pigs) can enjoy the diet they had before fertilizer companies offered farmers subsidies to use their fertilizer and grain and corn to replace grass as a natural cattle feed.

Corn is grown over 125,000 square miles in the U.S., about twice the size of New York state. Because it is so cheap, corn is fed routinely to cattle to fatten them up before sale. But it does damage to the digestive system because cattle have evolved to eat a grass diet. (24) Corn-fed beef is higher in saturated fats than grass-fed beef because corn metabolizes differently while heightening fat tissues. As the nation struggles to get rid of corn, taxpayers will subsidize $4 billion a year over 10 years for farmers to grow more of it, paying them $3 dollars a bushel to grow it while corn sells for $2 dollars a bushel, benefiting the large food and the high fructose corn syrup suppliers who use it in soft drinks and snack foods.

Lastly, A new awareness about animal welfare slaughtered for human consumption, cattle, chickens, turkeys and pigs, has activated reforms in holding pens and slaughterhouses. Cattle prods are out. Cages for chickens are a third larger than they used to be. Lighted entrances for cattle are standard. Sow stalls for pregnant pigs have been outlawed. People for the Ethical Treatment of Animals believe that if cats and dogs were treated like livestock animals producers would be subject to jail. Large outlets like McDonalds since 1997 have been requiring audits from livestock producers, raising its audits from 100 in 1999 to 500 in 2002, to conform to regulations developed by industry associations sensitive to animal-handling procedures. In concert with the Food Marketing Institute and National Council of Chain Restaurants, cooperating groups include the American Meat Institute, National Cattlemen's Beef Association, National Milk Producers Federation, National Chicken Council, National Pork Board, National Turkey Federation and United Egg Producers. Certified Human labels created by the Humane Farm Animal Care Association began appearing in stores in the fall, 2003.

The investigative reporting of a few journalists on the deaths and hundreds of sicknesses have alerted the general public about the lax regulations of the foods they eat. The processing and packaging of foods has made life simpler but has distanced us from the butchering, plucking, skinning and compromising with additives of our most cherished food, the beloved cow.

Milk and Cheese: Wisconsin and the West

If music is the drink of love, milk is the drink of life. And from milk all dairy products come, the buttermilk we ignore, the cheese we can't live without on the hamburger, and the ice cream we crave.

But like the meat industry, the U.S. Department of Agriculture has been monitoring the dairy industry for over a hundred years. For example, in 1870, want of proper care, proper feeding of cattle and even neglect of feeding were some of the mismanagement practices noted. Officials noted that farmers who made butter and cheese: "neglect to provide shelter for burning sun; (are) stinting of nutritious food in winter, after milking hard through summer month; the use of close, ill-ventilated stables and scantily littered stalls, and carelessness in eliminating noxious weeds from pastures, lead to the most serious evils and are frequent among farmers who cannot justly plead want of information as their excuse." (25)

American cheese factories originated in 1851 and thereafter cheese products doubled, from 103 million pounds to 240 million from 1860 to1869, and exports quadrupled, most shipped to England. The American Dairyman's Association reported in 1869 that there were 1,066 cheese and butter factories in the U.S. operating with 287,318 cows. The majority of factories were in New York. At the time only 31 were in Wisconsin. (26) By 1900 the Department of Agriculture was calling for laws regulating the production of butter, cheese, condensed milk and cream as rules were then nonexistent. (27) Making the conditions sanitary for animals has been a part of regulating the food processing industries for over a century.

Until very recently, Wisconsin was the leader in cheese and dairy production. If you haven't seen the cheese head hats at the Green Bay Packers football games, or on the heads of fans at the basketball games of the University of Wisconsin colleges, you've missed an entertaining view of sports regalia and an image of the state's most valuable agricultural commodity. Of the 8.25 billion pounds of cheese produced in the U.S. in 2000, about 25% came from Wisconsin, part of the $18 billion dollar dairy industry, where 85% of the state's milk is used to make cheese. But beginning in 2000 for the first time ever, western states produced more milk than north central states including Wisconsin with its 18,000 dairy farmers, imperiling that state's economy and public image. At the turn of the millennium, milk prices were in a downturn and the dairy industry was migrating from the north central states, consisting mostly of small farms, to the large-scale agricultural industries in western states headed by California. (28)

The dairy industry is industrializing like the beef industry and as it does it is shifting production location. The Wisconsin culture of preserving the family farm and resisting the corporate status of the dairy industry is hampering growth of its milk production and limiting cheese production, an expanding market in America. Raw milk is cheaper to produce in the west and hence the industry, headed by corporate dairy giants like Land O' Lakes, is moving west. The more efficient dairy farms in California, using the latest technology and production line processes, have cows that produce 21,000 pounds a milk per year compared to Wisconsin's 17,000 pounds. (29) About 2,000 large farms in California, Arizona, New Mexico and Idaho, averaging 650 cows per farm, can produce more milk than 18,000 small Wisconsin farms averaging 30-50 cows.

New England successfully lobbied Congress in 1997 to shift the national regulation of milk production to the states. The law was known as The Dairy Compact and was administered by a public commission composed of producers, processors and consumers. Its principal job is to establish a floor price for milk based on the regional economy. In New England there are roughly 1.3 million acres of space for dairy farming, down from 1.9 million acres in 1985. Suiza Foods based in Dallas purchased three-fourths of the milk producing plants in New England. (30)

Land and energy is generally cheaper in the west and there is less opposition from environment groups about the use of immigrant laborers. A few farmers in Wisconsin have experimented with larger milk producing farms without displacing families. But a few of the older farmers are getting tired of milking as family and worker replacements are harder to find. Wisconsin will still be a major milk and cheese producing state, but more milk and cheese will come from western sources.

Additionally, desired specialty cheeses like mascarpone, fromage blanc, ricotta and creme fraiche created at small creameries owned and managed by women in northern California are taking a bite out of Wisconsin's cheese industry. (31) The milk for the cheeses, from firms like Cowgirl Creamery, Bellwether Farms, Redwood Hill Farm and Point Reyes Farmstead Cheese Company, come from sheep and goats often mixed with cow's milk. California is now the country's largest dairy state and is expected soon to replace Wisconsin as the top cheese state. (32)

The Dairy Industry

Milk is the liquid of life from life and the most important cattle product besides meat. On a modest 6.5% of farmland, employing only 4.5% of farm labor, diary farms with 11 million milk cows, supports over 4 million employees, produce over 20% of the agricultural wealth in America. According to data from the National Agricultural Statistics Service, which are based on the 20 largest milk-producing states, cows are producing more milk on the average, up 1,440 pounds in 2001 from 1,343 in 1985, and from 1,185 pounds in 1991. In the decade from 1991 to 2001, milk production in pounds has risen 18 percent. One reason for increased milk production is the rise in advertising. The famous advertising tagline beginning in 1993, "Got Milk," is generally credited with the boost in domestic milk sales. (33) The advertising campaign, promoted by the California Milk Processor Board, the Dairy Management, Inc., and the Milk Processor Education Program, with its recognizable milk mustache, is reliably estimated to have cost $180 million nationwide. Milk had its brand name recognition.

One reason milk production has been increasing is because many dairy cows are being injected with recombinant bovine growth hormone (rBGH) a product manufactured by Monsanto, one of America's leading chemical producers. It is estimated that 30% of all dairy cows are injected with this growth hormone intended to produce faster growing cows that in turn produce increased milk supplies. What is unclear is what effect a growth hormone, sold as Posilac which has had approval by the Food and Drug Administration, has on humans who consume milk from injected cows.

Although the dairy industry advertises regularly to sell more of its product, nutrition experts know that calcium is a vital ingredient in the growth of healthy bones and skeletons in children and adolescents and that milk and cheese are the primary foods for absorbing calcium. The National Academy of Sciences recommends 800 milligrams of calcium daily for children 4 to 8 and 1,300 milligrams for teenagers 13 to 18. U. S. Department of Agriculture studies reveal that only 13 percent of girls and 36 percent of boys consumed the necessary dosage of calcium. What adolescents drank mostly were too many carbonated soft drinks and food juices.

Vegetarian activists, some physicians, and People for the Ethical Treatment of Animals (PETA) claim that there are other sources of calcium for strong bones besides milk. Kale, brussel sprouts, broccoli, tofu, Great Northern beans and soy milk are also sources of calcium and these products do not come with animal protein, cholesterol or saturated fat found in milk and milk products. (34) An average quart of milk will have 34 milligrams of cholesterol, 159 calories, 8 fat grams, and 4.9 grams of saturated fat. The adult bones of calcium-deficient individuals will fracture more easily and a host of other symptoms: osteoporosis, increases in blood limpid levels, hypertension, heightened blood pressure, cardiovascular problems, colon cancer are more likely. (35)

For those who do buy milk, economics plays a role.

Shoppers in the mid-west began getting seriously upset when the price of whole milk topped $3.69 a gallon in the summer of 2000 in Chicago, near one of the largest milk-producing regions in the country. Milwaukee, less than a hundred miles north, was selling milk for an average of 30% less. Many shoppers began taking their business to discount stores which were charging as much as $1 dollar less. In other parts of the country, however, milk prices had dropped by about 6%, or about 18 cents a gallon. (36) New York has a law against price-gouging in milk, the only such law in the nation as of this writing. The law does not permit retailers to sell for more than 200% of what farmers got paid for producing the food.

Economists offer one explanation for the accelerated price increase in milk—supermarket mergers. In 1995 there were ten top food chains in the U.S. that handled about 40% of all grocery business. By 2000 there were only five such major companies and mergers combined inventories. Economists explain that mergers are causing the price fluctuations that seem always to be in favor of the store and not the customer. (37) In fact, the price of raw milk has actually dropped since 1998 and there was a glut of milk on the market at the turn of the 21st century, while farmers are getting reimbursed the lowest price for their milk in decades, about 26% less since January, 1999, according to government surveys. Similarly, the cost of lower quality milk to make cheese has plunged 42% while the cost for such cheese in stores has remained stable. Large food chains have clearly decided that milk and dairy products are not going to be a loss leader for them but to rely on milk and dairy products to be profitable.

In spite of the declining numbers of dairy cattle, U.S. milk production has increased and dairying remains a vital industry, each year accounting for $75 billion in sales of dairy products. (38) Today's dairy farmers are still getting the same price per pound of milk as they did 20 years ago, about 12 cents. However, their gross income has increased yearly because every year the amount of milk produced per cow has increased.

A new formula for estimating a dairy herd's manure production and nitrogen content may help dairy farmers use less commercial fertilizer and make their farms cleaner and more efficient. When farmers can make accurate predictions about nitrogen content in herd manure, they can use animal waste more effectively, including replacing commercial fertilizers. Everywhere, governments are asking that farmers reduce waste runoff to protect waterways.

Tommy and Susan DeJong own the Rainbow Valley Dairy in Buckeye, Arizona, and their story is an example of the troubles that can come from cow waste. Tommy comes from a long line of dairy farmers dating to 17th century Holland. The DeJongs own a dairy herd of 3,300 cows and in 2000 had $10 million in annual sales.

By the 1980s it was clear that dairy farmers and the nation were aware that cow waste was polluting downstream water. DeJong had what he thought was a reasonable solution. He asked the U.S. Dept of Agriculture for assistance in helping design a system for water draining across his property. The Department's Soil Conservation System helped with the $15,000 cost. DeJong had to pay another $150,000 to implement the system that was completed in 1992. The system, designed to hold water from the 22 acres of cattle pens from running into the natural wash, consisted of seven ponds stretched across the open range. DeJong also bought pumps to send the overflow from the ponds back uphill with generators that worked even when electricity didn't, to purchase adjacent property and to install a mile-long pipe. The total cost for all these improvements was $600,000. Then the U.S. Environmental Agency (EPA) got involved. EPA thought the system was inadequate and took the DeJongs to court and a federal jury indicted him in 1996 on three counts of polluting the waters. In 2000, after an eight–year court case, a federal judge ruled in DeJong's favor, awarding him $205,000, the cost of his legal fees. The DeJongs were only the second defendant to win such a court battle in the history of this kind of legislation.

Other examples are not so rosy.

Experienced husband and wife investigative reporters Jane Akre and Steve Wilson working for a television station in Tampa, Florida (WTVT), spent three years beginning in 1996 researching how widespread the use of rBGH was given to dairy cows and much of that milk was sold through major food and grocery outlets. They investigated over three months and in five states. When it came time to air their program, which revealed that Monsanto, the dairy industry and food outlets had contrived to conceal information from the public, pressure from Monsanto lawyers caused the TV station to kill the story and the reporters were fired. The TV station then ran a sanitized version of the story. Akre and Wilson filed a whistleblower lawsuit against Fox Television according to Florida law and won $425,000 in damages in a court decision in 2000. Although Oprah Winfrey prevailed against the cattle industry, this was the first decision won by journalists against a news organization accused of illegally distorting the news.

Arrayed against independent and courageous investigative reporters, however, are the well-funded coalitions of drug, dairy and food industries who limit discussion in the press, intimidate editors and reporters, and hire experts and lawyers to disparage any criticism about food safety concerns the public is entitled to know. Consumer groups are aligned against such pro-industry representatives as: The Dairy Coalition, the International Food Information Council, the National Association of State Departments of Agriculture, the American Farm Bureau, and even the American Dietetic Association.

Safety measures can occur even if industry officials know the industry's product has unhealthy or unsafe ingredients in it, like the organ ingredients in certain health food products, especially when deceitful and misleading advertising is widely practiced. Consumers, unless they have access to reliable information about the relative health and safety risks in a product, are unable to make sensible choices. In matters of public food safety the public should demand the following: regulation, public access to information about products, and stiff penalties for non-compliance.

None of this discussion about food safety in the beef or dairy industries would be so ominous unless as a global village we had come to rely so comprehensively on the cow, our most reliable food source, for beef, cheese, butter and milk. If we had several alternatives food products we favored it would be relatively easy to switch protein diets. However, our eating habits have become integral to our culture. We are servants, not just to advertising campaigns, but to our palates and accustomed comfort foods. As it was millennia ago, the cow is our unacknowledged goddess. We pay obeisance to her by choice in the meat section of the supermarket, in our kitchens, at our favorite dining and restaurant tables, and at the pickup counters of the drive-through fast food outlets.

Finally, teaching the young ethics may be more difficult but instructive than teaching them animal husbandry.

At the 2002 Missouri State Fair, all 416 steers exhibited were meticulously analyzed by Don Lock, chief of the fingerprint section of the Missouri State Highway Patrol. Were the cows animal terrorists that they needed to be fingerprinted? Lock took fingerprints of each steer's nose, which also has a unique, individual pattern. He compared each print from a copy taken the previous spring at the farms where the cows originated to ensure that the same cow was entered. Too often in the past, cows had been switched since the cows that win first and second place prizes have increased sales value. Deception had become so extensive that modern detection techniques had to be called in to ferret out cheaters. (39)

The detections started in 1994 when 7 of 10 grand champions in Ohio were found to have been given illegal drugs, principally steroids to build muscle. That year in Ohio, 17 people were convicted of felonies committed with or to livestock and for violating pure food laws including injecting animals with corn oil, water, compressed air and padding implants in dairy cows to yield balanced udders. It was a sad but informative day when these crimes were revealed since so many involved junior Future Farmers of America. Fair stakes have not been evident at state fairs.

Learning about livestock has been an apprenticeship for the young since cows were domesticated. Cattle have been part of an individual's education in America since their importation first from Norway and then from Spain. A neglected American history lesson is that wherever European settlement was initiated or took root, in St. Augustine, Santa Fe, Plymouth, San Diego, cattle accompanied the first journeymen and settlers. These sites and communities would not have survived or thrived in successive years had it not been for the sustenance of cattle.

Meat and poultry constitute one-third of food spending in the U.S., though poultry consumption has increased at the expense of meat in the 1990s. The average American eats 66 pounds of beef annually, 51 pounds of pork, and 76 pounds of chicken. (40)

Years of warnings about high-fat content slowed purchases of beef products in the late 1990s. But by 2000, beef was back as the choice American food created by an aggressive advertising campaign by the beef industry spurred on by the catchy slogans, "Beef. Real Food for Real People," and "Beef. It's What's for Dinner." Together with a sleek advertising campaign, a robust economy at the time was also credited with the new meat-eating popularity. By the beginning of the 3rd millennium Americans spent over $26.2 billion on beef products, up over 8 percent. So-called precooked beef products, some cooked to cover rancidness and unpalatable material, were up over 67%. It is estimated that Americans will eat about 70 pounds of beef per person, up from 65 pounds in 1993. Even advocates for the American Dietary Association concede that some beef in moderation is good as an acceptable nutritional package or even for use in a diet plan. (41)

Western diets, compared to Asian diets, have always been heavy with meat consumption. But other nations are acquiring the meat-eating habit as meat consumption rises throughout the world, a testimony again to the theme that the cow culture is global. Increases in beef consumption were occurring not only because of rising human population but because of individual eating habits. Similar increases were occurring in Europe, Brazil, China, Nigeria and elsewhere. (42)

Growth in the organic foods industry, largely because of a voluntary seal of approval by the U.S. Department of Agriculture for all such crops and livestock, has been rising. All products must be 95% organic and this means that feed for cattle

a) may not contain plastic pellets, instead of roughage,
b) or urea, a synthetic protein,

c) or litter, a mixture of waste and bedding fed to cattle,

d) or manure and parts of other slaughtered animals.

Additionally, cattle may not receive growth drugs of hormones and must have access to outdoors, fresh air, exercise and freedom of movement.

According to a growing number of environmentalists and ecologists, cattle may be the best way of preserving cactuses and brush from fires that destroy so much of the natural environment in the arid southwest and intermountain west. A study at Colorado State University compared 93 ranch sites and found that ranches had at least as many bird, plant and animal species as found on protected wildlife reserves, countering the common assumption that ranching is detrimental to the total environment. (43)

In an unusual combination of food and energy sources, researchers have developed a new strain of alfalfa to feed cattle and help generate electricity. (44) In a project to be funded by the U.S. Department of Energy, the first generation of the new alfalfa type, grown and marketed by local farmers and produced strictly as animal feed, is taller, stronger, with thicker stems than alfalfa typically fed to cattle. The leaves can be ground into meal and fed to cattle, while the stems can fuel electric generating plants. A farmers' cooperative plans to build a generating unit and sell the electricity. The experimental program holds promise as both a high-protein feed source for dairy cattle and an environmentally friendly source to generate electricity.

The invisible beef byproducts industry, and the ultimate recycling industry, is rendering. The U.S. Department of Agriculture calls rendering a process of treating the leftovers of the slaughterhouse, the bone, fat, offal and carcass, with heat prior to selling it as pet food and supplement feed to chickens, cows, sheep and other animals. Each year about 12.5 million tons of material from dead animals becomes the protein supplement for other animals and in the products of gelatin and cosmetics.

Environmentally friendly ways to turn cattle hides, the most valuable co-product from the meat packing industry, into leather while ensuring better quality products have been developed by U.S. Agricultural Research Service scientists. (45) The U.S. produces about 35 million hides each year and exported hides bring more than $1 billion in foreign trade. About 63% of all finished hides are exported. The trade in finished tanned leather is worth about $4 billion. In addition, USDA scientists in 1998 turned hide-tanning waste into high-value commercial products, including adhesives and packaging films. From 100 pounds of cattle hides, a tanner gets half leather and half waste. Each year, about 60,000 metric tons of shavings end up as waste bound for landfills. (46) By using enzymes from a common laundry detergent, chemists modified the cleansing process to create a high-value grade of gelatin for industrial and agricultural use.

After years of selective breeding practices, raising cattle in America has become extremely efficient and scientific in breeding and feeding practices, nutrition, dietary supplements resulting in a cow with a proportionate weight gain and leaner meat. The most popular cattle brand to eat in America has been the Hereford, imported first into the U.S. from England by the Kentucky statesman, Henry Clay in 1817, while in Europe the most popular meat source is the Charolais, a large breed with heavy muscling originating in France. The high–end costs of raising cattle are in land, labor and feed. In pasture breeding, usually one bull is required for every 25 cows. When cows are artificially inseminated, the breeder can have one outstanding sire inseminate thousands of calves annually.

Today, technology exists to track cars, ships or airplanes anywhere in the world and can give directions to motorist or pilots when lost. Global Positioning Service or GPS is one tool allowing researchers and ranchers to study and learn why cattle make grazing choices. Collars have been attached to cows with special radio receivers. Using the coordinates from as many as 24–30 satellites, researchers can determine within a few meters when and where a cow is. GPS units can also monitor head movements, indicating whether cattle are eating, sleeping, or just walking. (47)

The history of food safety over the past few decades has been untrustworthy. Large drug and food industries spend millions on public relations to keep the public ignorant of potentially damaging ingredients and safety methods in the production and sale of their products. In general, sectors of the food industry will seek to remain unregulated, will hide evidence, hire experts, lawyers and respected organizations to testify in favor of the safety of their product, and even threaten to sue detractors who criticize their product. A significant number of states have passed statutes prohibiting the disparagement of perishable food products with stiff fines for offenders, a move which on the face of it appears to run counter to state and U.S. Constitution First Amendment rights. Clearly, some exploiters wish to silence all criticism legislatively where they cannot judicially. The mere existence of such statutes creates a chilling and somber reminder that vigilance in food safety ranks as high as vigilance in defense of all liberties.

Figure 5: Running with the Bulls Javea, Spain

CHAPTER TEN
CATTLE AND UNWANTED
ORGANISMS

The intricate symbiosis between humans and cattle means they can share common infections. Bacteria or viruses that can kill a cow can also kill a human.

Mad cow disease first appeared in the United States in December, 2003 and originated from a non-ambulatory, obviously sick Holstein cow that was nonetheless slaughtered in Mabton, Washington. Since mad cow disease is carried in the cow's nervous system, neck, spinal columns, brains, the passage of infected meat to humans is presumed to be low, although muscle meat mixed with other processed parts can include tissue that carries the disease because inspections are unevenly enforced. Because of the ubiquity of beef consumption, one case of mad cow almost automatically closes international markets. And because the U.S. meat industry has been able to lobby effectively against the imposition of tough legal restrictions on improved food safety measures, the mad cow disease that began in England and was never thought to land on American shores, exposed the weak side of consumer safety. It was difficult to say whether a hamburger was safe to eat after this scare but it certainly became less appealing.

This chapter explores how government regulations and industry collusion combine with food safety and disease to generate economic losses, worldwide panic and public response. All because of cows.

Like the holes in the brains of cows with mad cow disease, one case exposed the holes in the U.S. regulatory system, seemingly designed to protect the health and safety of the special interests of the beef industry and not the consumer. This non-ambulatory cow from Mabton, Washington was nevertheless slaughtered because meat processors have resisted banning neck bones and spinal columns from their meat recovery systems. What did become exposed was the circular food chain system of feeding feedlot trash from other animals like pigs and chickens into cow feed and then bovine blood to pigs and chickens. The efficient system of re-cycling waste and uneatable parts to other animals is not designed with food safety as its primary objective.

Within a week of the discovery, the USDA issued tougher, long overdue regulations that

1) disallowed cows tested for disease from having their meat distributed in the food supply
2) banned sick cows from being used in human consumption
3) restricted the use of mechanically separated meat
4) imposed tighter controls on meat recovery systems that might unintentionally let some spinal cord tissue slip into processed meat
5) created of a national tracking system that would help officials respond to any outbreak. (1)

In that same week in 2003 more than one million Americans were sickened by a food borne disease with increases in E.coli and listeria since 1996 when the Centers for Disease Control through its FoodNet program began active surveillance of these infections. Food borne diseases infect an estimated 76 million annually from which about a 100 die and 6000 require hospitalization. (2)

Regrettably, government responsiveness does not regularly match the accuracy of scientific findings. In the 1990s the mad cow crisis happened with an intensity no one could have imagined, creating a global public health crisis. Suddenly, questions for public health and safety rose quickly: how are animals raised? What foods do they eat? How are they slaughtered? How are diseases prevented in them during their short lives? How sanitary is their processing for human consumption? Humans and animals sharing some of the same germs and bacteria had been a common thread of coexistence. But because of the ubiquity of cattle and the global cultural trait of eating beef, when a cow gets a sniffle, metaphorically the human cough can be heard globally.

About 10,000 years ago crowded human societies came into contact with many infectious diseases, caused by evolving microbes from domesticated animals like cows and pigs. Ancient peoples lived intimately close to animals, their breath, urine, feces and sores. Since the measles virus is close to the pathogen that causes rinderpest in cattle, it's possible the virus transferred into a measles virus and evolved to infect humans. Similarly, tuberculosis and smallpox virus also likely comes from cattle. (3)

Nature has some strange inventions and the inter-connecting life forms that cattle and humans share is now better understood. For example, the fungus pilobolus lives in cow dung. To live in cow dung, the pilobolus fungus must first get into the cow. Cows eat the tough spores of pilobolus while grazing but since cows cannot digest them, they pass through the animal's digestive system and are excreted. To fulfill its life cycle, the pilobolus spores must land on vegetation. They perform this by shooting spore capsules upwards, away from the cow and onto the grass. Their gun is a stalk swollen with cell sap, bearing a black mass of spores on the top. Below the swollen tip is a light-sensitive area, the eye of the black capsule. The light sensing region affects the growth of pilobolus by causing it to face toward the sun, the region of greatest light. As the fungus matures, water pressure builds in the stalk in the tower of a stem until the tip explodes like a rocket shooting the spores, over a hundred thousand in one blast, into the daylight at heights of six feet or more and away from the cow. (4) The cows that made the dung will likely eat these spore-covered plants and begin the process again.

This cultural study of the cow includes how a society copes with the spread of diseases through invisible organisms and with educating the public about what potentially lethal cow diseases might be dangerous for humans. Anthrax can kill cows and humans but is not transferable from cows to humans. Bovine spongiform encephalopathy or BSE, commonly known as mad cow disease, a degenerative brain disease, is transferable to humans. The variant in humans is known as Creutzfeldt-Jakob disease, or CJDv, and is fatal. Hoof and mouth disease, when widespread in a country, can disrupt economies, the food supply, even the travel industry as Britain discovered to its dismay and chagrin.

Mad Cow Disease or BSE: From Deformity to Death

An epidemic was occurring in the highlands of Papua New Guinea in the 1960s among the Fore peoples. Women and children were experiencing dementia, staggering motions, followed by uncontrollable shakings and excessive fatigue. The disease came to be known as kuru, the shakes. Soon affected individuals were unable to stand without assistance. Within about nine months they died. Researchers, attempting to trace this disease, interviewed survivors to see what they had eaten or imbibed which might have triggered convulsions.

Health officials discovered that tribal members were cannibals who had eaten their dead relatives to obtain their power. The men had taken the best dead human flesh, the muscle, and the women and children had gotten the least desirable edible parts, the brain, liver and other organs. The women and children had thereafter become infected largely through the infected brains of their dead relatives. Cannibalism ended. But the search to discover what the mysterious disease was that had caused such deaths continued. It would appear in various forms two decades later. (5)

A disease affecting sheep known as scrapie, because the sheep scratch off a huge section of wool from the backside, had existed for over 200 years in Britain. It had never spread to other animals so shepherds and sheep farmers were not much concerned about its ripple effects because no one knew what caused it, nor had it had been very widespread.

But by the late 1980s, cows were beginning to experience strange and similar disorders of movement, muscle spasms, sudden fallings and laying down for long periods. In 1989 in Wisconsin mink farmers observed that thousands of mink were suddenly dying after experiencing listlessness and shaking. Investigating biologists noticed that mink had been fed "downer" cattle, or cows that had been laying down for prolonged periods, perhaps because they had BSE, or bovine spongiform encephalopathy. But after examining the carcasses and conducting autopsies on 6,100 minks, no BSE was ever found in minks.

But by 1996, 161,000 cows were affected in Britain. As BSE spread throughout Britain, autopsies of dead, infected cows revealed large potholes throughout cow brains, as if parts of the brain had itself been eaten. The alarm grew as the popular press ran screaming headlines. What caused this disease? Would it become an epidemic ruining the British beef industry? Could it spread to humans who ate beef and beef products? Not enough was known about this diseases and possible linkages with other diseases. European governments and inter-governmental agencies hustled to find out how many people might die from eating infected beef. There was no agreement on the incubation period; if it was long, this meant a potential epidemic was a decade or more distant.

British health officials scrambled to repair the damage to the British beef industry, tourism and Britain's wounded national pride. Epidemiologists told some health officials in 1996 that the meat simply had to be cooked thoroughly as heat destroys the virus. Later, it was found that BSE was not a virus at all. Prime Minister John Major postponed the destruction of British cows in order to calm public fears blaming the entire crisis on his political opponents. Even the National Farmers Union in Britain suggested that the public could be reassured if Britain ordered the destruction of thousands of older cattle. The Minister of Agriculture in England from 1995-97, with the improbable name of Douglas Hogg, announced new measures on July 23, 1996, to keep animal body parts, brains, spleens, spinal columns, out of the animal and human food chain. He was roundly denounced for his pathetic handling of the crisis but given an award by the vegetarian society for increasing its membership. (6)

Meanwhile, scientists and health researchers could find no bacterial traces and no virus. No response from the immune system signified that the organism had been invaded by a foreign agent. The concern became international news as more cattle became infected and videos of them stumbling and falling flashed on TV sets around the world. About the same time similar symptoms began to develop among British domestic cats that experienced patterns of coordination disability. But the biggest concern was whether or not humans were susceptible to this disease. The answer would come soon enough.

Steven Churchill in the late 1980s in Britain was not acting himself, and his family and younger sister were alarmed. As a normal teenager he was beginning to exhibit signs of a lack of coordination and excessive fatigue. Progressively, this turned into jerky movements, uncontrollable shaking and eventually a total loss of all coordination. In June 1995 when he was only 18 years old, he died. On March 20, 1996, the British government announced that 10 other young Brits had caught this debilitating disorder from eating beef. No older people were ever infected and none died. Demand for beef plummeted throughout Europe.

Steven had exhibited the classic symptoms of Creutzfeldt-Jakob disease, or CJDv, a variant of BSE. This disease generally affects some people over 60 and has a long period of incubation during which the victim feels no symptoms. The effects, however, are often fatal—the person dying within a couple of years. Once the CJDv is discovered, the victim shows signs of a failing memory and lack of coordination. Since thousands of cows in Britain were experiencing a similar disease, biologists were pondering whether or not there might have been a rare cross-species transfer of a disease that could be the beginnings of a human epidemic. By 1989, unwilling to take any chances on the disease spreading, the U.S. banned the importation of British beef.

A public health investigation into five BSE-related deaths in northern England between 1998 and 2000 concluded in 2001 that the cause was a local butchering method that permitted brain tissues from diseased animals (though unknown to be diseased by the butcher) to contaminate the meat. (7) Four local butchers cut the cow's carcass with the head still attached. From the 1940s to the mid-1980s English butchers had catered to a clientele who ate cow brains as a protein source. Butcher contact with cow brains was banned in 1989 three years after the disease was first reported. These five deaths were the largest cluster attributed to CJDv in one locale in an epidemic that had taken the lives of 90 people in Britain by 2000. By 1996, just a year after Steven Churchill's death, 10 more cases of CJDv appeared in Britain, all in young people who normally would not have this disease. A Frenchman died the same year from CJDv but it is thought his death was attributable to taking bovine-derived growth hormones for body-building.

Scientists soon discovered that British beef were being fed meat from dead sheep that may or may not have had scrapie. Could this sheep disease have leaped over the species chasm to cows and could it leap again to humans? Biologists knew that when a disease crosses into another species that the viral strain becomes more virulent, less resistant to inoculation, acquiring a longer host range making it more difficult to identity until it reaches a lethal stage.

Cow deaths kept occurring while international efforts were pursued to locate the cause of these symptoms. Scientists were hoping to pinpoint the disease before it became a human epidemic. They began with the knowledge that BSE is not like a bacteria nor a virus since the immune system, programmed to attack invader agents, had never been active in the presence of BSE. This appeared to violate the known rules of biology. Alcohol, usually effective, had no reaction on it, nor did burning or boiling. BSE did not conform to the regular pathogens of other biological agents and seemed to resist all attempts to destroy it. (8)

One investigator recalled that he had read an article published in 1967 pointing out how a protein could be an infectious agent. A protein? But proteins are the good guys in the body, necessary ingredients of all life, amino acids consisting of oxygen, carbon, hydrogen and nitrogen. A protein contains no DNA and no nucleic acid. How could it possibly replicate itself and grow undetected in an organism?

Written by mathematician, J. S. Griffith, the article noted that there was the possibility of two proteins, normal ones and rogue or abnormal ones, called prions, which have irregular shapes. (9) Prions can bind to normal proteins and cause the normal one to change into an abnormal protein, a prion. By this process, which is not strictly speaking replication, abnormal proteins take over parts of an organ like the brain and cause it to be abnormal. It was a profound, though according to critics, not entirely and original discovery. (10) Protein itself, the building block of all life, becomes an infectious agent, eventually lethal to the organism. The startling insight was that the process was a totally new way for a person to become disabled, not through bacterial or viral replication, but through biological conversion.

Europe had banned imports of British beef when mad cow disease became widespread in Britain. England had already slaughtered four and a half million cattle, effectively decimating the British beef industry, the greatest catastrophe ever to befall British farmers. (11) Having already banned British beef in 1989, in 1997, the U.S. Department of Agriculture banned all feeding of animal remains to cows. However, worse news was to follow.

After this discovery, European and American governments acted promptly. Yet the British government dawdled and denied. Prince Charles publicly ate British beef and proudly declared it "delicious." (12) A British government commission ordered a section of a report that related scrapie to BSE expunged, something no competent scientist would have suggested. Clearly, the British were attempting to protect the powerful beef, dairy and related dairy products industries like Nestle at the expense of public health.

In 1999 American doctors, endorsed by the Food and Drug Administration (FDA), banned blood donations from individuals who had been in Britain for at least six months, especially those who had visited in the 1980s, in order to prevent some form of BSE from arriving on American shores. During that time many BSE-infected cattle were sold in markets. The fear is that some people may have absorbed the BSE by eating contaminated beef and that this contamination might have spread into the blood used for transfusions. This would affect the total blood supply by about ten percent. After further review in 2001, a panel appointed by the FDA to study the effects of donated blood coming from Britain which might cause BSE barred blood donations from donors who may have been exposed to BSE while in Europe, those who had spent five years or more in Europe since 1980, or three months or more in Britain from 1980 to 1996. (13) Twenty-five percent of all blood in New York City was imported from donors in Europe. Ceasing the transfer of blood would definitely shorten hospital blood supply used in surgeries and trauma units. This decision was endorsed by the American Red Cross. (14)

Many people could become infected from cow beef if one infected person's blood was pooled with others. Conventional blood tests do not detect the presence of CJDv. Though no single case had been detected, there was a media scare sufficient to alarm the general public and hence politicians. The fear in the United States was that a variant of the mad cow disease could be incubating and not show up for years then spread quickly. Deer, elk and sheep had been infected with similar diseases.

The European Union (EU) scientific steering committee approved the renewal of British beef sales throughout Europe in 1999. European countries piously proclaimed they were 100% free of BSE disease from British or their own beef products. But an European Union report that year chillingly described the health risks of human exposure to BSE, noting that it is nearly impossible to detect until its terminal stages and may incubate for decades in humans. (15)

After a three-year investigation studying the mad cow problem, British officials released a 4,000-page report in October, 2000. (16) It was careful not to actually lay blame at the doorstep of individuals. But it did chronicle the missteps and misstatements made by nervous and guileful officials who were forced to concede, reluctantly, that when 4 million cows were slaughtered and 88 Britons died and seven others suffered from the disease that some form of BSE did get transferred to humans. The British government agreed to establish a fund to pay the medical costs of victims and to compensate the families for deaths. Farmers had already been reimbursed billions of English pounds for the loss of herds.

BSE in the Third Millennium

The problem would not go away. By November of 2000, fears of mad cow disease were rising in France, a country well noted for its culinary refinements. France had discovered 90 cases of the disease in its beef, triple the amount from 1999. Supermarkets recalled beef from a herd infected with BSE. School cafeterias in France removed beef from their menus and quickly followed by schools in Switzerland and Italy. The discovery of mad cow disease in France sent ripples of fear through the country that relies so heavily on foreign tourists, three million Americans annually, who come as much for French food as for its art and architecture. Suddenly, what was once just a health issue was infecting tourism, trade and economics. Beef prices plummeted and some countries rejected French beef. French officials ordered a temporary ban on livestock feed which contains beef products. All Europe reacted to the distressing signs that other beef may be infected. The court of the European Union in December 2001 said that France's continuing ban on British beef was illegal. (17) Spain reported its first case of BSE in late November, 2000. By early 2001, 23 cases had been detected in France, 20 in Germany, 15 in Spain. By the end of 2001, every European country except Sweden had at least one case of BSE.

There were complications surrounding which countries should set their health codes because there are natural cuisine and cultural traditions towards certain foods, citizen safety standards and international free trade issues involved. How should international associations and trade groups respond to these conflicting demands? For example, the Germans devour Kochmettwurst, a sausage made of cattle brains. Should the consumptions of such a natural delicacy be discontinued because of the potential health risks? Should warning labels be placed on products even if not packaged? How risky are certain cattle parts even if cooked thoroughly? By 2001, Germans were consuming 40% less beef products, including sausages. All animal parts are rendered for animal food consumption: sheep carcasses, road kill, plate waste, etc. Bone and animal meal fed to cows had by this time been banned by the European Commission.

Tackling the firmly entrenched agri-business industry in Europe, receiving $8 billion annually in government subsidies in Germany alone, is an enormous undertaking. The beef business was not aided by a United Nations Food and Agricultural Organization (FAO) report in early 2001 that warned that cattle in more than 100 countries may have been exposed to mad-cow disease. (18)

Dr. Thomas Pringle, a trained biochemist in Eugene, Oregon, was fortunate enough to receive an inheritance that allowed him to establish his own research foundation investigating mad cow disease. Since 1996, he has been collecting and disseminating information about how the prion gene family interacts within the Human Genome Project. (19)

Pringle became interested in mad cow disease from his work with wasting diseases among deer and elk. His dissemination of information on research involving mad cow, and the related human disease, Creutzfeldt-Jacob or CJDv disease, revived the perennial debate about what shared scientific information should remain in the public domain and which should stay among private pharmaceutical companies seeking medicinal patents. Investigators working with medical research companies hoping to develop an antidote to CJDv are not anxious to share their latest research.

Meanwhile, in the United States a mad cow disease scare erupted. In March, 2001, American agricultural department officials seized 234 sheep in Greensboro and Warren, Vermont after four sheep showed signs of mad cow disease. Larry and Linda Faillace had determinedly fought to keep their Vermont flock. But U.S. Department of Agriculture veterinarians suspected these sheep had been exposed to tainted animal feed brought from Belgium in 1996 during a time when import restrictions were lower. For the first time in the U.S. sheep were seized, slaughtered and their brains analyzed for the disease. (20) The tests were to take years as sheep brain matter was to be inserted experimentally into mice to see if they got the disease.

The most lingering question was the incubation period and scientists were cautious in estimating the numbers of how many people were infected, and whether or not there is any human-to-human transmission of the disease. (21) By late 2001, the U.S. Department of Agriculture, based on a Harvard study of mad cow disease, reported that the risk for American cattle contacting the disease was low and unlikely to pose a public health danger. (22)

Chronic wasting disease is a variant of mad cow disease. In 2002 it was identified in wild and captive herds of elk and deer throughout the Midwest, inter-mountain west and central Canada. A few hundred deer were diagnosed and thousands more killed for scientific analysis. (23) Many believe that the illegal practice of selling captured deer and elk has contributed to the spread of chronic wasting disease. As the mad cow scare demonstrated, England behaved badly in the safety of its people, even once the disease was publicly announced, in order to protect its agricultural industry. Exports of infected products led to widespread transmission of a potentially fatal disease in Europe. Beleaguered nations that had accepted contaminated agricultural exports unwittingly had to scramble to contain the spread of the infection.

Just when it seemed the curse affecting the meat industry had abated, mad cow disease was diagnosed in 2003 by an alert Canadian veterinarian in Alberta, Canada. This provincial veterinarian noticed that an eight-year old cow scheduled for slaughter was unusually thin and he suspected pneumonia. When the remains tested positive for mad cow disease, the U.S. immediately placed a ban on Canadian beef, a $1.1 billion dollar annual industry. Investors sold stock in McDonald's, Wendy's and other fast food outlets selling beef products. (24) Again, no one knew the origin of the infection or how many other cattle might be tainted.

Investigations by Max (2004) revealed that by 2004, nine people who ate beef at the Garden State Race Track since 1997 have died of Creutzfeldt-Jakob disease. In January, 2005, Human Rights Watch severely criticized the meatpacking industry for its inhumane working conditions noting that that workers are three times more likely to suffer injury than all other private industries.

The dominance of agribusinesses, the demise of the small farmer and the potential contamination of food products with suspect nutritional items like taste enhancers all contribute to the continuing national discourse about contaminated beef. The spread of common diseases also reveals the tight symbiosis between cattle and humans.

E. Coli Bacteria and Other Nasty Diseases

According to the Centers for Disease Control (CDC), the E.coli bacteria kill approximately 60 people annually and sicken about 73,000 (25) although scientists at the CDC and U.S. Department of Agriculture indicate that it is unlikely the bacteria will show up at the supermarket. More sensitive detection methods for discovering this difficult-to-eradicate bacterium have been perfected. Nevertheless, erring on the side of caution, the U.S. Department of Agriculture closed a beef processing plant in Greeley, Colorado owned by Swift and Co. that had recalled 18.6 million pounds of meat in July, 2002 when 28 people in seven states became ill because of meat containing E.coli. Inspectors were removed when they found evidence of fecal contamination. The plant with 2,500 workers had received a warning in the summer of 2002 and remained closed until the contamination was eradicated.

Similar diseases affect other edible animals that can cause humans worse than indigestion. Take the case of "Fully Cooked Barbecued Chicken" packaged for sale in August, 2000 and sent to several states. The Food Safety and Inspection Service of the U.S. Dept. of Agriculture said that 7,640 pounds of ready-to-eat or precooked chickens may have listeria monocytogenes, a rare but potentially lethal disease. (26) The food processor, the House of Raeford Farms in Raeford, NC, refused to recall the sale of more than three and a half tons of barbecued chicken suspected of contamination. Normally, a food processing company will warm distributors to return to destroy the merchandise. Rejecting a recall was never, until this time, an option.

What's the most common reaction to the announcement that you have an infectious disease? Ordinary people deny it could happen to them and may remain in a kind of psychological stupor for a long time. Mental denial often leads to verbal denial, and that is true for individuals, corporations and countries. For example, hoof and mouth disease, unlike HIV, E. Coli, and the human variant BSE, is not, as of this writing, known to be contagious to humans.

A minor flap occurred early in the Bush Administration in 2001 when a U.S. Agriculture Department official declared that salmonella testing in ground beef would be eliminated for school lunch programs. Red flags were posted by innumerable agencies and school groups. Clearly, the lack of testing for this disease would save the beef industry money and the implication was that the Administration was catering to the needs of industry at the expense of children's safety. Within days the Administration reversed its policy and decided to keep the testing policy in place, upholding regulations passed during the Clinton Administration. (28) Food safety must always fight with the powerful forces of the meat industry that consistently objects to testing for meat quality, even for potentially lethal diseases that could be transmitted to children.

In the 1970s most infectious diseases were thought to be under control globally. But a troubling display of lethal, not just annoying, diseases appeared which spread alarm throughout the world community. These included Lyme disease acquired from deer, AIDS, Legionaire's disease, the Ebola virus, and mad cow disease. Too many variables like sexual practices, changes in land use, population increases of people and animals, industrial food processing, overcrowding, and intravenous drug use, contributed to the spread of these new lethal diseases. The evolutionary resistance of micro-organisms was more rapid than the drugs used to prevent their spread.

According the Centers for Disease Control, salmonella causes 600 deaths annually. McDonalds and other fast food chains have recognized the potential hazards and have taken steps to prevent salmonella's appearance in its outlets. When the ground beef served in any restaurant can come from as many as a hundred different cows, every precaution has to be taken to insure protection against contamination. By mid-August 2002, the West Nile virus had infected nearly 300 people. This particular nasty disease had been unknown in North America prior to 1999. Shortly thereafter, it infected 29 species of mosquitoes, over a hundred species of birds, and killed a few people. (29)

Hoof or Foot-and-Mouth Disease

Hoof and mouth disease, which is not considered dangerous in humans, is an enemy virus that can ruin a livestock industry. The first documented case appeared in the United States in 1914 when 324,000 livestock in 22 states were killed. (30) It appeared again in the United States in 1929 and was thought to have been caused when ocean liners sold their garbage for pig food. Among the six cases reported in the 20th century, one occurred in 1924 near Berkeley, California and was introduced among pigs. Thousands of livestock had to be killed to prevent further contamination.

The virus afflicting British livestock was identified as Pan Asian Type O, one of at least seven different types and 80 sub-types of this virus. The concern of the agricultural industry is that the virus can contaminate any cloven hoofed animal, such as pigs, sheep and cattle. It does not affect horses, though horses can transmit the disease, and it can be carried on the wind quickly to distant locations. All animal related activities cease when the infection is discovered as officials who seek to contain its devastating contagion and rapid spread. Once an epidemic of hoof-and-mouth is declared as it was in England, there are ripple effects throughout the economy revealing that cattle and cattle byproducts have multiple uses in the economy.

February 21, 2001, was an inauspicious day in Britain. On that day Hoof-and-Mouth disease broke out in England, bad news for the British beef industry that was still reeling from recent outbreaks of mad cow disease and swine fever. The British government quarantined all cloven-hoofed animals, isolating farmers from their markets. As changes in the marketplace favored large agricultural producers, small British farmers were especially hard hit by the spread of this disease that effectively stopped all agricultural movement in the country. Hoof-and-mouth disease followed the mad cow disease scare a couple of years earlier crippling a British agricultural industry in which 51,000 people, nearly 10 percent of the farming work force, had left farming from 1998 to 2000. (31) By 2001, Britain had about 63 million livestock: 44 million sheep, 11.3 million cattle and 7.2 million pigs. (32)

The epidemic showed no signs of abatement as new farms were found daily to have the infection. (33) Britain had considered vaccination as a remedy but resisted it politically, believing it would harm exports. Moreover, tests are unable to determine an animal infected with the virus from a vaccinated one. The European Union had banned vaccination against hoof-and-mouth in the early 1990s. (34)

Within a five weeks, about a half million cows, sheep and pigs had been slaughtered and an additional 100,000 were targeted. (35) Within six weeks,

390,000 animals had been slaughtered, another 230,000 were waiting to be killed and 290,000 carcasses destroyed. Some believed the smoke from burning pyres was actually transmitting the disease through the air. The British army had by then been called in to supervise the logistics of burial in Cumbria county in England's north. Agricultural officials had widened the policy of containment whereby farm animals as far as two miles distant from where the disease had been detected were targeted for destruction. It was as if the Black Death, which killed an estimated one-third to one-half of the human population of Europe in 1348-50, was ravaging the cattle and livestock population. Consumers were initially confused and then chagrined. Farmers were outraged and dispirited, and a few committed suicide. Trade bans multiplied as agricultural markets plummeted. Just when everyone suspected globalization or terrorism was the chief evil in the world, a borderless virus asserted its supremacy.

Within weeks shortages in other industries showed how important cattle byproducts are to civilization. The manufacturers and retailers of luxury leather products, such as car seats, bomber jackets, handbags and shoes, began feeling the shortages of unavailable cattle skins. As people ate less beef because of the mad cow scare, less cattle were being slaughtered and hence less hides came on the market. The lack of European supplies of cow skins suddenly had global impact in the leather industry, as the price for a Texas steer hide rose from $55 dollars to nearly $80. (36)

The bad press from this disease affected tourism in the United Kingdom as hiking trails, country footpaths and county events like lambing days were closed or cancelled affecting bed and breakfast hotels, shops and restaurants. Travel restrictions on 76% of the English countryside were costing the British tourism industry as much as $150 million weekly. The entire rural infrastructure was brought to a halt. During the spring of 2001, as many as 15 million visitors who would normally come to the Lake District were reduced to a trickle. About 350 jobs were lost weekly and the district's losses calculated at $15 million a week. Most farmers, who had spent years cultivating pedigree herds, went bust. Farmers were scheduled to receive some government financial assistance, but not rural business that had been economically devastated. By the turn of the millennium, tourism in Britain was close to a $100 billion dollar industry. The farming industry was $21 billion. Big businesses were severely affected by agriculture.

By late 2001, 3.8 million animals had been slaughtered. British farmers had preferred culling to vaccination, even though vaccinations were available. The disadvantage of vaccinations is that it takes a few days for the animals to become immune. Farmers were fearful that large-scale vaccinations would ruin the livestock export market and the European Union prohibited the use of vaccination to combat hoof and mouth disease, though later Britain received a waiver. (37) The food giant Nestle, which exports powdered milk products to Europe, was opposed to vaccination because of the public scare from other countries. The delay, inspired by politics and not health, caused the outbreak to linger.

Had strict quantitative measures been established and immediately imposed, the foot and mouth disease could have been more effectively managed and restricted to a smaller section of England. Officials did not take a lesson from the mad cow disease. The epidemic was only stopped after road blocks were set up to inspect farmers' cars traveling from infected to uninfected areas. (38) The absence of a contingency plan widened the epidemic and helped spread the disease. No one could have believed it would ravage the countryside so devastatingly. At the time, government officials had no plan in place for such an emergency, especially for disposing of so many thousands of carcasses. One estimate put the figure as high as 50,000 dead animals awaiting disposal, some lying exposed for two to three weeks.

The risks were enormous: public health, environmental, and of course political. At the peak of the outbreak there were 30 to 40 new animals infected daily. Ministers and government officials shied away from giving evidence at inquiries as if the commissions themselves were carriers of some disease. The government refused to hold a full inquiry.

Three stories will illustrate the plight of ordinary farmers in the United Kingdom to the Foot-and-Mouth plague.

In Bradford, England, a city with a high Muslim population, the Saqib Halal Meat Shop (halal means "permissible" in Arabic), normally full for the purchase of lambs to begin the Muslim holiday (Eid-al-Adha, or Eid-al-Kabir, the Feast of the Sacrifice which commemorates Abraham's sacrifice of a ram instead of his son), had no lambs for sale. Licensed halal meat shops slaughter lambs in accordance with Islamic law. According the Koran, Muslim believers adhere to the injunction: "To thee have we granted the fount of abundance. Therefore, to thy Lord turn in prayer and sacrifice." Like other Muslim specialty and halal shops, Saqid's had been cut off from meat suppliers because of the spread of the disease.

The Jones family had been dairy farmers in Wales for three generations. The 220-acre farm had 228 pedigree Holstein dairy cows. All family members, including the ten-year old daughter, knew each cow by its name. One day in March, 2001, hoof-and-mouth disease was discovered by inspectors among the Jones' cows. Each had to be shot where it stood. Mrs. Jones lamented how everything was now gone for the family business and they would have to start over from scratch. She could not watch the shooting. "Watch, no way. I heard it, that was enough." (39) "I did watch them burn. I needed to be there for them," she noted with a trembling sadness in her voice.

In Scotland, the Gretna House Farm of 650 acres received the calamitous news that hoof-and-mouth had been discovered at a neighbor's farm, whose owner had unwittingly carried the disease to his place by transporting his sheep from a local market. The market at Longtown just happened to be the epicenter of the epidemic in Scotland. Within days a lame ewe with mouth lesions was found at Gretna Farm and the inevitable had occurred. (40) Even though the owners of Gretna Farm had taken all possible precautions, such as attempting to persuade the local authorities to close a local road unsuccessfully, covering the road end with disinfectant and straw and foregoing trips to the pub and visits to friends, a virus spread by the wind is stronger than static preventive measures.

The Gretna Farm family raised pedigree beef cattle. The Highland cattle, with their long horns and shaggy manes, were shot by marksmen because they are difficult to round up. The pedigreed and highly regarded Charolais and Simmental cattle, many special show animals, were all summarily dispatched together with the sheep. Just days before the outbreak, the family was notified that their Charolais had been rated in the top ten percent of breeding cattle in Britain.

Moving more quickly than electronic transfers, this livestock scare swept across Europe as country by country imposed strict measures to control its spread. As British farmers slaughtered herds to contain the infection, German officials confiscated parcels of food, meat and milk products from British tourists and hunting trophies and other skin or hides were placed on a warning list. European travelers from England had the welcome mat laid out, but it was a disinfectant mat to counteract possible transmission of the disease from the soles of shoes.

Hoof-and-mouth or foot in mouth was found among cattle at a dairy farm near Laval, France despite strict precautions. All livestock exports from England ceased and thousands of suspected infected animals slaughtered. The skies were dark with the smoke of burning carcass pyres. Within weeks, Portugal, Spain and Belgium had closed its borders to French meat. (41) Dutch farmers faced the threat stoically, some prepared to slaughter years of breeding practices. (42)

Rugby matches were cancelled because officials felt that fans could carry the disease home on their boots, soles of their shoes, or clothing. Footpaths were closed to hikers. Public zoos were closed. The annual dog show was postponed. Horse racing was cancelled by Britain's Jockey Club. Dublin cancelled the St. Patrick's Day parade. Every country feared the spread of the disease in its backyard. Ireland suspended all horse and greyhound racing. As Northern Ireland livestock was found to be infected, over a thousand soldiers were stationed at the border between Ireland and Northern Ireland to prevent animals from straying south. It did not prevent Irish sheep near the border from becoming infected and the disease was discovered in March 22nd, 2001. (43) Officials immediately began slaughtering Irish sheep and hoped to contain the disease to the region and not the country.

On March 14, 2001, the United States banned imports of animals and animal products from all 15 countries of the European Union. The following day Australia imposed a similar ban on European livestock. European beef had already been banned earlier because of mad cow disease. Cheeses were not affected because they are pasteurized thus killing any infection. Only strict quarantine and vigilant border inspections have kept this disease at bay from American livestock.

While Britain was mobilizing its army to bury over a half million livestock, in Hong Kong, a former British colony until July, 1997, hoof-and-mouth disease was more casually treated, where it was treated at all. (44) The disease has been rampant in Asia for decades. In 2000, there were 3,283 confirmed cases in Hong Kong compared to 1,061 in Britain in 2001. Hong Kong does not export any of its meat or pork so it does not have to satisfy the control requirements of its trading partners, and no restrictions are placed in the importation of meat or dairy products from other countries that might have the disease, and no animals have been slaughtered. Though farmers are supposed to leave animal carcasses at designated collection places for officials to haul away, when animals known to have the disease are simply dumped along the road by Hong Kong farmers, sanitation workers pile lime on top of the carcasses.

In Japan, Japanese consumers were skeptical when one case of mad cow disease appeared on the north side of Tokyo Bay in the early fall, 2001. (45) In the 1990s Japan's beef imports more than doubled with $1.8 billion worth coming from American beef. But the one mad cow disease case in Japan and conflicting government announcements caused about half of Japanese consumers to reduce their consumption of beef dramatically. Meanwhile, South Korea killed 350,000 cattle in 2000 and Taiwan slaughtered whole herds.

Europe had run the course of its livestock and cattle hoof-and-mouth disaster by late spring, 2001. But America was still anxious about the possibility of the arrival of hoof-and-mouth on its shores and was preparing for the worst. The ultimate fear was that the disease would impact the wildlife herds of deer, buffalo, caribou and antelope, the virus carried by the wind or through the feces of infected birds that would have come in contact with diseased animals. More than 22,000 deer were killed when the disease was rampant in 1922. If the disease were to affect them, some wildlife herds, like the caribou, might never recover and might be permanently exterminated in North America. What was most feared economically was a ban on meat exported from the U.S., a $50 billion dollar a year industry. (46)

The serious outbreak of Foot-and-Mouth disease in European, Latin American and Asian countries had prompted a global scare. Canadian authorities posted warning posters and pamphlets at immigration entry points urging travelers not to bring any meat, animal or dairy products into Canada, according to the Canadian Food Inspection Agency. Severe penalties awaited any violators. Entrants were recommended to stay away from Canadian farms, zoos and ranches for two weeks after arrival from any country that had reported Foot-and-Mouth disease.

If ever a confirmation was needed that there is a strong, permanent linkage between humans and cattle, the worldwide spread of BSE, and hoof-and-mouth disease and the accompanying global fear of contamination, cinched the relationship. But anthrax, despite its occurrence as a terrorist's poisonous use in the aftermath of the September 11, 2001 disaster, is also a continuing biological threat.

Anthrax

What was unthinkable prior to September 11, 2001 suddenly had to be deliberated by public health officials. Biowarfare, terrorist attacks with infectious diseases, hit the American psyche like a bad cold and a rip-roaring headache. And that was for people unaffected by a poisonous chemical. One or more terrorists mailed cultivated anthrax spores to notable members of the media, the U.S. Congress, the U.S. Supreme Court and the White House in October, 2001. The first to die was a photo editor for a national media outlet in Florida, quickly followed by two postal workers in Washington, DC handling congressional mail.

Anthrax spores are patient assassins living in the soil often for decades. They are uprooted by grazing animals like cows and bison, unearthed when they pull up a plant then breathe in the microbes which multiply endlessly, gorging on the animal's nutrients until it dies. These bacteria then revert to spores, pour out through the dead animal's liquids and return to the dormant state in the soil again, lying in wait for the next disturbance to awaken them. (47)

Along the old cattle western trails from Texas to Wyoming or Kansas in the years 1865-1890, drovers left infected animals to die. When carcasses had swollen and rotted away the anthrax spores migrated back into the ground where the spores emerged again under certain propitious weather conditions, like heavy rains when they percolate to the surface, or when animals uprooted them while grazing.

More than 1,600 cattle died of anthrax in Texas in the summer of 2001, the most severe outbreak since 1987. Nobody knows how many deer were infected but the Texas deer population had decreased. Ranchers who raise deer for hunters are reluctant to talk about anthrax contamination for fear of hurting the business. (48) Anthrax is an occupational hazard for Texas ranchers. Anthrax can be eradicated if infected dead animals are burned, not buried, and if stock are vaccinated as most ranchers now do.

Humans are infected with anthrax cutaneously or through inhalation. Inside a body, the spores are absorbed by macrophages, cells designed to handle such impurities, where each spore turns into an anthrax bacterium, multiplying until the macrophage bursts, then the anthrax bacteria get released into the bloodstream. Released toxins bind to other cells in the body eventually destroying it. If detected early after the infection, an individual can be treated with antibiotics. Many rural Russians are exposed on a regular basis to the anthrax threat because of exposure from farm animals that carry the disease. (49)

But cattle may also save lives, and not just as a food source. In one of science's continuing working miracles, 12 scientists reported in an online publication that cloned calves can produce immune globulin or gamma globulin found as antibodies in humans. (50) The possibility increased that such antibodies coming from cows could be given to human patients infected with certain diseases for which they are now immune deficient making their recovery.

In the near future cattle could be a source for treatment of certain diseases at present only derived from antibodies produced in human blood, always in short supply. A cow carrying human antibody genes could be immunized against a particular disease and within a couple of months such antibodies could be collected and given to infected humans, thus saving lives. (51)

Mark Taliaferro, a rancher for 25 years, runs a cattle ranch on the outskirts of Dupuyer, Montana. A weird occurrence was ravishing his herd: the unnatural deaths of a select few of his cows by facial and genital mutilation. Some cows have had their face masks neatly incised from the carcass, and not by bite marks. There are no vehicle, animal tracks or footprints near the dead cows. Space aliens, satanic cultists and people inhabiting UFOs have been ruled out as culprits. But the mysterious deaths continue to puzzle animal scientists, veterinarians and forensic experts. (52)

Microscopic organisms are not fussy about which organisms they infect. Upon entering a biological system they proceed to wreck havoc by rapidly multiplying and then consuming the unsuspecting host. HIV was transferred from a chimpanzee in Africa to a human. Monkey pox, related to smallpox, appeared in the U.S. in 2003 among dozens of people who became infected by contact with pet prairie dogs infected by other small animals imported from Africa and sold as pets. SARS killed hundreds in Asia in 2003. The nipah virus discovered in 1999 was transferred from fox bats to pigs to humans in Malaysia and killed over a hundred who died within a couple of days after infection.

Agencies and bureaucracies are ill-equipped to monitor all these activities in these huge enterprises given all the variables for technical and human error. But the hunt continues for the origin of the infection from usually tropical animal species emerging from wider contact with animals or people. The infection in most cases can be fatal unless stopped with antibiotics if an antibiotic has been developed. Although all organisms share common ancestors from 600 million years ago, cows and humans do share respiratory and circulatory systems and digestive tracts, both susceptible to some of the same diseases that plague the planet.

In these remaining three chapters we examine the culture of the cow in Africa, India and Asia to illustrate the important role cattle make in the lives of peoples in differing parts of the world.

Figure 6: The Norse Myth of Gefion with her Four Sons
as Bulls Plowing A Fountain near the Harbor in Copenhagen, Denmark

CHAPTER ELEVEN
AFRICAN CATTLE: WEALTH AND INSURANCE

John Danforth, former senator from Missouri, in 2001 was appointed by President George W. Bush special envoy to the Sudan in an effort to step up relief as a step towards improving relations with the U.S. While being escorted in southern Sudan, villagers killed a cow in his honor that he had to gingerly step over as a part of the ritual avoiding the pool of blood leaking from the fatal neck wound. Nothing in his formal education prepared him for this cultural peculiarity, but knowing the value of the cow in all cultures, especially in Africa, he would perceive how important are the ritualistic acknowledgements of each culture to killing a cow in one's honor.

Africa is, according to scholars, the original home of all humans, but also the habitat of the animals the first humans hunted. The broad plains of east Africa, encompassing Kenya, Tanzania and Uganda, and the vast Serengeti Plain and preserve in north Tanzania, is home to some of the most wild, revered, and protected animals on earth. Most protected animals of all are the domesticated animals: the cattle of the various tribes in the region who regard them as the ultimate form of capital. Beef is sacred food, not just meat.

East Africa was once thick with green vegetation, the beauty of the acacia trees, undulating hills, plantations of cotton, maize, coffee, and sugar cane, herds of cattle and goats, great herds of wild animals, the giraffe, zebras, wildebeests, antelopes, impalas. But today much of sub-Saharan Africa is beset with civil war, genocide, human and animal over-population, and diseases like the HIV and Ebola viruses. Apart from over-crowded cities like Lagos, Nigeria and Nairobi in Kenya, life in the rural bush and plains relies on livestock, and especially cattle, as a means of livelihood.

Archaeological excavations and analyses have yielded insights into prehistoric pastoral peoples in central east Africa. (1) Bone analysis shows that humped cattle existed in this area over 2,000 years ago. Ethno-archaeological studies show similar living conditions to present-day Maasi peoples; the subsistence economy has changed little over the millennia and still revolves around cattle and livestock. Tribal members think in terms of number of head and not the money value of cattle. (2)

Africa is home to two kinds of cattle: the sanga, or forward-humped cattle which first evolved in east Africa, and the zebu that came from Arabia. The zebu cattle are somewhat small with short horns, short-legged with loose skin and fatty humps. There has been much inter-breeding between the two groups so intermediate breeds exist. The peoples who cultivate cattle are likewise two general kinds of tribal groups: the hamitic and semitic peoples who are nomadic and cattle-herding pastoralists, and the negro or Bantu in southern Africa who are settled agriculturalists. Most of the tribal, cattle-herding groups in east Africa are nomadic.

The so-called Big Five animals of South Africa are the lion, tiger, elephant, rhinoceros, and the water buffalo. Though wild throughout much of Africa where it is not domesticated for farm use, water buffalo are maintained in South Africa's well-kept game reserves. I saw herds of them in Krugerland National Park near the border with Mozambique from the comfort and safety of a large van in 1988 and again in 1998. They are formidable adversaries in the wild when threatened, not the docile creatures easily tamed for domestic farm work as they are in Asia. One knowledgeable observer, a native South African, told me that if threatened with harm they gallop off as a herd only to swing around and charge the attacker from the rear, almost like a military strategy. When initially encountered, they stare unblinkingly at the intruder, alert, and ready to move forward with horns lowered, or retreat to safety.

Like all wild animals in Africa, water buffalo forage and graze in the safety of herds and not singly to avoid predation. Apparently only the very sick, elderly and disabled are taken by lionesses, tigers, or hyenas. Their size and strength, their viciousness in defense, and the sharpness of their horns, give even voracious predators pause. Hence, its ranking as one of South Africa's Big Five is justified because of its fierceness as a wild animal, not for its domestic usefulness.

Survival comes first in Africa for man and beast, then adaptation to the environment. Human migration effects cattle, and tribal members move from dry to wet areas as the need dictates. The arid Saharan in the north and tropical climate in the equatorial and southern part of the continent can be cruel on humans and animals alike. The hunt is always for food and water and this frequently means migration. Cattle in Africa are not as large as in northern climes and thus have lower milk yields and shorter lactation periods for the calf's needs. (3)

Pastoralism is the most sustained way of a traditional economy in much of Africa in a grassland environment. The governments of Sudan, Somalia and Kenya recently tried to establish enclosures to encourage pastoralists to settle, ranch, avoid migrations, and to promote meat production, programs which actually began during the early 20th century with years of colonial administration. Nomadic peoples were relegated to poorer lands and many would have had to exchange a pastoral life for wage employment. The quest to incorporate disparate groups of ethnic peoples into a national economic scheme is presently destroying older ways of subsistence and may well result in the elimination of cultural identity and pastoral ways for certain groups. Successful breeding practices have resulted in numerical increases in animals, especially when viewed as capital currency among tribes like the Maasi. This puts a strain on grazing areas. Seasonal changes involve issues of grazing, animal diseases, water and water pollution, soil erosion, and even rain patterns. (4)

The land of the pastoralist does not always sustain a settled agriculture. The land is often unsuitable for the support of farming, as it is too sandy in Africa or the Middle East, or too rocky in western Ireland. Pastoral production is based on the livelihood from the care of large animal herds. Constant grazing puts limits on grasslands. When they move, pastoral peoples in Africa carry all their possessions with them or relocate to seasonal camps. For tribes like the Maasi, this means the cattle move with the people too. Sometimes, a move must be made every four to seven days as grazing areas become depleted during the wet season. Some animals, like camels or oxen where available, become pack animals. An eruption of the active volcano like Oldoinyo Lengai in 1967 in Africa forced a quick Maasi migration in a search for better pastures.

Five tribal groups will serve as examples of peoples in this chapter who have an intimate relationship with the land and African cattle: the Wachagga who live on the slopes of Kilimanjaro, the Barabaig of Tanzania, the Dassanetch of Ethiopia, the Maasi of Kenya and Tanzania, and the Cattle-Herding Himba Tribe of Namibia.

The Wachagga

Towering nearly 20,000 feet just three degrees south of the equator is the world's largest volcano and the tallest free-standing mountain in the world. Kilimanjaro is a word that means "mountain of spring waters" in the language of the Wachagga, the people who live on its lower slopes. Whether you are a romantic who has read Ernest Hemingway's *The Snows of Kilimanjaro*, an adventurer who would like to climb over 25 miles in six days through a rise of 15,000 feet to the glaciers near the top, or one who wants to study tropical wildlife up close and personal, this mountain is a magnet. The vegetation holds the rainfall and provides a protected sanctuary for tropical wildlife and a closed canopy for some of the earth's most exotic birds. And because there is negligible erosion, there are perpetual free-flowing streams providing a never-ending and pure water source for drinking and irrigation to the Wachagga who live downstream. The mountain slopes provide some of the thickest and most luxuriant vegetation anywhere in the world, receiving 78 inches of rainfall annually, more than London. No other place on earth offers such a diversity of wildlife.

I recall vividly an airline flight directly over Kilimanjaro during an early morning sunrise, and recalling how awesome the view was down into its half-mile wide crater pit of smoldering hot ash. The gigantic mountain that dominates the east Africa plain is in fact three distinct volcanoes that have merged over one million years.

The tribe own hundreds of thousands of cattle, but live in a region with sparse grazing lands, even though there are abundant tree crops for cultivation. Kilimanjaro has little pasturage and the nearby countryside is becoming thickly populated with people and livestock. To complicate matters, the region is infected with the tsetse fly. As a result, tribal owners keep half their cattle herd in huts and feed them with cut grasses. A cow will eat about 100 lbs. in 2-3 days so the women of the village must descend further down the mountain slope, cut grass, then pack it uphill to feed the herd. (5) The women do not trust the diligence of the men to do this work because they must rely on the milk and fat of the cow as a basic food source. No milk is sold until all relatives have had their share.

As the tribal members have many cattle, they often deposit them with relatives and friends who are only too pleased to keep the cows in exchange for the milk. The owner thus avoids possible confiscation of some of his cows by his enemies and creditors. Only when an owner dies, not having revealed the whereabouts of all his cattle, does a dispute arise about ownership.

Bulls are kept in separate huts where they are fed bananas to fatten them for slaughter at feasts. A man could have several such bulls. Those with many bulls can have many feasts, and, because he invites so many guests, can attend many feasts of his friends with large numbers of bulls in reciprocity for his generosity. (6)

The Barabaig of Tanzania

The 25,000 members of the Barabaig people live in Tanzania and are devoted to cattle raising and herding. Their cow of choice is the zebu, humped at the shoulders, standing about 5 feet, with one to one-and-a-half foot horns. The climate they inhabit is uncompromising and unforgiving. To survive the Barabaig must migrate frequently with their herds to find new grasses and watering holes. Yet, these animals thrive in semiarid conditions, but long walks to find good grazing frequently result in exhaustion and poor lactation. The Barabaig have an emotional as well as a livelihood attachment to cattle and all products of the cow are used in ordinary life. (7)

Deceit is required to get the cows milked. The milker holds a calf between his legs so the mother cow thinks the calf is sucking while the milker milks the udder into a gourd, then he lets the calf suckle. If a calf dies, the milker stuffs straw into the dead animal's skin, lets the cow lick the skin to begin lactating and then is milked.

The milk supplies are sporadic, even during the rainy season when grazing is more abundant. Milk storage is not possible since there is no refrigeration. Men prefer to drink raw milk; curdled milk is the women's choice. A form of liquid butter, called ghee, is used to rub on leather products like harnesses to keep them supple, and even on themselves as a skin ointment and moisturizer. Milk is sometimes mixed with cow's blood to produce a paste.

The skin of the dead or butchered cattle makes a variety of leather goods. Women use the calf's skin, making it supple by first scraping off hair and flesh, stretching it out, and then letting it soak for a few days in human urine. Cow urine is used as an antiseptic to wounds.

Here are a few of the topical concerns in understanding how the Barabaig build their herds.

- *Land*: Unless the tribe migrates, the cattle will starve or die of thirst, and new lands must contain both grasses and water.

- *Water:* In their part of the world, there is less than 20 inches of rain per year and no rivers. The cattle must survive on pools of water accumulated after a rain, and the cattle must compete with wild animals for the few water sources.

- *Grazing Grasses*: The cattle must also compete with wild animals, like zebras, gazelles and wildebeest, for grassland foraging. Accidental or man-caused fires can destroy these grasslands and eliminate or seriously reduce this food source for cattle. Army worms can also reduce grasslands.

- *Diseases*: There are many African diseases that can deplete a tribe's cattle, among them bovine pleuropneumonia, rinderpest, hoof and mouth disease, rift valley fever, anthrax and sleeping sickness.

The Tsetse fly is a small, two-winged insect inhabiting the bush in central west and east Africa and deadly because it carries lethal viruses. The mosquito carrying malaria is bad enough, but the tsetse fly carries a sleeping sickness, trypanosomiasis, fatal to cattle whose blood it sucks thus transmitting the disease. The trypanosome is the genus name for parasitic flagellate, a protozoan, which infests the blood of vertebrates transmitted though the bite of an insect. Trypanosomiasis, a form of sleeping sickness, is the resulting infection. The zebu tolerance is very low, and the cattle of the N'Dama and the small dwarf humpless shorthorns of West Africa also have a high tolerance for the bite and disease of this insect. Immunization programs help, as do efforts to control the insect itself.

Here are other issues in appreciating the relevance of cattle to the lives of these people:

- *Foraging*: Long walks across the savannah means diminished lactation and a lessening of the milk supply. This also means the meat for human consumption is tougher and less satisfactory.

- *Cattle Theft*: All tribes take precautions against cattle theft, though theft is infrequent. Because of their similar color, cattle can easily be hidden among other cattle in a herd from another tribe.

- *Social Prestige*: All marriages are established on the economic basis of cattle and livestock. The cattle give sustenance to the husband and wife, and the cattle usually come from the husband's family. The journey of the newlyweds to their new home is interrupted by gifts of cattle by the father, mother, and brother, if married. The wife will know the qualities of the animals, and will probably receive on average about a dozen cattle. The range of cattle received varies from two to twenty.

The traditional gift of a prospective suitor in all of Africa helps determine whether he will win the hand of the potential bride and receive the blessing of relatives and ancestors is cattle. In rural communities, five to ten cows is been the going price for a bride. (8) But as salaries have stagnated and the local currency inflated, the price of a single cow in Zimbabwe, for example, has risen from about $54 a head to over $200 in 2004 prices. Some families want payment in American dollars. The exorbitant prices demanded by poor families are straining the demands of poor lovers. It isn't only the cows that are getting milked.

The Dassanetch of Ethiopia

In the southern corner of Ethiopia, just north of Kenya and Lake Turkana and next to the border with Sudan, lies an area inhabited by the

Dassanetch, a tribal group little known to the outside world. Their population is relatively small, at about 20,000, and they practice animal husbandry, limited fishing and small-scale agriculture. Though in the not-too-distant past they may have been full time pastoralists, today they subsist on a combination of agriculture and livestock management to survive. The cultivation of crops, primarily a second growth of maize and a third growth of sorghum, is for human consumption and for animal grazing. Like the Maasi and other pastoralists, they herd cattle and other small livestock and their values, status, and social relationships are built around their animals. (9) The few Africans who fish full time for sustenance are believed by the Dassanetch to have a lower social status. The 30 or so elders who preside over the affairs of the tribe are called "bulls." In fact, the animal bulls technically are the property of the human "bulls," and no decision can be made about the animals without the permission of the elder bulls. (10)

Like the Maasi, the Dassanetch rely on rainfall to determine where and when they will graze their cattle and seek water. During the dry season they will lead the cattle to water, to wells, the nearby Omo river, or the lake, one day early in the morning and return later that evening to begin grazing the following day. Cattle are only watered every second day in the dry season during which time they are slaughtered more or less continually. Some scientific observers thought they were witnessing ceremonial feasts, when in fact they were only witnessing what is a common experience of the tribe. There are varying patterns of consumption of the beef: by the men exclusively, by an age-group, by a family, by females, or for the old and sick depending on their number. Naturally, meat feasts also occur on special occasions such as births, circumcisions, marriages, or deaths.

The women and wives control the milk but not the cows that produce it, nor can they decide where the cows will graze. Likewise the man may dispose of the animals in any way he pleases, subject to the rights of the elders and custom. For example, if a first male child is born, the husband must allocate some animals for the child's personal herd, though they will remain under his management. He cannot use these animals for his own purposes, or even to slaughter one to entertain his son's friends without incurring bad faith. Bachelors and married husbands who are childless cannot own a bull.

The second important event in the life of a household of the Dassanetch next to the birth of a son is the creation of bride-wealth, a dowry. Two or three years before the girl's breasts "come out" a man begins the transfer from his own herd of about 12 bride's-wealth cattle, the beginning of a long period of creating bonds and affinities between different tribal members some of whom will become part of his household and extended family. He may have to borrow cattle from friends, or even from the herd in his own household to create a dowry, known as dimi, the official coming-out ceremony for the young girl, and again when he gives away his daughter to her new husband. The marriage of a daughter may mean that new cattle come into the household's herd, but the marriage of a son means the cattle leave for good. When the head of a household dies, all the bulls and rams, plus one heifer, must be slaughtered immediately.

The family divides its responsibilities to the herd among several family members in different locations. The stock is typically divided among those kept at the settlement camp, where the wives live and milch cows are maintained, and looked after by the adults, and that part of the herd looked after by boys or young men at stock camps. Milk and manpower, stock and grain, are shifted as needed between these camps. These camps relocate depending on the season, as during the dry season there is less grain and milk since the cows are watered only every other day.

> *What is land? Land is nothing. Today I cultivate here, tomorrow I cultivate in another place. Today land exists tomorrow it vanishes. Cattle stay forever. (Dassanetch Elder) (11)*

The Maasi

The entire lives of the Maasi tribe, about 250,000 in Kenya and 100,000 in Tanzania, revolve around cattle and some sheep and goats. The Maasi tribe is known as the "people of cattle." (12) Maasi cattle are sturdy and for the most part disease-resistant. (13) The people subsist on cattle blood, meat and milk, and occasional kills of other animals such as the eland, an antelope they believe is a wild cow. Cattle manure is a vital source for some crops, and serve as fuel and plaster for the construction of huts. In recent years the Maasi have traded cattle for grain products as a dietary supplement. Occasionally, they use other tribal groups as sharecroppers to build vegetable farms near their camps, often less than an acre.

Bonds of intimacy between a husband and wife and between a father and his children are created and sustained through the sharing of wealth, which means in central east Africa giving and receiving cattle. Cattle, the main source of wealth in the nomadic pastoral communities of Africa, are a vehicle for building and keeping social bonds to the family, the extended family, the clan, the tribe and members of neighboring tribes.

There is no stable ratio of human to cattle population, but a ratio that fluctuates according to conditions. Several studies have attempted to calculate the proper ratio of cattle per head or per household to sustain this pastoral economy, from a low estimate of five cattle per adult head to ten. What must also be factored in is the size of the small stock herd and this may vary from 25-50 percent of the total livestock holdings.

For example, there was a systematic decrease in the ratio of cattle per head from 1960, from about 13 cattle per head, to 6 per head by 1978. During the same time, there was an increase in the size of the small livestock herd, mainly goats and sheep. It appeared as if the small stock were gradually replacing the cattle in the Maasi economy, a possible sign of impoverishment. Cattle began to increase in the 1980s, and this fluctuation of the ratio of human population to cattle was attributed to severe drought. Clearly, the weather and environmental conditions determine the fate of mammals as rain and droughts predict the amount of vegetative fodder for cattle and hence the livelihood of the pastoral people.

The dietary and nutritional standards vary according to the amount of cattle holdings, and this depends in turn on the amount of rainfall and grazing conditions. The wet season provides the highest amount of caloric intake, and the dry season the lowest as there is less grain and less milk yield. The diet combines a balance of milk, meat and grain. Food standards are highest when cattle prosperity is highest and when excess cattle can be exchanged for grains. In general, the men consume more milk than either the women or children and the high protein intake may account for longevity.

Women do not control the sale of cattle or livestock. Only the men perform that duty and they own the animals even if the wife brought them in a dowry. The women are responsible for the feeding of the family and they alone have control over the milk distribution in the household.

Wild animals have an impact on the livelihood of Africans and their cattle. In the latter part of the 20th century there was a huge increase in the wildebeest population, doubling from 1965 to 1980. Wildebeest graze on the same grasses as cattle thus reducing available forage. Mingling with the wildebeest herds is discouraged because the wildebeest calves during the birthing season carry a malignant catarrhal fever to the cattle. As a result, the tribe moves quickly away from wildebeest herds. Droughts and disease to the cattle are the chief threats and thus cattle are always kept as insurance against epidemics or the loss of grazing lands.

Diseases are frequent and often fatal to the herd. For example, East coast fever is carried by a tick. Foot and moth disease and anthrax are prevalent and widespread. Roughly 8-10 % of a livestock herd can succumb to a variety of such diseases and early mortality. Veterinary services are rudimentary and the supply of cleansers used in dipping irregular. Regular dipping, as done with sheep in Scotland, for example, would easily control the tick that disseminates east coast fever.

The Maasi drink the blood of their live cattle as a protein supplement without killing the animal, a practice that would be anathema to many ethnic and religious groups. Leather straps are placed around the neck and the jugular vein is pinched with the fingers. After enough blood has been extracted, the leather strap is released.

Among the Maasi, the man appointed to perform a circumcision wears a coat made of cowhide. He will receive a bull if he performs a circumcision on a man who marries before getting circumcised, or if he operates on a handicapped man.

The Maasi remain great hunters, especially of the lion, a constant threat to their animals, and hunt to maintain their manly status and skills, though they generally do not hunt to eat wildlife. (14)

The plaster covering the pole and slat huts bound with bark strips is cow dung. Maasi renew it regularly as it decays. The accumulation of heaps of cow dung in the corrals (kraal, the Dutch equivalent) is one reason why these pastoralists move periodically.

There are mainly two seasons for the people and the cattle: the dry and wet seasons. During the dry season where the permanent camps are the tribe remain at the higher elevations where they can find grasses for the cattle and where water is piped into villages. These camps may be near springs or streams. Wet season camps, temporary and dependent on the rains, coming between November and April and then only in a few, heavy downpours, are near the water sources in the lowlands when flushes of grasses begin to fill the plains. During the wet season, the tribe descends to the plains leaving only the elderly women, small children and the milking cows in the camps in the highlands.

For the Maasi in northern Tanzania settled on national parks and game preserves, nomadic and roaming homelands have essentially been usurped by the government to satisfy the needs of international wildlife tourism. Maasi of the Serengeti were expelled when game preserves were established in the 1950s and 1960s. They have been relegated to marginal lands and may continue to live in and around the preserves but cannot call the land theirs as they once could. One such place is the Ngorongoro Conservation Area in northern Tanzania, renown for its physical beauty and cultural legacies. Its 8,000 square kilometers to the east of the Serengeti encompasses the famed Olduvai Gorge, site of the discoveries of some of the oldest humanoid remains and home to a spectacular abundance of African wildlife is also home to about 15,000 pastoral Maasi.

The social, cultural and political life of the Maasi is adaptable to the environment and their cattle. If scarcity of land or water occurs, modifications are made in social, clan or family herding units as the tribe makes whatever adjustments are necessary on behalf of their chief investment, cattle. On the other hand, many government officials believe that environmental conservation is not compatible with the needs of a cattle and livestock economy. The fate of the pastoral peoples with cattle will always be a balance between conservation of dwindling land resources and development of natural spaces, and the economic needs of government as opposed to the needs of a subsistence people. As an example, Tanzania experimented with large-scale state-owned ranches, farms and plantations raising beef cattle under a socialist government in the 1970s. At the expense of the pastoral people, these ranches did not produce the amount of beef needed for the urban market nor the profits needed to keep it running.

Moreover, attempts decades ago to pipe water into villages met with limited success as hastily constructed dams were breached and even gravity flow pipes were never repaired. Consequently, the Maasi have had to rely on traditional, scarce means of obtaining water for themselves and their cattle and livestock.

Recent development aid programs have met with modest degrees of success, but just as often with disappointment as local administrators failed to distinguish between the differing needs of pastoralists from those people of settled agriculture, or failed to make adequate plans, for example, for relocation to settled villages or for the provision of grain in settled villages. For a people whose sole sense of economic security lies in its cattle and livestock to succumb to nationally administered bureaucratic inefficiencies, poor planning and new administrative structures, like ministries and conservation authorities, to their cultural seniority decision-making process is to undermine not only their economic security but their political autonomy.

A combination of factors has contributed to a lowering of life standards, including a deterioration in the food supply as pastoral households own less cattle than in previous decades and hence have less milk. This leads to a decline in health and living standards. The people are more dependent on supplies of purchased grain, for which they must sell cattle, a market strategy they cannot control. Even grain is insufficient to meet all the energy needs of the families and households.

Governmental inadequacies and inefficiencies in land and water management crucial to survival of African native peoples contribute to a diminution of quality of life. No crop cultivation is permitted in conservation areas. Some people have been evicted from certain areas they used to roam for grazing of their cattle and are excluded from certain salt licks and water holes and can no longer collect resin in the forests, or set fires to control pastures. The wild animals are protected and allowed to roam free but not the Maasi people. Wild animals compete with Maasi cattle for the same grazing lands.

There is a widening gap between the rich and poor among the pastoral community and a rising antagonism with government officials who see the Maasi as obstacles to wildlife management. If the Maasi leave or become assimilated in the larger life of the nation, and the subsistence life of the pastoralists fades, so goes the traditional cattle culture of Africa. People like the Maasi in central east Africa, and in other parts of Africa and Asia, are threatened by government regulations for conservation, tourism, and land and water management. Although diseases, droughts and natural catastrophes occur periodically threatening both human and livestock life, it is the man-made decisions which might ultimately decide the destiny of these pastoral peoples and the cattle that sustain them.

Survival in a milk-based subsistence culture is not consistent with the cost-benefit analysis of economists. For example, development efforts promoted national beef production at the expense of survival among the Maasi tribes. We could say that one goal is economic, and the other anthropological or sociological, but the academic distinction blurs the idea of how a survivalist economy like the Maasi whose cattle is its capital is supposed to make the transition to mass production based on national goals.

The problem in recent times for pastoral peoples clearly is the diminution of a land base and grazing lands for herds. Many livestock development projects in east Africa have failed because of a poor understanding of the internal dynamics of pastoralists. (15) African governments do not support the concept of many acknowledged rights, like land rights, water rights, human rights. There is no tenure to lands.

East Africa attempts to make optimum use of resources, but it must exploit its animals, which can devastate the land through indiscriminate grazing. The key to survival is not just the slaughter of animals but the sustained yield that can support a burgeoning human population. The increase in both human and animal population throughout sub-Saharan Africa has exacerbated the ability of governments, where they have chosen to be involved at all, to manage natural resources. Here is a strident and potent message for civilization concerning the relationship between humans, domesticated animals and the environment.

The Cattle Herding Himba Tribe of Namibia

When the cattle are threatened or killed by stalkers, the warders become hunters of the predators. It's as necessary to protect one's currency and livelihood as it is to obtain it. A Himba man is a warrior. A woman's life is part ritual and part domestic. Lives of all members of the tribe revolve around cattle.

Kuwiya is a Himba cattle herder from northern Namibia. Recently, two of his cows had calved as he drove his small herd from a dry season pasture to a place where the grass was plentiful. He and his nephew built a sturdy makeshift enclosure of thorn branches about waist high and slept. During the night, a predator had breached the thorny barrier and eaten half of one of the new calves. Believing the night thief was a leopard, Kuwiya enlisted the support of others to help him capture and kill the intruder, the cleverest, dangerous and most cunning of culprits.

Selecting a deep hole near the enclosure, the men enlarged it and covered it with leaves and branches as a trap pit. Securing a piece of fresh goat meat as bait, they placed it carefully in the center to lure the leopard. They hoped to conceal the pit, get the animal to fall in then to spear it to death. That night they watched breathlessly as the leopard cautiously approached, sniffed the meat, looked around carefully, extended a paw to test the rim, then backed away. It returned moments later and slowly advanced on the goat meat bait, and quickly fell into the pit. The men ran to the dark hole and threw their spears in, heard the shrieks of the leopard and warily retreated. They approached the pit again moments later and, when the moon appeared from behind clouds, they found the pit empty. At midmorning the next day a boy from the village reported that a leopard had been found in dense undergrowth near the trap. It had a spear in its rump and another in its rib cage. (16)

Similar stories pervade the Himba cultural lore of men protecting their cattle.

Katwerwa, is also a Himba man, tall and formidable-looking, though he is by nature gentle and unpretentious. Vigorous looking, he did not enjoy walking long distances when herding cattle. His left thigh still showed the deep scar acquired as a young man defending his cattle from wild predators.

As a youth, he had lingered near the well as the cattle grazed all day in the bush awaiting their return to drink before nightfall. As it neared twilight and the cattle did not return he grew anxious. When he saw vultures circling by the foothills the following morning he asked his sons to inspect where the cattle had grazed. They returned to say that an ox had been killed and suspected a lion. A few of the young men in the village gathered and vowed to hunt the lion, following its trail through the grasses. Suddenly, someone spotted the lion looking at them resting in the shade. As they approached, it moved deeper into the brush. Within moments, the lion had lunged toward the group and bit one in the shoulder then fled. Katwerwa threw his spear but the lion swerved and the lance fell amiss. The lion charged again and while Karwerwa grabbed one of the lion's ears and rode its back, one youth slit one of its hind legs with his knife. As he lost hold of the ear, the lion then turned and ripped at one of his legs with his powerful jaws and teeth. Another youth seized the lion's mane and held it while a third cut open its back and choked it until he dropped its hold on Katwerwa's leg. The others then quickly dispatched it with their spears.

When a young Himba girl comes of age, a ceremony surrounds her coming out. Kakara had such a ceremony for her. While singing and dancing continued around the ancestral fire, she emerged from her homestead and was led to the fire while her father placed three large gourds of milk in her hands. This milk had come from her grandfather's sacred herd, from cows that represented departed ancestors and is only drunk at a girl's first menstruation, marriage and after each birth.

Among the Himba, visiting the graves of the fathers is a time of remembrance and celebration, a time of hope when the heifer will be well received by the deceased father, that ancestors will know they are not forgotten.

The graveyard is covered with the skulls of slaughtered cattle adorning wooden stakes. About 60 to 70 people had gathered for this remembrance ceremony of selecting a heifer as an ancestral cow. While the elderly man, Wamesepa's father, watched from a nearby grove of ironwood trees, a small herd was released into the graveyard. A golden brown heifer edged closer to a particular headstone and while chewing steadily suddenly laid its nose on the headstone. Cautiously inching forward, Wamesepa touched the heifer's flank and identified this heifer as the one chosen for his father. As others moved forward to restrain the animal, a grandson, Wandisa, slit the heifer's ear to mark it as belonging to his grandfather. As the heifer scampered to rejoin the herd, he let the ear cuttings fall on the grave as he announced the heifer belonged to his grandfather and the ceremony was concluded.

The joining of cows in the life of ownership, belonging, kinship, ancestry and spirit worship, is a powerful combination of cultural forces uniting religion, tradition and daily nourishment. Africa may well be the ancestral home of humans, but it is equally the home of the peculiar reverence for cattle that rules every African society. Yet despite the reliance on cattle and their importance to the cultural and dietary life of so many Africans, over-population, over-grazing, civil war, mismanagement, government corruption, deforestation, the inexorable spread of HIV and AIDS, and widespread famine still plague the continent.

CHAPTER TWELVE
INDIA, LAND OF THE SACRED
BULL AND COW

India. The name conjures up exotic beliefs, myths, curvaceous stone carvings, a vast landscape filled with heavily populated cities and villages, people with unpronounceable names, and sacred cows. To a westerner the idea that cows, those placid animals that seem to be everywhere in the countryside, could be deities or have a kind of divine essence, appears naive. But to an Indian, reared on the belief that all living things are manifestations of the divine presence, such an idea seems natural and reasonable and is moreover prescribed in sacred texts.

How does one begin to understand India and its reverence for cows? It isn't just its one billion plus people, its pantheon of gods and goddesses, its class structure, its history begging for boundaries. In India sanctity conquers sanitation, and incantation and ritual subdue science. India's collective soul is a mystique, an undefined but constant quality. (1)

One scene captured part of India's bovine psyche.

I stared at the man and woman on the street in Vrindavan in India as they tried to keep steady on their heads between them in their small, oval basket a large load of dried cow dung patties they were carrying to roadside restaurants. They jostled past crowds of people making their way through the narrow street. This delivery scene is common in India's city streets because it symbolizes in this ancient and sacred culture how even the excreta of the cow is used in daily life, and how this animal like the elephant has been incorporated deep into religious beliefs. Lord Krishna, Hinduism's cowherd symbol and god of love, is worshipped by 800 million Hindus.

India has only 3% of the earth's land mass but 33% of its cattle population, largely the zebu but also the bachaur, the grey-white short-horned, and the water buffalo (*bubalus bubalus*). The cattle in India constitute one-sixth of all the cattle in the world, outnumbering all other Indian livestock. Half of the water buffalo in the world are in India. About 85 percent of cows and 90 percent of water buffalo are kept for milk and diary byproducts while the others work as draught animals.

And nowhere do man and animal come together as intricately as in India. Even today, traditional Hindu religious practices clash with the cow as a source of food in a country with a billion population and with steady economic development. Humans and cows have existed side by side in India for millennia, at least since excavations have revealed their coexistence and the domestication of cattle in the third millennium BCE. All of the cultural characteristics regarding the bull cult, representing masculinity, virility and brute strength, the bull's position in the hierarchy of deities and the cow as a symbol of fertility were all features of Middle Eastern and Egyptian civilizations.

Village life is two-thirds of India's population and so among the billion people rural life involves cattle as the primary work animal and the source of the village's dairy products. The cattle not only provide milk to the people, most of whom are vegetarians, but their draught work includes pulling ploughs and trampling hay with their hooves while hitched together and walking four abreast around a pole, pulverizing hay for cattle consumption.

The tradition and images of the cow in India are also true of Hindu society. I have seen cattle grazing indiscriminately in fields, on lawns and in the backyards of Hindu residents. An unchecked cow, free to roam while it wants, can do serious damage to a field of crops or to a backyard vegetable or flower garden. This is a major problem for poor countries where there is often insufficient food for the human population. Nevertheless, the cow is a sacred animal worshipped even by the gods. According to the great modern leader of non-violence and political independence, Mohandas Gandhi, the cow is more than just a symbol of Hinduism. It represents Hindu life. The cow is not just an animal to Hindus. It represents an identity all Hindus have with both the natural and spiritual world, and protecting cows is supporting Hindu religious belief.

How did it all begin? Like the cow culture worldwide, India's essential cow character emanates from origins in domestication.

Neolithic Cattle in India

The Neolithic period or New Stone Age in human development is characterized by the domestication of cereals and animals, the beginnings of settlements, and the origin of crafts like pottery and weaving. It roughly dates from 8,000–5,500 BCE. Scientists have traced the archaeological record of Neolithic peoples into India from excavations into ash mounds, protuberances built up from the burning of cow dung. There are over 50 such sites in central India. Neolithic people entered central India in a series of waves from northeastern Iran prior to 2,200 BCE. They had herds of *Bos Indicus*, the humped Indian cattle. One scientist surmises that these mounds are the former sites of cattle pens, rimmed with thorn fences, where cattle dung was accumulated and then systematically burned. He suggests that these burnings were not the result of catastrophes or of carelessness but of the clan ridding itself of trampled dung accumulated over a set period of time, either a few grazing seasons or a few years. He further estimates, based on the size of the mounds, the herd size to be between 500–800 head. A cattle herd will leave between one and two feet of dung or more accumulated in a corral over a grazing season. (2)

Excavations of settlements near the pens yielded pottery, stone blades and animal bone fragments, revealing that beef was a large and regular part of the diet and the mainstay of the economy. Cattle bones also mean that Hindu beliefs in the sacredness of the cow had not yet been firmly established and thus was a later cultural and religious development.

During the monsoon season the cattle could be penned near the principal water source, usually a river or stream. During the dry season the community had to move into forested regions and to build fences of thorns or timber to protect the cattle and the herders from wild animals. Calcified hoof prints and human remains show that these early herders cohabited with their cattle at night inside corrals at temporary cattle stations. As timber became scarcer, it is probable that the cow dung was a continual fuel source resulting in the piles of ash mounds. Terra cotta models of cows and bulls were also discovered among the remains. Painted bulls on rock walls occur throughout parts of central India from this Neolithic period and repeatedly show humped bulls with exaggerated long horns.

Throughout India there are seasonal festivals that celebrate fertility and ward off disease. In many of these festivals, cattle and fire are always associated, the cattle as symbols of fertility and the sustenance of life, and the fire to ward off diseases. In some cases, cow dung is used as the fuel for these fires. It is possible that the burning of the cow dung in Neolithic times may also have been an early religious-like festival associated with the pastoral life, seasonal changes and migrations and the benefits of cattle and livestock.

A 1920s researcher documented a remarkable tribe in southern India known as the Kuravers who existed largely by stealing cattle. But as it is true of so many books of the early 20th century, the author reveals more of the morals and customs of the colonial British, who ruled India for nearly 200 years, than objectivity or sympathy for the customs of the people he investigates. (3)

Hinduism: The Sanctity of the Cow

As a good cow to him who milks, we call the doer of fair deeds,
To our assistance day by day.
Come thou to our libations, drink of Soma; Soma-drinker thou!
The rich One's rapture gives kine. (4)

India's oldest writings are called the *Vedas,* a term which means simply "knowledge," canonical collections of hymns, prayers and liturgical formulas that together form the basis for Hindu beliefs. The earliest texts of the *Vedas* date from the 14th century BCE, or about 600 years before the earliest texts of the Bible were written. The *Upanishads* date from about 1,000–500 BCE. These texts contain the oldest literature and religious expressions in humankind together with a complicated cosmology and intricate theology. The *Vedas* mention the cow and bull more than any other animal. Even Buddha, the leader of one of the ancient world's great belief systems beginning in the 6th century BCE, was born under the astrological sign of Taurus.

The belief in the holiness of the cow emanates from Hinduism, a complex religion without a founder or a dogma, whose beliefs have simply accumulated, superimposed one upon another over the centuries. (5) Hinduism, with its various sects of non-violence, cow worship, renunciation, and meditation, defines Indian society. Understanding a few of its primary principles and key words will help place India's cow worship in perspective.

Hindu is a Persian word that means Indian, and Hinduism, using the sacred language of Sanskrit, is the belief in the eternal law of existence or *dharma.* Belief in a god is not required. Individual karma is a part of the karma of the universe which one needs to conform to in order to be saved. Conforming to one's karma is similar to conforming to God's will. As the *Upanishads* say, "When they depart this life, the wise, who have realized Brahman as the Self in all beings, become immortal." (6) But when such souls depart they might return in another form as the great cycle of life is determined by one's actions in a previous life.

KEY WORDS IN HINDU BELIEF

dharma:	a natural law and order
karma:	one's individual actions, or the universe's actions
samsara:	the wheel or revolution
moksha	(or mukti) escape, release, liberation
brahman:	sacred (state of liberated soul or the eternal holy)

The Lord of Lords in Hindu religious belief is Vishnu, often displayed with several attendant deities. But the more energetic god is Shiva represented in stone and bronze carvings at the center of a whirling circle of cosmic activity showing five symbolic aspects of eternal energy: creation, preservation, destruction, favor and concealment.

A luminous bronze carving of Shiva stands in the center of a hall of Indian artifacts in the Victoria and Albert museum in London where he is shown with a thin waist, a smiling Lord of the Dance. I was transfixed, staring at this marvelous carving one afternoon not just because of its artistic beauty but because it represented the essence of Hindu thought. Shiva balances on his right leg while hooking his left across his body revealing a sense of motion, a characteristic of all Indian sculpture. A cobra curls from one arm and in one upturned palm he holds fire. His dreadlocks fan out from his head like flames or scattered musical scores. He dances inside a ring of fire that is also the circle of life. A nearby drum represents the sound of creation with flames, the sign of destruction and release. Shiva is the destroyer and creator of life. He frequently treads, as he does in this sculpture, on the symbol of ignorance symbolized by a prostrate dwarf known as *Apasmara*.

In other representations, he is shown with his consort Parvati, also known as Uma, and they are the primordial couple. In these depictions Vishnu holds a trident in one hand, the symbol of his active life, and in the other hand a rosary, sign of his contemplative nature. At their feet are the bull and the lion, symbols of their domination over nature.

Jainism, which probably originated among the Brahmin caste, is an Indian religious faith that believes the soul can be freed with right knowledge, faith and conduct. Jainism declined with the rise of Hinduism and after the conquest of northern provinces in India under the Muslims about the 11th century CE. The cow is a life form for the Jain believer that demands the respect all living things require. For the Hindu, however, the cow is more like a god to be venerated. The concept of non-violence and non-injury to living things including cattle, or *ahimsa*, has a persistent hold on Indian beliefs with a long history and high prestige status, adopted intact by the Hindus, but not the Moslems. (7)

The Moslem rulers of India were not opposed to the slaughter and sacrifice of the bull or of beef eating as were the Hindus a thousand years ago. Various Moslem rulers (Babar, Akbar, Humayan) issued edicts against the killing of cow to appease the native population rather than from any personal religious belief. This singular experience with the cow and bull, linked as it is so intricately with religious belief, is a source of tension even today where Moslems and Hindus live in close proximity to each other. Ironically, the Moslems still practice the sacrifice of the bull, but also the less costly lamb. On religious feasts they follow the most ancient traditions of appeasing God.

If the culture of the West is centered in materialism, individualism and trade and commerce, that of India revolves around the spiritual, the empire of the dollar contrasted to the land of the soul. India's goal is non-violence, meditation, austerity and spiritual enlightenment. It is impossible to understand the land of the sacred cow without understanding this cultural meaning in India. In a land where all life is sacred, the cow, literally the backbone and driver of the rural economy, is regarded as the embodiment of the divine spirit of Shiva and his embodiment, Krishna.

Belief in the special sanctity of the cow is relatively recent in India's history, only within the past two thousand years or so. Asoka (273–232 BCE) who unified northern India through war, a great king who personified moral leadership and became a follower of the Buddha, proposed the rule of *Ahimsa*, or right conduct, equivalent in the West to the concept of virtue. *Ahimsa* is the foremost virtue and results in the worship and reverence for the cow, the reluctance to kill any living thing, and in the practice of vegetarianism, all solidly entrenched in Hindu society. (8) This is a reversal of Asoka's earlier propensity to war and conflict, which he later repudiated. He engraved his set of beliefs of compassion, truthfulness, non-violence, non-acquisitiveness and non-injury to animals on stone columns throughout the kingdom and founded hospitals for humans and animals including cattle. Non-violence, he believed, should be extended to all living things. Hereafter, conquest was to be by *dharma*, or the principles of right life. The Brahmin, the upper caste in Indian society, interpreted and disseminated Asoka's ideas until they became an integral part of Indian Hindu spiritual belief.

But it is not just a moral principle that drives the belief in the sanctity of cattle. Gods have taken on the role of cowherd protector. Krishna is the incarnation of the great god Vishnu, the preserver of the world. Krishna's myth began to gain momentum about two thousand years ago and is today the most popular in the Hindu pantheon. Krishna is the playful one, the hero who has inspired poetry, literature and philosophy, much like Dionysius of Greek fable. As Christ is symbolically referred to as the divine shepherd, Krisna is the Divine Cowherd. The cow and Krishna have become inseparable in Hindu mentality.

Cows definitely have economic importance, and Hindus share repugnance for slaughter. These factors have combined in Hindu society to project a religious significance at least as great as that of Mesopotamian societies. But the moral and religious principle has been extended to become a moral code in itself. *Ahimsa* is a sacred pact and one deviates from it at one's peril and undergoes social isolation from the community.

Nandi, Shiva's Bull

In the city of Thanjavur (ancient Tangore) on India's southeastern coast stands a Hindu temple. More than a thousand years old, it is covered with carvings of lions, elephants and other animals. But under an open air covering in the courtyard is the 20-foot high stone statute of the cow goddess, Nandi, the sacred transportation vehicle of the chief god, Shiva. In keeping with the ancient belief that farming and the domestication of animals occurred in human history more or less simultaneously, Hindus believe that Nandi also makes rain and crops grow.

Every two weeks, about 50,000 people gather to watch this stone statue receive a bath. Standing on wooden scaffolding, priests bathe the statue of the cow goddess in a sweet mixture of curried milk, coconut water, vegetable oil, honey and sugarcane juice, ladling the orange mixture onto the cow's stone head from huge vats filled by hoses from the courtyard floor. The faithful clap during this ceremony and pray for favors while their feet slowly get soaked from the tumeric-smelling, curry-colored ceremonial mixture flooding the floor and coating their shoeless feet. (9) For Hindus, Nandi is Shiva's bull, an extension of his benevolence to all life. Nandi is the symbol of Shiva's power that can give life force to humanity.

Krishna may be depicted as a cowherd god, and Surabhi the mother of all cows, but Nandi, a carved stone image of a white bull reclining on a raised platform facing the entrance to any shrine, resides in every Shiva temple compound in India. There are over 3,000 such temples alone in the city of Benares along the Ganges River. An enormous stone statue of Nandi dating from the 8th century CE can still be seen in Kanchipuram in India.

At one of Benares' sacred shrines in the inner city, votaries enter during festival time bearing cups of water they have collected from the Ganges nearby. They enter past the statue of Nandi and struggle through the crowd to pour an offering of milk over a large black stone jutting out of a stone receptacle. The basalt stone represents the phallic symbol of Shiva and the power of life. The milk offering also represents the power of life that this substance gives the people. This ancient and powerful ritual commemorates natural forces that hold the material world together, keep life intact and further the work of the spirit.

Vegetarianism

Vegetarianism is not so much a commandment to eat vegetables, fruits, grains, nuts and some dairy products as it is a prohibition against eating meat. Because living things are sacred, meat is also sacred and not to be slaughtered. Not even meat products like gravy should be consumed. So powerful is this belief that entire cities, like the large northwestern city of Hardiwar along the Ganges, has no meat, not even eggs, for sale in the city limits. When I stayed as an honored guest with a family living in the city along the Ganges at the family's request, I was fed nothing but a mixture of grains and vegetables. At my host's request, I also had to bathe in the sacred waters of the Ganges, which I indicated might take considerable time, as the purpose was to cleanse all my sins.

Marco Polo was the first to record in the West that the people of Maabar in India worshipped the ox, refused to eat meat, and would not kill any living thing. Subsequent travelers, like Ralph Fitch in the 17th century, confirmed this belief. (10) For as long as records have existed, Indians have followed this belief. To the western mind, such practices seem unusual and perhaps simplistic. If animal spirits can be the sites of former human spirits, perhaps even former ancestors, then looking after animals in a caring manner does not appear extraordinary.

Objections to the meat-eating culture have been raised by scientists, nutrition experts and vegetarians. Why eat meat at all? One response is ingrained custom, but another reason is that meat tastes good. We like it for its fat. Of all the agricultural land in the U.S., 45 percent of the total land mass is used to raise animals. More than half the water consumed in the U.S. is used to raise animals for food. Vegetarians claim that a vegetarian diet requires only 300 gallons of water daily while a meat-eating diet needs more than 4,000 gallons daily.

Animals raised for food cause water and ground pollution. More than one-third of all raw materials and fossil fuels are used to raise animals for food. Raising cattle often means deforestation in many developing countries. Cattle receive more than 80 percent of all the corn grown in the U.S. and more than 95 percent of all oats. In addition, according to vegetarian groups, animal activists and organizations like People for the Ethical Treatment of Animals (PETA), vegetarians avoid meat and dairy products for health not religious reasons and have the lowest rate of coronary heart disease, a fraction of the heart attacks, and only 40 percent of the cancer rate. They outlive the average person by six years.

Assisted Living for Cows

Three thousand years ago, India had *goshalas*, literally "places for cows," not so much sheds or pens as grazing lands for disabled cattle. These were, and still are today, animal homes. The more elaborate *pinjrapoles* were also places for cattle but attached to temples and served as hospitals for sick and weak cattle. (11) Both of these types of homes for cattle still exist today in India, though the words are often used interchangeably, and the existence of such places explains much about the close relationship between people and cattle in India. One additional reason for the *pinjrapoles* is a refuge for protection of cattle against the potential deprecation of these animals by Moslems.

In the United States, animal shelters contain domestic pets for adoption but not cows. In India, the vast majority of animals cared for are cattle, but also horses, sheep, goats, and even rabbits and deer. One animal shelter in the heart of Delhi, the Jain Bird Hospital, is devoted exclusively to sheltering weak and injured birds.

The acquisition of land by many of these animal shelters has greatly benefited them economically. Rent from pasturelands for sharecropping or crop cultivation is an additional source of revenue. Commercial cattle breeding, for a small number of cattle shelters, constitutes another form of economic resource enhancement.

One field study in the west central province of Gujarat found that over 70 percent of the animal population in one *pinjrapole* were cattle including calves. Some of the animals are taken to these shelters and animal hospitals because their owners can no longer manage them and not because the cattle were disabled, a small number of the total population in such shelters. Salvageable animals are often reclaimed by other owners who find ways of making the cattle productive.

The most notable *goshala* or cattle shelters are those found at temples. While paying homage to the divinities, worshippers can also pay tribute to the cows themselves in the temple compound. The milk collected from these cows is used to make ghee, a kind of milk paste or curd which is used to fuel the temple lamps, and to make a milky nectar which is distributed to participants at temple ceremonies. The cow thus not only figures prominently in sanctification but in the rituals associated with the temple activities.

The sale of milk and dung cakes as cooking fuel constitutes a good percentage of the revenue for these animal refuges. Hence, while honoring the religious devotion to the cow, the cow's economic viability is not over-looked as a supportive benefit. Many *pinjrapoles* have a commercial dairy on the property, though free milk is often distributed to the poor. Most will have a stud bull in attendance and several available cows not too old or disabled. Male calves are usually sold or traded. Female calves are kept to enrich the herd, especially if they are to be milk–producing. If a milk herd has been established, it perpetuates itself and uses the proceeds as revenue to support the institution.

Fuel for the temple and compound for cooking is cow dung, patted into cakes, and left out to dry, collected each day by the women employed by the temple, often the wives of cowherds. Later, excess dung cakes are collected, placed in a basket, and taken into the village to be sold to benefit the temple *goshala* as a supplement to government funds received to operate these animal hospitals. One temple complex uses cow dung from its *pinjrapole* to light the temple by converting the dung into methane at its gas plant.

The presence of these animal shelters is noteworthy especially during calamitous times when, for example, during prolonged droughts, farmers are unable to feed their cattle they can bring their cows to the animal shelters. It is ecologically correct to re-cycle animals during inconvenient or adverse times, and such animals can later be released to individuals who can prove they have need of them. Poor pastoralists or farmers can thus place their animals in shelters when they cannot support them, during periods of drought for instance, and retrieve them at a later, more promising time.

Lodrick theorizes that these animal shelters, assisted living for disabled cattle, are religious institutions. (12) They are animal sheltering institutions, like ancient temples and shrines, operating on religious principles. Moreover, these shelters are not economically viable because it costs more to care for these animals than is returned to the institution financially from the sale of their products. These shelters preserve animal life. The religious and human value of life supersedes that of economic efficiency in the maintenance of these institutions, a value at variance with the operation of most of the global economy. Hence, animal homes must rely upon government grants, donations and gifts, land and building rents and the sale of some milk to sustain their costs. The Jainism non-injury concept of *ahimsa* extended to all living things, and its religious adoption into Hinduism, takes precedence in India over economic viability.

Sometimes in the autumn, as calculated by a lunar cycle, there occurs a unique festival involving cows at a *goshala* like Sri Nathji Goshala in Nathdwara in western India. It is known as *Gocharan*, meaning "a grazing of the cows." It commemorates a time when the youthful Krishna took his foster-father's cattle out to graze. The festival, which draws attendees from miles around, begins with the washing of the cows, which are then decorated with peacock feathers on their silver painted horns, daubed with orange dots or hand prints on their flanks, and fed special foods like molasses and sweet mash. Several cows have garlands slung over their thick necks. About a hundred of these decorated and highly select cows are paraded through the city streets.

Then, the chief of the nearby temple leads the crowd in the cow-worshipping ceremony that may last until evening. It is attended by all social strata of Indian society who bring gifts of food to offer on the makeshift table of ritual in the *goshala*. Participants walk clockwise around the bull, crouch to walk under its belly and touch its tail to their foreheads. This ancient and unique Indian festival must be somewhat similar to the veneration shown the bull or the cow in ancient Middle East civilizations. The strong link between the cow as source of nourishment and as the embodiment of divinity and symbol of a god's presence is never more apparent than in the celebration of this Hindu festival.

Although this activity is performed outside the temple, other ceremonies may accompany this cow worship. Often, a cow dung image of Krishna, the cowherd god, with a flowing scarf, legs crossed, and pendants hanging from his neck, outlined with white chalk, may be constructed on the temple floor for devotees to admire. At the conclusion of all ceremonies, a calf is allowed to walk over the image to destroy it, thereby losing its divine power, as no person could destroy this reproduction, as the calf is the special image of Krishna himself.

The highlight of the festival is a bull-fighting event, not between a man and a bull, but between two bulls. While guests are seated on circular stone platforms, two bulls are led into the inner courtyard of the *goshala*. Each is branded with special insignias, a trident, symbolizing the god Shiva on one flank, and the circle of the sun god on the other flank. The bulls butt each other, lock horns, and exhibit their massive strength until one bull flees pursued by the other bull which must be distracted by cowherds as the bulls are not allowed to harm each other. It was the same kind of spectator sport I had observed in the United Arab Emirates imported there by Portuguese sailors.

In the early autumn of 2000, the beleaguered President of Indonesia was agitated because violent car bombings were occurring in downtown Jakarta, the capitol. Ex-President Suharto was about to come to trial on corruption charges and many assumed his military supporters were trying to stop the trial. In a fit of pique reported in the local and popular press, then President Wahid, weakened by bouts of violence across the country, demanded that police arrest any suspects. "There are no sacred cows in this country," he is reported to have said to a cabinet secretary. It was a loose, unthinking remark, motivated by a need for action, but a slight to the Hindu minority in the country. There are sacred cows on the island of Bali where a majority of Hindus reside. But for Hindus everywhere all cows are sacred.

Africa represents the currency of the cow. India symbolizes its sacredness. Latin America, which we visit next, is the emblem of the cow as principal food source.

218

Figure 7: A Suckling Calf From Egypt in the Ashmoleum Museum in Oxford

CHAPTER THIRTEEN
LATIN AMERICAN BEEF-EATING CULTURE

Beginning Thursday, October 11, 1492, and in journal entries throughout the rest of that month, Christopher Columbus, a man of uncanny navigational skill but filled with religious fervor, materialistic acquisitiveness, evangelical zeal, and callous disregard of the American natives, was awestruck with the natural beauty of the flora and fauna of the islands and lands he encountered. (1) The small communities he established would soon need more than the delights of local fruit to provide nourishment. Ships with cows were soon to follow.

All Latin American cattle can be traced to the few hundred imported by the Spanish from Andalusia and the Canary Islands in the years of conquest and settlement that followed Columbus. Like explorers and colonists, the cows landed in Hispaniola (Haiti and the Dominican Republic) in 1493 after Columbus landed there the previous year, and then were transported to the settlements in New Spain (Mexico) once base stocks were secured. Royal villas were established as early as 1498 in order to raise cattle for breeding stock and for cowhides as the leather of clothing necessity.

Some of these early cattle ranchers became more renowned for their explorations. Hernan Cortes, a successful cattle man on Hispaniola, financed his explorations and conquests in New Spain with the sale of cattle and properties. By 1520 cattle were self-supporting in Puerto Rico and throughout the established colonies of the West Indies, then reaching New Spain's Tampico in 1521 and Veracruz in 1522. From this foundation herd, Spanish cattle populated all of Latin and South America numbering in the millions by the 18th century. Spanish cattle first came to Florida in 1565. (2)

These cattle were the only domesticated bovines in the hemisphere in the 16th century. The descendants of this special breed, known as criollo cattle, interbred with British breeds over many generations, have mostly disappeared from the continent. One branch of criollo descendants is the Longhorn of Texas. Cattle came to Brazil courtesy of the Portuguese who brought a similar criollo cattle via the Cape Verde Islands to Sao Vicente by 1532.

Tropical cattlemen realized that a larger calf resulted from cross-breeding the criollo with a zebu about 1870 and, since the progeny, the

Santa Gertrudis, produced a better tasting meat, they continued the experiment. The disadvantage was that the offspring did not adapt as easily to the inhospitable tropical environment. The small size of remaining criollo cattle, such as among those in Bolivia, weigh only about 700 pounds, the small size the consequence of living at higher elevations and several generations of eating the sparse vegetation at that altitude. Through selective crossbreeding, natural adaptation and genetic evolution, the criollo introduced into the Americas has become as extinct as the Spanish conquistadors who imported them.

Following quickly on the heels of the Spanish and Portuguese armies in the 1530s were the ministering clerics and missionaries, the Jesuits in South America and the Franciscans in Mexico and the American southwest. Cattle came with the soldiers and priests to South as they did to North America as the principal protein for the colonizers. Within a century, wild cattle had proliferated over most of the verdant grasslands of southern South America, in Chile, Argentina, Brazil, Uruguay, and plentiful beef had become the primary food, eaten three times a day, creating a beef-eating culture that exists today. The riches of South America were measured in cattle and wealthy landowners, as they were in Texas and the West, and created oligarchies that controlled both cattle and the pastures that sustained them.

As the Spanish developed the silver mines, like the rich Potosi veins in what is now Bolivia beginning in 1544, cattle were located in the northern pampas of Argentina to feed the burgeoning mining population. A mint was established in 1672 to coin silver and 86 churches erected to accommodate the devotions of the 200,000 then living in the city, the largest and wealthiest in Latin America at the time. When the silver ran out, the people ran out and the larger cattle herds migrated to southern Argentina where they became wild and accustomed to the rangeland in Paraguay and Uruguay through the 18th century. Trade in these wild cattle constituted the principal livestock trade in this region until modern times. By the middle of the 19th century, the value of cattle in Latin and South America exceeded all the treasure of silver and gold taken by the Spanish from Andean mines, illustrating the point that renewable resources are the best investments. Brazil today has 150 million head of cattle, Argentina 56 million. In comparison, the U.S. has 105 million.

The social structure of the landowners in the countryside was formed in the middle of the 19th century when an enterprising elite emerged to gain control of land used for the raising of commercial livestock and then political power through their exercise of wealth with exported raw materials, in which the beef industry played a leading role. (3) Political clout faded for these aristocratic landowners who owned cattle as the power shifted to urban manufacturing interests. Neither raw economic wealth nor political power alone accounts for the cultural attachment to cattle. Beef eating of beef in Latin America is a sign that the love affair Latinos have with beef shows no symptoms of decline.

I enjoyed such a typical beef feast with friends in an outdoor restaurant in the capitol Brasilia one balmy evening. As I sat at the rectangular table, wandering waiters with large beef shanks sliced off limitless portions on my plate until I told them to stop. Brazilian restaurants in larger U.S. cities feature meat of all kinds, beef, pork and lamb, rarely chicken and never fish. From this and similar meals, I needed no statistics to understand the profundity of the beef culture in South America, though I was to learn soon enough, as I report here, of the economic implications of the trade. Follow with me as I untangle this skein of data to leave an impression of the cattle wealth of these continent, a singular feature of the global economy as Africa's is of the subsistence economy.

The Economics of Beef Production

I have repeatedly referred to the cultural impact of the bull and cow on civilization. But unlike the south Indians, cattle were not a spiritual problem. Capitalism ruled the life of cows in all Latin America then and now. The economics of cattle in the continent relies excessively, and in some regions of Argentina exclusively, on cattle as food and source of wealth. Since Latin America is more integrated economically in the world's production of goods than Africa, the economics of cattle production in the continent and heavy dependence on this one agricultural product fluctuates with a host of indicators, a few of which I examine here. The numbers I use are from 2005 and will obviously fluctuate yearly but the trends will not substantively vary.

By the late 1970s agriculture was the largest export sector for all of Latin America, more than 52 percent of 12 countries. Yet beef products were only about seven percent of agricultural exports for the region, but over 50 percent for countries like Argentina, Uruguay and Brazil. By the 1980s about half of all Latin American countries were beef exporters. As human population increases in Latin America, disproportionate to Europe and North America, beef production is growing more slowly than demand. Unless beef production increases at least as fast as Latin America's human population, or unless Latinos turn to other food sources or more beef is imported, shortages or escalating prices may result. A concern about the health risks and an increase in the incidence of heart disease because of the concentration of cholesterol led to a reduction of beef consumption in the 1990s. The collapse of the Argentine economy at the turn of the 21st century placed an additional burden on a middle class nation that once enjoyed the highest per capita income in South America.

Latin America produces about 12 percent of the world's meat production, measured in metric tons, and has 22 percent of the world's cattle with about 8 percent reserved for milk production. Comparatively, the U.S. and Canada together produce about 20 percent, western Europe another 20 percent, and eastern Europe another 8 percent. About 70 percent of beef consumption occurs in the developed countries, primarily Europe and North America. Latin America and the Caribbean currently have a cattle population of over 350 million head. More than 60% are raised in three countries, Brazil, Argentina and Colombia. About 60 million head of cattle are slaughtered per year in the region. Despite the overall low average production indices of the Latin American beef industries, trade in beef and derived products provides substantial net earnings of $1.3 billion per year for the region.

The implications of these figures are two-fold. First, that Latin America, the Caribbean, Central and South America, are, like cattle, a continuation of western culture imported from Europe, primarily the Iberian peninsula. Second, that Latin America is more or less equal with Europe and North America as a cattle rearing and meat-producing continent.

But the other conclusion, as the following figures show, is that Latinos eat much of the beef they produce. Beef consumption in the 1980s was over 70% in several countries like Argentina, Uruguay and Columbia, and about 60% in much of the rest of Central and South America. (4) By contrast in the United States and Canada, beef consumption was about 55%, pork about 25%, poultry 21% and sheep or goat at one percent. By the 1990s in the U.S., chicken replaced beef as the number one meat product consumed for a variety of reasons such as real of presumed health hazards like high levels of cholesterol, contaminated meat and bad publicity about slaughtering practices, and by the promotion of less-fatty chicken in fast food outlets. But beef consumption is rising in South America about one percent per year for a variety of reasons, most notably population increases, a tendency towards urbanization, migration and job creation in the mega-cities of the southern hemisphere.

Throughout much of the economically developing world, eating beef is a luxury activity. But in Latin America eating beef is a regular part of the dietary intake and provides high nutritional value, a major source of protein for all levels of society, except in rural areas where the consumption of a whole cow is problematic and transport and storage difficult. Hence, rural families tend to eat smaller livestock. But in the mega-cities of the southern hemisphere like Sao Paulo and Buenos Aires, beef is the largest food acquisition of Latin Americans, estimated at about 15 percent of total expenditures and 30 percent of food expenditures for a family.

Government policies in many Latin American countries tend to encourage beef consumption at the expense of production, the opposite of what is needed, at a time when new technologies, available but not always implemented, should be favoring beef production to feed a growing local human population, and the acceleration of the beef export trade, a source of much needed income.

According to Public Citizen, a public watch group, "Fast Track" exacerbates the damage to The North American Free Trade Agreement's (NAFTA) to U.S. ranchers and farmers raising beef. (5) In 2002, President George W. Bush asked Congress for powers to expand to 31 countries in Central America and the Caribbean through a proposed Free Trade Area of the Americas (CAFTA), legislation that was passed in the summer of 2005. The U.S. already had growing meat trade deficits with Argentina, Brazil and Uruguay. The trade agreement was expected to increase Latin American beef exports to the U.S., according to the U.S. Department of Agriculture (USDA), granting foreign producers new import rights into the U.S. consumer market.

The U.S. beef trade deficit with Argentina was more than $100 million annually during the 1990s. The U.S. beef deficit with Brazil has grown 1,400% since 1991, from $6 million to $91 million. In the U.S., the cattle and beef sectors' $21 million surplus in 1995 had become a $152 million deficit by 1999. The dairy trade deficit nearly doubled from $416 million in 1995 to $796 million in 1999. According to the USDA, the beef deficit with Uruguay has increased by 75% since 1991 from $26 million in 1991 to $46 million in 2000. U.S. ranchers would face stiffer competition from imports but would not gain in Latin American and Caribbean export markets.

Beginning in 1993 over trade agreement debates with South America, U.S. ranchers were promised that the agreement would provide a path to lasting economic success through rising exports. Consumers were promised lower food prices. Neither materialized. Farm income declined and consumer prices rose while some agribusinesses, which lobbied hard for NAFTA, had record profits. U.S. agribusinesses took advantage of NAFTA to invest in Mexico's low-wage, minimal regulatory environment. For instance, Cargill Corporation purchased a beef and chicken production plant in Saltillo, Mexico, thus securing access to lower wages and lower regulatory standards.

Beef ranching in Latin America has traditionally been associated with large ranches using extensive land pastures and a few cowboys to produce high quality beef. However, small and medium-sized ranches with just a few hectares (one hectare = 2.47 acres) make significant contributions. Any minimal increase in cattle production on a small farm, in traction, milk, breeding or fattening, and especially if cattle culture is combined with agriculture, has a direct impact on the smallholder's income. One reason for lowered productivity is the unknown amount of forage and grazing land available for cattle during any given season, a problem technology can only partially solve, especially because of uncertainty in weather conditions.

A similar movement to assist cattle in the U.S. to finish with grass and not grains is promoted by the U.S. Dept. of Agriculture. U.S. ranchers, like Latin American counterparts who put their beef cattle out to pasture, can produce animals that are ready for market because adding more grass to cattle's diet will produce high-quality beef.

Romosinuano cattle, a South American breed, may pass along more traits that are ideal for cattle production in the southeastern United States. They mature quickly, resist insect pests and heat, and have consistent carcass quality and good temperament. The U.S. Agricultural Research Service brought 143 Romosinuano embryos from Venezuela to Florida where researchers implanted them in surrogate cows. The purpose was to evaluate the breed's qualities most desirable for cattle herds in the southeastern U.S. Because of U.S. import restrictions, it took five years and the cooperation of the Venezuelan government to arrange this embryo transfer. The only U.S. herd with Romosinuano genes is highly inbred and therefore not the best predictor of the breed's true potential.

So Latinos produce lots of cattle and eat lots of beef. So what do the cattle eat? Agricultural scientists have perfected some grains for cattle consumption.

Findings from a comparative 3-year study show beef cattle can be fed as efficiently on grass pastures, with some grain, as they can with mostly grain. Agricultural scientists experimented using wheat pasture and perennial grass pastures, such as Old World Bluestem, millions of acres of which grow in the southern Great Plains region. They stocked the grass pastures with twice as many cattle as they would normally to ensure that most of the grass would be consumed thus decreasing the amount of grain. A high-energy diet composed mostly of corn was provided in a covered feeder to give cattle additional energy for fattening. In the grain-on-grass system, cattle make their own dietary choices, deciding how much grain they need based on the grass supply. As the grass supply dwindled, the cattle ate more of the high-grain diet.

Cattle fed grass plus grain needed less feed to reach market weight than herd-mates fed in the feedlot. Less feed means lower production costs with savings about $25 per animal. With four animals per acre, that's more dollars per acre than could be anticipated from other uses of the grass. From an ecological viewpoint, the system of using grain-on-grass reduces animal waste and allows producers to finish cattle without incurring the added cost of waste disposal.

But, by one of those ironies in nature, what food cattle prefer is also a factor, and, as it turns out, cattle hanker for tropical corn from Mexico and Central America. This corn could become an alternative cattle feed to sorghum in the southern United States. Yielding about 87 percent more dry matter than sorghum, thereby making each acre more productive, dairy cows and steers alike seem to prefer this corn over sorghum, based on studies by the U.S. Agricultural Research Service. Dairy cows ate so much more tropical corn silage that their milk production increased 10 to 20% over eating sorghum. Tests of its nutritional value showed it to be slightly less digestible than forage-type sorghum. Tropical corn is a good alternative because it grows well in heat and tolerates insects well. Alternating tropical corn and the winter crop would protect the ground from erosion and give cows two quality feeds.

Cattle producers retain their assets as long as they think that retaining cattle is more valuable than slaughtering them. When prices are too low, and grazing lands (or animal feed) do not meet expectations, there is a glut of beef (or milk) on the market, or there is a weak credit market. An increase in the price of beef will make keeping heifers longer more profitable. An older cow may be kept for good milk production when milk production drops, and become a candidate for slaughter as beef. The average cost of beef production is lower in Argentina than it is the U.S., even though the average age of a steer for slaughter is higher than in the U.S.

The profitability of the cattle industry, wherever located and the weather notwithstanding, depends on analyzing an integrated cattle system that includes breeding technologies, fattening practices, calving rates, control of parasites, land management, transportation and marketing costs and domestic government policies and export demands. Latin America could be faced with unattractive dilemmas, like retaining beef for local consumption and foregoing beef for export that receives higher profits, if developed countries increase imports because of surpluses.

Because beef production and consumption is such a prized and financially stable part of Latin American economics, wild fluctuations in beef prices, sometimes effected by the weather but more often by government policies in producing countries, can also effect the economy as a whole causing disruption in labor costs and consumer incomes, even leading to inflation. Like other economic sectors, cattle production is cyclic and variable depending on a variety of factors, including price fluctuations, trade restrictions, quotas and tariffs in other beef producing countries which seek to protect their own domestic cattle producers.

Like other commodities, beef markets are unstable for a variety of reasons. The chief reason is that among the world's trading nations there is variability in price, partly explained by government intervention, usually by injecting import quotas and subsidies in the market to protect local cattle raisers and consumers and to maintain stability in local markets. Latin beef export countries could lose valuable foreign exchange earnings if beef importing countries like America would only eat less beef.

Encouraging short-term slaughter in order to raise beef prices domestically in Argentina is a defeatist policy because it creates shortages of beef in the long term by reducing the stock. Because beef is consumed in Latin America by all income levels, price stability is essential for social harmony. Higher prices for beef have the advantage of encouraging production and exports. But low prices improve the welfare of consumers by restraining wages and lowering inflation tendencies. All governments face this economic and historic dilemma, and not just over cattle prices. Both income and population growth exacerbate this situation and make government policies choices more difficult.

What is true of the economics of beef is also true of milk production. World prices for milk fluctuates as do those for beef and milk subsidies as dairy farmers attempt to maintain a just price for their milk production without damaging the importation of cheaper milk in times of short supplies. Milk production has greater risks than meat as fresh milk must be sold daily. Milking facilities and refrigerated trucks are essential and transportation and processing facilities indispensable. Cheese, because of limited markets in Latin America and developing countries, and problems of quality production and storage, is an even more difficult commodity to produce regularly for a profit.

So is it possible to have your beef and eat it too? Not if you want to make some money for the country by exporting it. Local farmers can use dairy cows and a few cattle for family food and domestic consumption. But because the beef business is of industrial strength in Latin America, its reach is global not just domestic. Let Argentina serve as an illustration of the economics of the cattle industry.

An Argentinean Example

Named after the Latin word for silver, Argentina is still trying to live up to its wealthy connotation. With 40 million people and 50 million cattle, Argentina is a country with a highly literate population, rich in natural resources with a diversified industrial and agricultural base. But in 1989 when debts increased and inflation reached 200%, the country made a series of regrettable decisions, like pegging the peso to the dollar that exacerbated the financial situation. By the mid-1990s, economic setbacks in Mexico and Russia, a failure to resolve the crisis in the banking system, and a massive default on its international loans deepened Argentina's economic woes. By January 2002 when the peso was released from the dollar, inflation increased, the unemployment rate ran to 25%. As a result, 60% of Argentineans, double the number from 2001, were below the poverty line of less than $220 dollars a month for a family of four. (6) But by 2005 the country had surprisingly rebounded and gained some economic stability.

This economic situation conditions an understanding of reliance on the cattle culture. Meat is closely identified with the country's national identity. The most conspicuous ceremony of Argentine and Brazilian life is the *asado*, the eating of different layers of cooked beef cuts from the skewer and grill. Argentina was earning more than $600 million in exports profits from the cattle industry.

Argentina produces about 4 percent of the world's beef, slightly less than Brazil, the two largest cattle producing countries in Latin America. Though poultry, pork and chicken are stable meat products on the continent, beef surpasses all these foods combined in production by weight and consumption. When milk is added to this food equation, cattle are by far the biggest commercial food value.

The organized cattle industry in Argentina is linked to the saladero, a rudimentary beef packing plant begun in 1810 in Buenos Aires. Cattle were held in corrals next to the processing facilities, killed, and their hides stripped and salted for shipment. The meat was cut in strips, soaked in a brine, salted, and packed in barrels. The fat was rendered into tallow.

Like the beef shipped from the U.S., the British too would consume Argentinean beef if marbled with sufficient fat. Accordingly, by the 1870s British investors came to Argentina and introduced shorthorn bulls, Angus and Herefords, to cross-breed with the criollo cows. Soon the pampas was dotted with palatable crossbreeds for shipping in refrigerated cabins to England. Within a few generations the original Spanish cow had disappeared from the South American landscape.

Even today the economics of the beef industry in Argentina is linked to its politics. The National Meat Board was established in the 1930s and since then there has been state control of the marketing system. Argentina attempts to avoid uncertainties in the market for beef and to cushion against cyclical swings of the industry. Real and hidden subsidies or tax incentives to beef producers distort international trade since they do not reduce domestic consumption. Smith traced the historical development of the politics of Argentine beef and has shown how the grievances of special interest groups transferred their complaints to the political system, bypassing the courts, and turning a economic grievance into a political issue. (7)

In addition, the Food and Agricultural Organization (FAO) of the United Nations devises laws and regulations to protect the health of consumers and insure fair trade practices. Programs promote coordination of standards in food safety, pesticide residues, food additives, and hygienic practices for fresh meat. (8)

Argentine beef is highly prized in the U.S. because the cattle are fed from the grasses of the humid pampas whose temperate climate permits the cattle to graze year-round which obviates the need for growth steroids. These cattle produce a steak that has less cholesterol and fat, though it tastes a bit tougher and even gamy. For some, grass-fed animals is a return to a simpler way of raising cattle whose meat tastes better, has less fat and fewer calories, and is less prone to diseases like E.coli. The public may not be totally convinced of the overall switch from grass-fed to corn-fed. But as consumers become more concerned about the overall health benefits, like organic vegetables, they are responding in an upward trajectory to grass-fed beef.

On April 18, 1996, the Animal and Plant Inspection Service of the U.S. Dept. of Agriculture proposed a change in federal regulations regarding the possible importation from Argentina of diseases effecting animals and people. The agency is always concerned about hoof-and-mouth disease if cattle are imported. Nevertheless, the agency permitted, under certain conditions, the importation of fresh, chilled and frozen beef from Argentina. (9) Argentinean beef constitutes a major import as some American fast food chains rely on its beef for hamburgers.

This U.S. federal action, or any such international food trade action such as from the World Trade organization (WTO), takes into consideration both consumers and producers. The quota of 20,000 metric tons of Argentinean beef would mean American consumers would save about $90 million annually in lower beef prices, if, indeed, the lower costs were actually passed on to the consumer. However, these lower prices would also mean an increased cost of $40 million to the beef and dairy industries.

Argentina is the source of large-scale beef and milk production for all of Latin America and its beef products are exported to America. This accounts for the fact that beef is the most important meat and protein source consumed in all of Latin America, slightly less than in developed countries but far ahead of developing nations.

Hoof and Mouth in South America

The world of beefeaters and beef producers has been alarmed because of cattle diseases. The diseases which can transfer to humans are especially worrisome, but diseases that do not spread to humans are also vexing because the deaths or incapacity of large numbers of cattle makes beef more expensive, perhaps milk undrinkable, and can ruin smaller stakeholders.

Policies in export countries, particularly North America and Europe, place a high premium on health standards for beef consumption and until recently some Latin American countries did not maintain or adhere to such strict standards, especially in not controlling for hoof-and-mouth disease. Columbia as of this writing does not have certification for controlling hoof and mouth disease, which exports beef to neighboring countries like Venezuela, also without such health restrictions.

The U.S. had approved the import of Argentine beef in 1997, believing the hoof-and-mouth disease to be under control, as long as the country maintained its vaccination program. Then in the late 1990s the mad cow disease galvanized international attention further damaging the meat industry. The Argentines believed that hoof-and-mouth disease initially was imported from illegal crossings of cattle from Paraguay. U.S. officials immediately banned beef imports from Argentina. (10)

In August, 2000, the hoof-and-mouth virus was detected in local cattle and the announcement by Argentine authorities was swift: they shot more than 3,000 cattle and thought they had contained the problem. When an outbreak occurred a few months later, officials made no public announcement, and then, in collusion with the meat industry, lied about its presence when word leaked out. (11) Argentine agricultural officials began secretly vaccinating cows months earlier. Animal sanitation authorities and ministers of agriculture were forced to resign in March, 2001. New officials promised a totally transparent policy of information. To return to profitability, Argentine beef producers were going to have to tighten control along the poorly supervised borders and enhance sanitary requirements in the global market. Regrettably, that didn't happen. The disease infected Brazilian cattle across the border within weeks.

In early May, 2001, the first case of hoof-and-mouth disease appeared on a ranch in the region of southern Alegrete around the town of Santana do Livramento. By June, 1,100 occurrences of hoof-and-mouth in its cattle were discovered in the region. Within days, Brazilian police were summoned to kill the animals. Health workers buried the cattle slain by the police on a ranch in Rio Grande do Sul. Another 1,200 cases were reported in Uruguay.

Brazilian cattle ranchers, and those in neighboring Uruguay and Paraguay, were quick to blame Argentinean ranchers of covering up an outbreak of hoof-and-mouth disease in 2000–2001. But it was too late to assign blame if officials hoped to stop the spread of the crippling disease. Brazilians authorities quickly banned the movement of animals north of Rio Grande do Sul and ordered the vaccination of 13 million head of cattle. Hoof-and-mouth disease, well known among cattle ranchers, nonetheless ran headlong into Brazilian cultural belief. Brazilians tend to deify their cattle, symbolically if not formally, and challenge any attempt at government intervention.

Brazil, the only industrialized economy in South America, is the world's largest producer of coffee and has a diversified agricultural base with the largest commercial cattle herd in the world clustered in southern zones. Yet it only exports about six percent of its total production. Only the United States and Australia export more beef. (12) Its cattle are grass-fed and disease free, the Brazilians proclaim, so when mad cow scares arise and threats of hoof and mouth disease threaten, it can advertise its cattle and beef products for maximum sales, as it did in 2001 when the European beef industry fell under a cloud when hoof-and-mouth and mad cow disease arose in England and quickly spread to the European continent. But because Canada was unconvinced that Brazilian beef was free from mad cow disease, it banned beef imports. Global beef consumers were understandably alarmed over the dangers of eating potentially infected beef and are unconcerned where the beef originates. Thus Brazilian beef sales at the time remained relatively stagnant.

Hunger, Rain Forests and Beef

Another cultural aspect between cattle and civilization, besides economics, is the impact of the environment, nowhere more evident than in tropical zones. The large-scale destruction of tropical rain forests and the permanent loss of complex eco-systems has become one of the most globally significant environmental issues. Some of this rain forest exists in parts of Central America, but the majority of it is in the Amazon Basin of Brazil, the location of the greatest source of biodiversity on earth. The rise and expansion of cattle ranching is cited as one of the chief causes of this deforestation. Small farmers who have cattle and other livestock can earn a reasonable living, better than their other choices, with the possible exception of coffee plantations, Brazil's chief money winner. Although some pastures and soils have become degraded in the region, some believe this is due to mismanagement. But cattle ranching can be positive in reducing poverty and in fostering agricultural growth, the backbone of any economy. (13)

Limited scientific evidence suggests that the tropical rain forest does regenerate and that secondary growths can turn into a mature forest again if left to its own resilient resources. There is no evidence, only speculation, that there will be mega-extinction of species. There is evidence that once forested land is cleared that productive agriculture can be achieved. When lush forest vegetation is lost the poor and weathered soil is unproductive, though useful for cattle raising. (14) Most farmers in the Amazon Basin have cattle for milk and beef production.

Brazil encouraged farmers and the unemployed rural poor whose job losses occurred because of technology advances in large-scale agriculture like dairying in the 1970s and 1980s to colonize the Amazon Basin to reduce crowded cities. The population of the Amazon was 1.8 million in 1960 but had increased eight-fold in 40 years to 16 million in 2000, or 10% of Brazil's total, creating a regional market for meat and dairy products. This political move was viewed by the government as integral to overall economic development, a plan as we saw that was similar to the wholesale migration and settlement of the American West.

But deforestation is the direct result of government policies, not poor farmers seeking a bit of land and income from cattle. The Brazilian government built roads into the interior of the Amazon in the 1960s without any scientific investigation of the environmental impact, did nothing to halt illegal migration, and then subsidized questionable large-scale economic development ventures. Cattle raising and milk production arrived in the Amazon to provide the food supply to the exploding population seeking free land to exploit.

How Brazil and neighboring countries connected to the Amazon will protect the remaining Basin and its rich environment is a speculative but controversial political issue. The record thus far has been inconsistent and deplorable. How nations protect the deforested land already occupied and in production is equally critical, but the preservation of cattle production is the most viable alternative. If there are positive arguments for agricultural use of deforested land, the optimal use of the cleared land is for cattle to provide meat and milk products for the growing inhabitants in the region.

According to some writers and environmental groups, beef production causes hunger and poverty by diverting grain and cropland to support livestock instead of people. (15) These groups claim that beef production in developing countries perpetuates poverty, especially if beef or livestock feed is produced for export. Cattle consume almost twice as much grain as is eaten by Americans. Globally, about 600 million tons of grain are fed to livestock, much of it to cattle. If agricultural production were shifted from livestock feed to food grains for human consumption, more than a billion people, it is estimated, could be fed, the approximate number currently suffering from hunger and malnourishment.

Others assert that scientific evidence does not support these views. Cattle production need not pollute the environment or damage the land. Careful land management can improve pasture and rangeland, often land that is marginal as cattle graze primarily on poorer lands that cannot be used for crops. Grazing globally more than doubles the land area used to produce food. Manure can be a valuable fertilizer rather than a damaging pollutant.

Reducing beef production would lead to a decrease in feed grain, but this does not mean that grain would necessarily become available to the world's hungriest people. It is more likely farmers would not grow grains because of lack of market demand for it. Grains are only a partial diet of cattle. The remainder consists of grasses and other vegetative material humans and non-ruminant animals cannot digest. Since transforming these plant materials into food for human consumption is not possible, they become ruminant food.

Further, contrary to certain allegations, there is negligible relationship between fast-food hamburger consumption in the U.S. and the destruction of rain forests in Central and South America. Less than one percent of the beef consumed in the U.S. comes from Latin American rain forest areas. As we have seen it is consumed locally, and much of that which is imported to the U.S. is pre-cooked and canned products rather than as ground beef.

In Brazil, 4.5 percent of the landowners own 81 percent of the farmland. The Brazilian government estimated that 38 percent of all the rain forest destroyed was attributable to large-scale cattle development. The average rain forest cattle ranch employs one person per 2,000 head of cattle, or about one person per twelve square miles. By contrast, peasant agriculture can often sustain a hundred people per square mile.

Costa Rica: Ranching and Deforestation

Costa Rica is a small central American country seventy-five miles wide at its narrowest, possessing an extraordinary diversity of physical features and biological habitats, with Atlantic and the Pacific coasts, mountain ranges, volcanoes, swamps, marshes, and arid deserts. It has three times as many species of birds as all of North America.

First visited by Columbus on his voyage in 1502–03 from Honduras to Darien (Panama), it was named the "Rich Coast." Although signs of gold were detected, this label characterizes Columbus' avaricious propensity for seeking gold anywhere to solidify his ambitions, repay his royal sponsors and support his fledgling communities. But in fact the strip along the Atlantic shoreline belonging to Costa Rica, as I have personally seen, consists largely of swampy lagoons filled with mosquitoes and alligators. Even today it is unfit for human habitation except for escapees of civilization and vacation seekers of exotic adventures.

But Costa Rica, responding to global imperatives, expanded cattle ranching as an economic method for the obtaining crucial foreign exchange currency. The demand for beef in the United States was extraordinarily high and dollar-strapped nations like Costa Rica unwisely chose deforestation to earn dollars, though its forests are a principal tourist destination. Latin America in general lost 11% of its rain forests to cattle ranching alone. But cattle interests in Costa Rica cleared 80% of the tropical forests in just 20 years, turning half the arable land into cattle pastures.

Costa Rica denudes nearly 57,000 hectares of rain forest, among the world's richest ecosystems, each year, the highest total in the Western Hemisphere. In 1969, the government established thirty-two national parks and reserves, approximately 10% of Costa Rica's national territory. One of these parks, Corcovado, contains 161 square miles of rain forest. Within this small area, about the size of Washington, DC metropolitan area, there are 285 species of birds, 139 species of mammals, 116 species of amphibians and 5,000-10,000 species of insects. If the logging of rain forests continues at its present rate, the rain forests in Costa Rica will disappear before the end of the 21st century.

Today, about 2,000 ranching families own over half the grazing land in Costa Rica with 2 million cattle. The Costa Rican government established specialized exchange rates and credit instruments in order to help cattle ranchers expand beef exports or attract new investors. Beef exports increased nearly 500% from the 1960's to the late early 1980's as pasture land increased from 27% to 54% of total land mass. Deforestation and the increase in pastures in Costa Rica marks one the most expansive environmental alterations as the chart shows.

LATIN AMERICAN DEFORESTATION IN THE 1980's

	Acres	Percent
Brazil	22,500,000	1.8
Colombia	2,225,000	1.7
Costa Rica	310,000	6.9
Ecuador	850,000	2.3

Approximately 2.2 billion metric tons of topsoil have in the aggregate been eroded in Costa Rica because of beef exports to the United States. Ironically, Costa Rica today receives greater tourist earnings from the preservation of its rain forests than from its exploitation and destruction of its land.

A similar process is occurring north of Costa Rica in southern Mexico.

The growth of poverty is linked to the reduction of rain forest in the Montes Azules Biosphere Reserve in Chiapas, a region of once proud cedar and mahogany trees. (16) Is it better to save lives or save trees? The subdivision of land by the peasants, who grow traditional crops of corn and beans, around the reserve has limited the amount available for the new farming generations. Peasants believe they are being starved out by ecological policies that preserve the forest at their expense. They cut trees, burn the undergrowth, plant corn, and when the soil is exhausted let their cattle graze until rain washes away the thin topsoil unable to support grass for cattle consumption. The evidence is that the regeneration of forested land occurs much more quickly after small-scale farmers abandon moderately used deforested land than was believed. The recuperative power of the forest is impressive, though heavily degraded land recovers less rapidly, acquiring nutrients, growing new shoots and replenishing the soil and developing over time into a mature forest again.

The trade-off in the debate is between deforestation and small-scale farming as populations increase and small farmers seek land and work to sustain modest incomes. Large-scale farming and land speculators, on the other hand, with capital supplied by agribusinesses or as subsidies from the government, can destroy large swathes of forest and exact a long-term effect on soil, air quality, species eradication, and timber.

One universal purpose in the days of early civilization was the sacrifice of the bull to propitiate the gods for intended blessings and to prevent unwanted disasters. If the cow is god in India, in Latin America the cow is queen. And if religious ritual were to be revived as an integral part of cultural life as it was in ancient times, the cow would be a goddess. However, if money and economics is a kind of god, then cattle are still kingly, accounting for a sizeable proportion of all worldwide agricultural products.

And if we think that bull sacrifice is a primitive form of ancient cultures, let's take a contemporary example from Brazil. The Brazilian football club Nautico believed that its fortunes were not because of the quality of its players, its training program, or its playing strategies. Some divine intervention was necessary. The manager decided to call upon the services of an Umbanda priest, Father Edu, to change the club's fortunes.

Umbanda is an animistic religion brought to Brazil by African slaves and has fairly strong public support from all ranks of Brazilian society. Father Edu won acclaim when he discovered a skeleton under the stadium of another football club in Sao Paulo and then exorcised its spirit. The football team went on to win its league championship after 21 years of dismal performance. Father Edu had accurately predicted the poor performance of individual players whom he said had acquired evil spirits around them that needed to be expunged. He spent seven days in ceremonies for the Recife team that included sacrificing a bull, four goats, eight cocks, and various potions including a bottle of whiskey. It is not known how the team later performed after these ceremonies and ablutions.

But if religious practices were imported from Africa, other forms of entertainment have come from America. The rodeo has transcended the American west and found a home in Brazil. (17) It is the green and yellow flag fluttering over the stands that lets you realize that you are not in the USA. In fact, you are at the event billed as "The World's Biggest Rodeo Festival" in Barretos, Brazil, a city of about 100,000 about 500 miles west of Rio de Janeiro in cattle country. For 11 days in August, the 35,000 seat stadium is filled daily with spectators from the U.S., Canada, and Australia who come to watch the rodeo competition.

Once the terminus of long cattle drives and the restless cowboys (caubois) who drove them and still the place of Brazil's major slaughterhouse, Barretos has boasted of a rodeo since 1956 as an entertainment to keep cowboys from getting drunk and causing disturbances. Today, the rodeo industry is a billion dollar enterprise. There are 1,200 other rodeos throughout Brazil with sales of over 24 million tickets sold annually. Fans come to see a winner on a bucking horse or bull ride. There are no ties—someone wins or loses every eight seconds on a bull ride as the rider either stays atop the bull or is bucked off.

The rodeo may provide entertainment in Brazil as it does in the American West, but beef is the lifeline to the people and the culture. It's hard to imagine what the world would be like without beef to enhance a meal. There are a few places in the world, the Middle East is one, where, because of the heat beef cattle cannot comfortably exist, lamb is the typical meat of choice, and its slaughter for Jewish and Islamic religious feasts is deeply ingrained. But the Spanish and Portuguese colonized Latin America with their languages, culture and cattle, importing the meat that defined their eating habits, their principal industry and their lifestyle.

Figure 8: The Bull Rython or Drinking Cup From the Ashmoleum Museum, Oxford

CHAPTER FOURTEEN
THE ASIAN COW:
WATER BUFFALO AND YAK

Saudi Arabia has the largest integrated dairy farm in the world. Certified by the Guinness Book of World Records, this dairy farm in Al Kharj about 100 miles south of the capitol Riyadh, began in 1973 when the Saudis realized their vulnerability to a possible food embargo, the Achilles heel of a desert kingdom, at the same time the industrialized world realized its economic exposure with the 1973 oil embargo. The farm has 29,000 skinny dairy cows bred from a British dairy herd and produces 122,000 gallons of milk daily. These cows never see the sun or grass. The farm grows its own sorghum for silage for the cows to eat, and makes *creme caramel* and strawberry laban, a yoghurt drink from its milk. The cows linger in pens cooled by water spraying fans lowering the temperature from about 120 degrees down to near 80. (1)

So although Asia is home to all bovine species like the zebu and yak, but most commonly the water buffalo, the workhorse of the continent, cattle are nurtured in Saudi Arabia where they would never survive in the harsh, natural desert environment like a camel.

But if cattle can be industrialized in a forbidding climate, individuals like Ahmed can also exist with a few, hardy cattle in a habitat where they would not normally thrive.

Ahmed bin Mohammed bin Zaharo is in his mid-60s and lives near Fujairah in the United Arab Emirates in a place where his family has lived for centuries. The harsh rocky desert mountains have only a few trees and plants that grow in the spring after the rains have subsided and they provide him with his medicinal cures. Like his Bedouin father and grandfather before him he was born here and will live out his days in this place. His four sons and one daughter live in cities nearby, are married, and have jobs in local businesses and civil service. Ahmed lives in a thatched hut and exists on the milk and meat he gets from his cattle, descendants of the cows brought by the Portuguese explorers when they commanded the water routes from Europe to Asia over 500 years ago.

Let these two desert examples from the Middle East serve as an introduction to the adaptability of cattle, like people, to difficult Asian environments, and the resilience in co-existing with mankind everywhere in the world. Let's begin with the water buffalo.

Uses of the Water Buffalo

Water buffalo, literally the work–horse of the small farmer, helping feed about half the world's population, exist in over 40 countries throughout the world where they provide the work-energy for rural farmers. The global water buffalo population approximates 200 million, half in India where they contribute about 60% of the total milk produced, another 20% in China, and 10% in Pakistan. Oxen and water buffalo outnumber all mules, donkeys, horses and camels throughout the world. (2) I have seen water buffalo at work in the fields and farms of Japan, Malaysia, Borneo, India, Nepal, Thailand, China and southern Italy and saw the extensive uses of the water buffalo in rice fields, in haulage and forest logging, touring terrain where agriculture is the only economy.

The water buffalo is gray at birth turning to a slate blue at maturity. Its horns grow horizontally and laterally when young before forming a semi-circle as it matures. At maturity it is stocky, with a large belly and a long neck, a short face and powerful shoulders but inferior hindquarters. When harnessed, it ploughs, harrows the field, pulls carts and hauls timber. It is slow to pull weight but when it does it pulls heavy objects like logs and loaded carts relatively easily. Its work habits vary depending on country and agricultural needs. In rice producing countries it will work to plough and harrow or tenderize the soil for planting and again at harvesting, haul equipment, harvest crops and transport goods and people.

When not working, it prefers to wallow in shallow water, for which farmers provide ponds if not available nearby. Water buffalo have adapted well to hot and humid tropical climates but need water for heat reduction. The "water" part of the designation of this animal helps define its need for the proximity of water it has adapted to in tropical and sub-tropical climes. It will graze while wallowing, partly or fully submerged, and finds water fronds and aquatic plants particularly delectable for their high water content. The daily wallow, which cakes the skin for hours, further protects from insect bites and the tropical sun.

It trains easily, though rates vary from country to country and training begins when it is about 3 years of age. A week to ten days is sufficient for one person to train it to walk straight, not to raise its head while plowing like a cow does, and to respond to commands to go left, right or stop. Its inherent docility helps in training once its becomes accustomed to the sounds and movements of humans and then it becomes extremely tractable. Unless alarmed or frightened, it rarely attacks or threatens humans and is apprehensive of strangers. It is frequently led, fed, and cared for by the elderly and young in a farming family who regard it as a gentle provider.

A common misconception is that the water buffalo exists only in Asia. In fact it is found in more than 40 countries, in the marshes of southern Iraq, in Russia, along the alluvial flat lands of the Nile river in Egypt, northern Australia and the riverine drainage of the Amazon river in Brazil, as well as in China, India, and in all the countries of south and southeast Asia and several European countries. A second misconception is that the water buffalo is a symbol of primitive societies. But it is not just an animal adored from the past, but a contemporary that still has powerful economic implications for the future. It has ecological importance as well since it consumes huge amounts of fibrous plant material, like straw and rice stalks, that have no human use.

In other ways, the water buffalo depends intimately on human beings, for, as the human population increases, so too must that of the buffalo whose fortunes are tied to human management and care, and, to some extent, the survival of subsistence farmers in developing countries. Recommendations by the Food and Agricultural Organization (FAO) of the United Nations and the International Atomic Energy Agency suggested a combination of vaccines and de-worming increase calf survival and is the easiest way to increase the buffalo population. Protein dietary supplements are encouraged for milking cows. The low reproductive rate of buffaloes is a problem that has not been satisfactorily resolved. (3) The buffalo population will need to keep pace proportionately with the human population if even minimal economic production rates are to retain equivalency over time.

Because of both human and buffalo population increases throughout Asia, especially in highly dense population areas like the island of Java in Indonesia where buffalo is the major draught animal, grazing land is becoming scarce. Buffalo are often reduced to grazing along roadside verges. Among farmers in Asia, who, on the average farm only a few acres or hectares, rice straw may be the only food source for the water buffalo. Farmers in Bangladesh, Indonesia or Sri Lanka often cannot afford high quality feedstuffs. Thus, the buffalo population is being further reduced by under-nourishment, which leads to poor ovarian cycles, and by excessive buffalo slaughter for the saleable meat to supplement agricultural income.

If it is sold when it is older and disabled or less able to work well, its meat is tougher and less tasty. If raised for meat production and not draught work, its meat would be more palatable and of a higher quality to consumers than beef. (4) In The Philippines, the water buffalo, known locally as the *carabao*, is the meat of choice over beef. (5)

Urea is a solid nitrogen compound extracted from mammal urine. Together with molasses, urea is highly recommended as a dietary supplement to forage food intake of the water buffalo, which largely consists of crop residues and wasteland grasses. (6) When diets are improved, females tend to calve earlier in their cycle, 18 months instead of every 2 years. When diets are inadequate, the ovarian function almost ceases to exist. Smallholder farmers are frequently un-knowledgeable in the management of their buffaloes or do not have the resources to provide necessary dietary supplements. Lastly, the buffalo is subject to the variety of viral, bacterial and parasitic infections of any animal, even a ruminant, which has its nose on the ground. Where infections exist, mortality rates increase.

In Italy, one of the delicacies of the region south of Naples is "mozzarella di bufalo," cheese made from the milk of the water buffalo. It's deliciously different, more liquid than regular mozzarella. The water buffalo's origin in Italy dates back to the 13th century when they were brought to Europe by returning Crusaders. Thereafter, water buffalo breeding increased. The name "mozzo" in Italian means cut or chopped off, and derives from the fresh cheese cut from the full-cream milk. It is porcelain in color and elastic. Drops of a white, fatty liquid emerge when the cheese is cut.

After walking along the southern Italian coast in the summer of 2001 in the ancient city of Paestum, a city founded by the Greeks and only abandoned in the Middle Ages and still possessing some of the most remarkably preserved temples, I lounged in a nearby cafe with a glass of white wine and a lettuce salad laced with tomatoes and buffalo mozzarella, a delightful and memorable luncheon and my first experience of this wonderful cheese. Mozzarella of the water buffalo can be smoked, but even then its distinctive taste can be easily identified and distinguished from the mozzarella of the beef cattle. Apart from the Italian cultivation of the water buffalo for its cheese, the draught ox has nearly disappeared in Europe though it exists in some parts of eastern Europe. A few days later I ordered it again in a restaurant in Capri and have been partial to its delicacy every since.

The water buffalo has an economic advantage over mechanization such as a tractor. Its grazing food, where available, is less expensive than fossil fuels which would have to be purchased at a distance from the farm, transported and then stored. Tractor repairs might be more expensive that veterinarian fees. It does not pollute the air with carbon monoxide. Its presence in a family means that children can be easily taught animal care and husbandry. It has a low milk yield, but its milk can be a family nutritional supplement, especially in countries with poor protein and calcium selections. With selective breeding, the water buffalo can reproduce itself and hence contribute to the wealth of the family farmer.

In rice producing countries, tractors would easily become stuck in the thick mud of the flooded rice field, whereas the buffalo can work belly deep in the padi. Indeed, the water buffalo is known as the living tractor. Its slow, plodding movements and traction in mud and wet soil make it ideal for work performance in rural areas where roads are not paved, irrigation ditches are plentiful and transport vehicles scarce. In western China, where I saw entire hillsides terraced for rice growing, buffalo are efficient at working the narrow and steep terraced slopes. A tractor would become quickly immobilized in such conditions, if it could even reach the higher elevations without destroying terraces.

As a contrast, high mechanization and a minimum of manpower without draught animals is used in developed countries and among global food producers for high agricultural yields. For example, in the Sutter Basin in California there are extensive growing areas for rice. The soil is harrowed by disk before it is flooded. Rice seeds are soaked for a day with a small dose of chlorine to prevent fungus growth before planting and to begin germination. An airplane seeds the field from an altitude of about 25 feet, and the heavier seeds find their way into the ordered troughs. A combine harvests the rice. (7) In California, there is no need for the work of the water buffalo.

The future development of the water buffalo in the developing countries of Asia will depend on small-scale farmer adoption of modern methods of reproduction and the application of improved technology. As crop methods improve, the use of the buffalo may decline as a puller of ploughs and carts, but increase as a meat and milk producer. Among poor farmers in Asia, inadequate reproductive management in its estrus detection, lost semen and infertility, has kept the water buffalo population relatively stable to what it could be with improved techniques.

Water Buffalo in China

As was true in ancient Middle Eastern civilizations, the Chinese had a God of Stockbreeding in their mythology about three thousand years ago. (8) *Niu-wang*, the so-called King of the Kine or protector of cattle, is the Chinese personification of the buffalo. *Niu Wang's* image, over 16 feet high, had horns, and the mouth and ears of a buffalo.

In Chinese folklore, oxen, which included water oxen, existed only in heaven and not on earth. People did not live very well because there was little to eat and no animal to help them with the work. Consequently, the Emperor of Heaven sent the ox down from the sky to tell the people they should eat at least once every three days. The ox got the message confused and told the people that the Emperor says they should eat three times a day.

Because of this grave mistake, the Emperor sent the ox to earth permanently to work for humankind by plowing. Originally found only in heaven, oxen are now on earth to help people. This is the cattle equivalent of original sin. It is not known if the quality of their draught work qualifies oxen for re-entry to heaven.

The work of water buffalo in China is similar to that in other Asian countries. Buffalo plough fields, puddle and harrow the flooded rice fields, grind millet, grain or corn, dig for wells and raise water from wells for irrigation, even mine the earth's surface for minerals like silver by plowing. Water buffalo universally throughout Asia are used for rural transportation, but routinely from home to field, if at a distance, and back again. The water buffalo might produce a very good source of meat and protein, complementing pork, chicken and duck, for the Chinese diet and for export. However, agricultural research and large-scale productivity of the water buffalo has lagged behind in economic development until very recently. Still, China needs a wide supply of food and protein products for its growing 1 billion plus population.

A comparable example comes from an unlikely source: the ostrich.

While in Hong Kong in the summer of 1997 as a journalist to cover the ceremonies of the British transferring the territory back to China, I was following a food story about the importation into China of ostrich. The ostrich, a 6–foot high flightless bird with valuable feathers, seems an unlikely source of popular food. It is the largest bird on earth, each one often weighing over 300 pounds. Its long neck and legs, coupled with a bad temper and disposition, make it unacceptable as a household pet. But the taste, more like beef than chicken, make the ostrich an attractive protein source for international palates. Ostrich breed in hot, dry climates, and can mature in 3 years and produce as many as 30 chicks annually, making them an ideal protein source and a relatively inexpensive product to feed and bring to maturity. (9)

The same situation is not true for the water buffalo despite its equal potential as a high protein food source for China with 25% of the world's population to feed. Could this be because the Chinese culture and rural farmer views the water buffalo as a chief draught worker of the farm, almost a member of the family, and not as an animal to be eaten because of its labor-intensive productivity? The water buffalo is one of the most docile of all animals, does not bellow or low like a cow, and, like a dog, will come when called by name. It has a working life of about 25 years, about the same as a person. The pig, the duck, the goose, chicken—none of these farm animals provides the farmer with such docile, productive labor as does the water buffalo. Although the Chinese will eat just about anything as food, Chinese reluctance to eat the water buffalo is understandable.

As befits a country with the world's largest population, the Chinese meat sector is massive, with over one-half of the pork consumed in the world. But since 1985 the size of the pork pie chart has been reduced from 86% of the Chinese market to 66%, and poultry and beef have both increased as a percentage of meat consumed. (10) Although not a country usually associated with beef, China is emerging as a giant in the world beef market as it makes the transformation from a socialist, centrally planned economy to a market-driven economy.

Buffalo Races

In the village of Kakkoor in a district of Kerala in southern India a bullock race can occur at any time in a paddy field. Two bullocks race together linked by a wooden pole roped so that two loin-clothed men can race alongside and keep it and the bulls straight. Men and bulls splash muddy water in all directions as they race towards a pre-established goal.

As owners of buffalo prize their animals dearly, it is not surprising that they would use them at play as well as at work. Some countries (Laos, for one) have forbidden fights between bulls because of excessive gambling. But throughout Southeast Asia, bullfights—sometimes just extended shoving matches, but often to the death—occur on festival days and are well attended.

Zebu bulls, a bovine animal with a humped back, on the island of Madura north of Java in Indonesia are bred for racing purposes. These animals are traditionally used for farming among the rice paddies and agricultural plains of the island. But once a year they are raced for prize money. The race occurs annually but preparations continue throughout the year as the top prize money is about $20,000 dollars, a fortune to native islanders. The bulls are bred on a cocktail of eggs, chocolate and beer, among other ingredients, to fatten them and make them strong and fast. Two zebu bulls are yoked together and a rider straddles them, holding on with each hand to a tail, ostensibly for steering but also to maintain balance during the bumpy ride. The rider stands on a wooden platform similar to a short ladder with a place to hook his legs and bounces up and down as the bulls attain a speed of nearly 40 miles an hour. There are heats to the races and finally only the fastest are raced in the final event of the all-day affair on the island.

On nearby Bali island, which has a predominate Hindu population, the sacred water buffalo sacred are bedecked with fanciful headdresses while enormous bells, almost three feet across and a foot and a half high, are hung from their necks during the annual New Year's festival race. Two buffaloes are hitched to a two-wheeled cart, much like a Roman chariot. After the bulls have been matched in color and size, two carts race with two buffaloes and one driver each race along a dirt track. The prize is often four times what the winning buffalo is worth on the open market. (11).

One unintended race drew no bettors. The following tale is a true story, seemingly unbelievable, like a story from *The Arabian Nights*. Unmarked cargo planes come and go at the Dubai airport in the United Arab Emirates on the Persian Gulf coast. They take off and land with commodities to and from Asia. These planes are often commissioned through intermediaries who hire the planes and crews, usually Russian or Ukrainian pilots who are usually much cheaper than western crews.

On one such trip, a Russian crew was returning from South Asia with a light load and refueling at a small airport in an undisclosed South Asian country. The crew waited on the tarmac. The crews claim they never get enough to eat on these long-haul trips. Spotting cows grazing in a field next to the airport the pilots decided to grab a cow and take it back to the Gulf and have a butcher, for a share of the meat, cut it into portions for them. They hit the cow over the head with a wrench and, by lowering the back of the airplane's ramp, drug it into the back of the cargo hold and took off.

Once airborne, and while the plane was over the Gulf waters, the cow recovered consciousness and, wildly excited, began thrashing through the cargo area. The crew didn't want the goods damaged and unable to contain the revitalized cow, lowered the back ramp and the cow dropped out. Descending at over a hundred miles an hour, it hit an unfortunately located fishing boat and sank it. The astonished fishermen were rescued by a nearby boat and taken to Dubai where they reported to the police that their boat had been hit by a flying cow and wanted to claim insurance for their loss. Unwilling to believe such a fantastic story, and agreeing that the fishermen were attempting insurance fraud, the police threw them in jail.

A few days later at a diplomatic reception, when the chief of police was relating this story to the Russian Ambassador about the fishermen whose boat had been sunk by a flying cow, the ambassador, who was familiar with the other side of the story from the Russian pilots, suggested that the chief should investigate further. The Russian Ambassador later sent a letter to the chief explaining that indeed it was possible a cow sank the fishing boat as on the same day in question a cow had been ejected from a high-flying but unknown airplane. The fishermen were released. But there is no indication of whether or not they received any compensation for the loss of their boat or the catch of the day.

Buffalo Cloning: Until the Cows Come Home

Bessie, of Sioux City, Iowa, in November 2000 became the first cow to give birth to a cloned ox, known as a gaur. Bessie was the surrogate mother who carried the gaur fetus known as Noah. Scientists had known of the possibility of one species giving birth to an implanted embryo of a similar species. Agricultural specialists took some DNA from one of Bessie's eggs, artificially induced to begin dividing without fertilization, and fused it with a skin cell of a living gaur. Noah was cloned from a single cell taken from a dead gaur, and hence Noah has no father. The cell was nurtured in a laboratory where, fused to Bessie's eggs, it grew into an embryo and was then transferred to Bessie's womb. Bessie's immune system accepted this transmission. (12) Bessie was one of 32 cows implanted with fused eggs but the only one to bring the embryo to term. Eight of the cows retained embryos; five had miscarriages and two were removed for early fetus analysis.

This cloning technique to seen as a measure to help preserve endangered species like Pandas, or even extinct species like a woolly mammoth, a 20,000 year-old specimen recently found in Siberia. Cloning extinct species might mean that select animals will return in a cloned form to a habitat that also doesn't exist in the way it did when they lived earlier. This historic cloning of a gaur in a surrogate cow is another example of how cattle save civilization...as well as each other.

The Yak

The domestic yak (*bos grunniens*), also affectionately known as the grunting ox, has its home on the high Tibetan plateau that stretches from the Himalayas to the Gobi desert. If you want to have friends in high places, or just escape summer heat and see the yak in action, the Tibetan plateau in western China is the place to go. The average elevation is about 10,000 feet above ground level. About 85 percent of the world's estimated 15 million yak live in this part of China, in Qinghai province and westward to the border with Kazakhstan. Others are scattered about south Asia, Tibet, Afghanistan and Russia.

Like the water buffalo among people in the tropics and subtropics, yak are crucially vital to the lives of the people who live at these high elevations with their meat, milk, hair, hides, dung and other animal products, as neither cattle nor buffalo would survive during winter at these altitudes. An average bull will weigh from 300–500 pounds, and the female somewhat less.

The first thing that strikes you when you see a yak is that it looks like it has every worn out rug draped over its back. Its shaggy long hair reaching nearly to the ground is of course just the insulation it needs in the severe cold of this region during winter months. Its hair is also a valued commodity. Yak are shorn once a year and both the wool and long coarse hair is used for weaving blankets. Their thick coats are normally black or dark brown, but can also be red or tawny, or even silver-gray. Its appearance can be deceiving because with surprising agility it can carry a 100 pound load for more than 20 miles along a steep, narrow path in the high mountains and plateaus of central Asia. Its usefulness as a pack animal is legendary and indeed absolutely necessary in this climate and topography.

When I first visited this part of the world in 1988, I traveled with Chinese hosts from the provincial capitol city of Xining in Qinghai province in western China along the dirt road heading west into the setting sun, into a vast, desolate interior of the province, situated high on the plateau where the ground was relatively level but the altitude was 9,000 feet. It was summer in the western hemisphere but it was cold. The average annual temperature here is about 30 degrees Fahrenheit, or zero degrees Celsius. We had in our van a 50-gallon drum of gasoline as there are no gas stations along the way and none at any destination. In the distance along the route to Qinghai lake, a large fresh water lake without an outlet, we could see small bands of grazing yak tended by the nomadic Tibetan herdsman, the only permanent inhabitants, together with their yaks, in this forbidding landscape.

Besides the value of yak meat, the most prized food the yak delivers is its milk which yields a strong, pungent butter tea, yogurt, and an even more powerful fermented drink which can rock you back on your heels, or your backside if not seated. In this country and climate, it is an indispensable beverage, another sign of vital human nourishment from a form of cow few with have encountered and a drink fewer still can stomach.

There are some restrictions among these Buddhist believers about killing animals, but not if they have been killed by accident or others. Hence, a cottage industry has arisen of yak killers, a class of Tibetan butchers who slaughter enough yak once a year for the season.

Fresh milk from the yak is not the usual local beverage for the Tibetans of central Asia or the Sherpas of Nepal. But the curd and buttermilk is extensive as butter is used both for domestic and ritual use. The buttermilk is often boiled until it becomes solid like a hard cheese and then can be eaten or carried in a leather pouch for consumption while traveling.

Finally, you shouldn't be insulted if you visit the Matsuzaka Beef Farm in Japan if you are politely asked to remove your shoes on entering the cattle barn. Mr. Kubo, the owner, insists it is to protect the dissemination of hoof – and-mouth disease, one of the reasons Japanese beef was banned in 2000 for shipment to the U.S. Market. But the real reason is because Matsuzaka beef is prized meat for Japanese palates, selling at upwards of $60 a pound in markets, when you can find it, and for about $144 for a grilled steak in a Tokyo restaurant, when you can find it. (13) Removing your visiting shoes protects the cow, not you.

Like the more famous Kobe beef, Matsuzaka cows are pampered, massaged and fed a high fiber diet of wheat, hay, corn, bran and soy bean washed down with copious amounts of beer. The cow's head is held high and is fed beer through a bottle just like an ordinary guzzler. Exercise is discouraged. Cows are kept in this plush environment for at least three years, a long time by beef cultivation standards before they are slaughtered.

I ate my first Kobe steak in Japan in 1964. It was more tender than an ordinary beefsteak. The fat was sweeter and the red meat seemed to retain the juices more because it was marbled. Everyone said that the massaging of the cows daily with a stiff brush, together with the high fiber diet which stretched the stomach to allowed the cow to eat more, made Kobe beef a special culinary treat.

Figure 9: Sargon II as a sculpted winged bull also from the British Museum

EPILOGUE

I asked myself in 1986 in Aleppo why I was seeing so many cow and bull sculptures from societies where civilization originated and what their significance was. I realize now that although there is a time lag between bovine museum sculptures and mad cow disease that the difference is less meaningful than the global permanence and continuity of cattle culture. Cattle, like relatives, are an indispensable part of civilized life. But the cow is not the myth; myths have accumulated around the cow like barnacles on a docked ship. The key ingredients for understanding the human context of the cow are civilization and culture, twin concepts which link humans to each other and the resources they need to sustain themselves.

Civilization comes from the same Latin word as city, citizen, civic and civility and refers to a high level of social and technological accomplishment. Writing is the chief example among less thrilling but no less important advances such as etchings on pottery and improvement of ploughs. But because of the domestication of animals for food and the improvement of methods for seeding and harvesting, hunting and gathering practices were minimized and settled communities and cities matured into robust civilizations. What began as tribal groupings expanded over centuries into pastoral settlements, evolved into towns and cities where people gathered to sell, trade and form social structures, and grew into empires of wealth and power with kings, armies and divinities for protection. Civilization is a union of soil and soul, the resources of the earth transformed by desire and maintained by social and legal discipline. A constant factor throughout the evolution of all the world's cultures is a reliance on the cow as the most valuable animal resource.

Similarly, culture, like the constituent atoms of a molecule, constitutes all the behavioral and historical elements of the way a people live. Art, painting, architecture, sculpture, literature, language and social actions define and shape every culture in countless expressions of uniqueness. The physical form, a flag, a crucifix, a classical Greek frieze, a menorah, an idiomatic saying, a food or clothing choice, all have significance within an elaborate cultural web. Cattle have been contributors to all civilizations and all societies, not just as food, but as suppliers of commodities, entertainment and even religion, making them peerless cross-cultural objects.

The use of the cow as food, drink, shelter, clothing and other consumable products parallels the development of both civilization and culture. As societies populate and disperse and individuals compete in an information age, people have less direct experience with food sources and literally have lost sight of agriculture's crucial importance to human sustenance. (1) In the past half century we have learned how tight the inter-species relationship is. (2) We share two common traits with cattle. One is a similar mammalian evolution, each originating millions of years ago, and a second is a relationship of cultural affiliation and dependency originating about 10,000 years ago. The typical belief is that humans are distinctly different from all other animals. However, mounting scientific evidence, from archaeology, paleontology, genetics, zoology and the life sciences, has compellingly demonstrated that humans come from a common ancestor and that all living organisms share a similar genetic structure.

Take mice for example. Humans and many other mammals descended from a common ancestor about the size of a small rat from 75 to 125 million years ago. (3) Mice and humans each have about 30,000 protein-coding genes and share 80 percent of the same genes; only 300 are unique to either organism. About 99 percent of genes in humans have related counterparts in the mouse and 80 percent have identical, one-to-one counterparts. It is possible to recognize regions in the two species that have descended relatively intact from a common ancestor. A genome is all the genetic material in an organism, unique to every plant or animal and about 40 percent of the two genomes are directly aligned. The proportion of unique mouse genes not detectable in the human genome (and vice versa) is less than one percent.

The mouse and the dog are the only mammals after the human whose genomes have been sequenced as of this writing. (4) Scientists now are working on genetic blueprints of the rat, cow, and chimpanzee. The mouse provides a special lens through which we can view ourselves and other living organisms. As the leading mammal for research over the past century, the mouse has provided a model for human physiology and disease, leading to major discoveries in such fields as immunology and metabolism.

As evidence accumulates of genetic similarities among mammals large and small, it is becoming harder to identify and understand the functional differences that make each species absolutely and not just relatively unique.

No, we don't eat mice and their size and taste argues against them as a future major food source. But when we come to larger animals, food choice does make a difference, and cattle have been humanity's primary selection for the dinner menu since before history. Although I do not have evidence nor the expertise to conduct such scientific research that supports a similar biological case for cattle, I have argued that the history and cultural evidence of humanity's association with cattle has been cumulative over millennia, just not fully documented.

Cattle are really everywhere but we don't always see them. By the turn of the 3rd millennium, there were 97 million cattle and calves in the U.S., including 33 million beef cows and 9 million milk cows, a number that has been stable since 1992. (5) The population of cattle in the U.S. is approximately one-third that of the human population. Nearly 3 million graze on over 306 million acres of public land in 11 western states, over 40 percent of the land of the West.

Over 70% of all grain produced in the U.S. is today fed to livestock, primarily to fatten cattle. Growing grain to feed animals and not humans is one of the major agricultural transformations in civilization and might never had occurred were it not for the insatiable appetites for marbled and fatty beef. But as we have seen, it is not just Americans who have a love affair with beef. Except for devout Hindus and other vegetarians, the whole world has adopted beef as a main source of protein. Humanity now seems genetically predisposed to its texture, taste, and fat, its usefulness as a work animal, and for long periods of human history as an object of veneration.

If we fast forward to the third millennium we can see the timely relevance of cattle not just on cultures, but on psyches and well as stomachs.

Control of beef production in the U.S. is in the hands of the powerful beef industry whose primary business assumptions and aims are industrial efficiency. Congress passed *The Beef Promotion Act* (1986) authorizing a program of research and promotion directed and funded by the cattle industry with oversight by the Secretary of Agriculture. An assessment of one dollar on all cattle sold in the U.S. and on imported beef and beef by-products goes into the coffers of a special board that promotes beef. The law, ignored more in the breach than the observance, prohibits this money from influencing government policy. (6) This is but one example of how meat industry lobbying has sought and found favorable legislation for its interests.

The meat industry, beef, chicken, turkey and pork producers, beginning in the 1990s began building feedlots in isolated rural areas. The industry captured the waste of cattle and hogs and sprayed them in nearby fields, operations that did not require permits as strict as those for the disposal of human waste. As a result, according to the Environmental Protection Agency, over 35,000 miles of rivers have been polluted. Citizen groups have filed numerous lawsuits to curtail or control the waste problem or eliminate these feedlots altogether. Feedlot operators say they already have nutrient operating plans. (7) Environmentalists claim that the standards are too loose, and do not allow for public review.

Some have seen the beef industry as a danger to the environment, of more people dying from heart disease, of forests depleting because of cut trees, of methane gases polluting the atmosphere, of ground water contaminated by cattle waste. The continued use of 270 million acres of federal public land for domestic livestock production is endangering species, effecting arid and semiarid lands and riparian areas, and disrupting ecosystems at unprecedented rates. (8)

Federal grazing permit holders include banks, large corporations, grazing associations, wealthy individuals and small family operations. Only a minority of permit holders have been in the business for more than a generation. Fewer than 23,000 livestock producers (of 1 million nationwide) possess federal grazing permits. For some, the ranch is a hobby or a tax deduction. The majority are small operators who depend on other, non-ranch income to support themselves. Public-land livestock grazing continues because federal agencies are reluctant to contest powerful western livestock interests as they have been for over a century. Few western communities are dependent economically on public land grazing. Agriculture as a whole comprises a fraction of the economies of most western states and livestock production an even smaller part. (9) While the coffers of corporate food giants have risen, farmers, idealized in political statements, have witnessed an unprecedented decline: 36 percent of all American farms, over one million, from 1970 to 1992. (10)

The U.S. Dept. of Agriculture approved $762 million to livestock producers in 2002. The Bush Administration approved grazing on tens of thousands of acres in the Conservation Reserve Program, on land set aside for wildlife habitats. Congress had been considering legislation that would exempt ranchers who lease public land from conforming to regulations of the National Environmental Act. Environmental and conservation groups accuse the ranchers of getting money for problems they themselves created.

Some see it differently. Grazing, because there are so many rules and environmental standards regulating its use, does not seriously damage the ecology and compete with the wildlife. (11) Ranchers and cattlemen belong on the grazing lands as much as anyone and with judicious use of land and water they can graze their herds and maintain a sense of western dignity as long as no new designated wilderness areas are opened for grazing. Ranch managers predict that no private cattle will be running on public land soon because the costs and risks are too high and the profit margins too low.

Clearly, cattle, like people, leave footprints where they live and travel. And unquestionably, people might be healthier if they ate only rice, fruits, nuts and vegetables. But the amount of food consumed throughout the world would not likely diminish, and therefore its production and processing would also leave environmental tracks unless everything disposable was also biodegradable or able to be re-cycled.

The grasses of the Great Plains in the 19th century fed the cattle and the cattle fed the nation, helping to create the romance and magic of the western cowboy, the embodiment of independence and hard work, a lifestyle as virile and unrelenting as the landscape.

The migration of so many people westward in the 19th century imported a culture as it resettled new land. The massive movement reduced physical and intellectual boundaries and expanded frontier intercourse among a plurality of peoples. In 1787 the states created the national government; a hundred years later the national government used land to create new states. The advantages of the land created a robust new mentality of rugged individualism, practical expediency, restless energy, an exuberance for freedom, a struggle for survival coupled with a profound self-interest. Food was available from plentiful wild game. But cattle came with colonizers and settlers, supplanted the native bison, increased exponentially, helped make fortunes, then exhausted the land, and stayed to become a part of the culture of food and income in the West. The continuing presence of so many cattle in the West is a living testimonial to the cow's preservation as humankind's chief animal.

Yet never before in human history has such large masses of people been so divorced from the land and the direct processing of their food. Perhaps this phenomenon has contributed to our alienation from nature as the natural environment has been subdued and recreated as subdivisions, business plazas, parking lots, shopping malls and highways. Increased population threatens further ecological damage. When food and its processing are left totally in the hands of agribusiness companies so we can conveniently purchase food already packaged, and when contamination occurs, only then do we realize the danger of a hopeless loss of control over the food we eat, unless, as in a few community vegetable plots in England, or in victory gardens I recall tending in the backyard as a boy during 1942–45, as everyone participated in growing food for the war effort.

The consolidation of the beef industry in the United States has benefited consumers with a plentiful supply of meat while posing greater dangers of infectious contamination, as beef is concentrated in more economically and politically powerful integrated industries which use migrant, unskilled labor and skirt safety regulations in order to cut costs. So universal is the reach of the beef industry that the discovery of E. coli, hoof-and-mouth and mad cow disease have global economic implications and the potential for the rapid spread of epidemics. The latest global panics of cow diseases are examples of how integral and interwoven is our symbiosis with animals throughout the world and how deep is the biological relationship in our food supply and cultural habits, such as dining at fast food outlets.

The cow is dismembered and repackaged for purchasing and preparation convenience. In our culture we have distanced ourselves from the distractions of nurturing, feeding, weaning and killing our animals for family food. As a child and youth I remember no one in the family buying meat from a butcher. Each year, my grandfather and father killed a deer, often an elk, occasionally a moose, once a bear. We hunted pheasants, ducks and geese with trained cocker spaniels. We fished for salmon, trout, bass and white fish. We only ate the meat of animals we killed in the forest, field, farmyard, rivers and streams like Native Americans and early settlers. Sadly, such wildlife largesse no longer exists and the animal bounty we enjoyed then has passed into the hands of mass-produced meat product companies, trout farms and rural, out-of-sight processing plants. As ancient icons of human fertility and virility, the cow and bull have been hidden in fields, behind feedlots and processing plants, artificially inseminated, force-fed chemicals, and turned into packaged meat. In the haste to buy foods conveniently and machine-processed, we have lost touch with the close identification and affiliation humanity has had with the cow.

Industrial automation began, not with the automobile, but with cattle. One form of automation is cloning. The rush is now on to challenge the arguments against cloning cows. In the spring of 2001 leading scientists testified before Congress that cloned animals would have too many defects to be acceptable. They agreed that it would be immoral to press on with cloning. Other scientists view this challenge as preventing experiments in human cloning. Scientists at Advanced Cell Technology in Worcester, Massachusetts have cloned 24 cows that grew to adulthood, subjecting them to every medical and scientific test devisable, and they all are normal in every manner. The company created 500 embryos from which 50 survived birth and 24 matured. Skin cells from fetal calves were transferred to cow eggs whose genetic material had been removed. These cloned embryos were then transplanted to cows serving as surrogate mothers. (12) Your next hamburger could come from a genetically engineered cow.

The hamburger is the new cultural icon. There are nearly 600,000 fast food restaurants in the U.S. feeding 100 million Americans daily and serving over 40% of all meat consumed. These food industries employing eight million, the largest employer of any industry in the U.S. Combined with animal fat to ensure firmness, beef has monopolized the meat market globally and made eating out cheap and quick, without even the need for utensils.

Fast food restaurants using beef are able to offer tasty food, but of questionable nutritional value and with added substances such as flavor enhancers and stabilizers. (13) A "Big Mac," for example, is 7.6 ounces with 590 calories, 34 grams of fat, 11 grams of saturated fat, and 1,070 grams of sodium. An ordinary cheeseburger, usually 4.2 ounces, has about two-thirds less of everything. Eating a Hardees Six Dollar Burger will cause you to have to burn off 949 calories with 62 grams of fat, 25 grams of saturated fat and 1,685 grams of sodium, unless of course you are in a weight enhancement program. These caloric and fat counts are exclusive of whatever else a client purchases and eats, like fries made in beef fat, sugar drinks and desserts, during the same meal, typically totaling more than half the caloric intake for an average individual's needs of 2,000 calories a day, and containing negligible or minuscule amounts of the enzymes that fresh fruit and vegetables provide for nutritional balance. (14)

Because more than 60% of American adults and a growing number of children and adolescents are overweight or obese, a national public health crisis has loomed, which, according to the Surgeon General, is costing Americans $117 billion a year in hypertension, diabetes and heart disease. (15) Suits filed against food giants by individuals who claim they have grown obese from eating fast food products may appear frivolous, but unless ingredients are properly labeled, and nutrition breakdowns are not easy to find even on the web sites, the exact composition of products is unknown to the consumer and potentially unhealthy, quite apart from whether there is injurious bacterium or virus included. (16)

As the guest of a governmental agency in Almaty, Kazakhstan in the mid-1990s, I was honored at a dinner where a whole roasted lamb was served, including the head. As was customary among the Kazakhs, I was given a slice of the ear to eat so that I might more carefully hear the injunctions of my elders, and one whole eye so that I might see things more clearly. Luckily, I had consumed enough vodka and brandy by then to tolerate such ethnic and culinary delicacies. What was unusual is that serving the whole animal is rare except in those cultures, like the Hawaiian roasted pig, where the practice is traditional.

The history of the cow and bull is as old as civilization. Though the dog and horse, which we do not normally eat, are domesticated for our use, no other animal has so dominated in so many ways in a human relationship as the cow. The earliest artistic depictions reveal an awe, reverence and fear of the power and work potential of the bull and the efficacious benefits of the cow. It's no wonder that ancient peoples deified the bull and cow as symbols of a divine spirit indispensable to human survival and created cultic myths around its status and preservation.

The most persistent cultural trait that we are reminded of daily without questioning are our food choices. Wherever we see it, the image of a cooked steak makes us salivate. Acquired tastes and experimentation with a variety of ethnic and national foods partially dictate what we eat. But genetic inheritances also determine food choices. Since humankind has been chiefly beefeaters for millennia it would be difficult to resist such strong biologically determined inclinations.

Of course it is possible to make conscious changes in our dietary habits for health or religious reasons, for example switching from meat to fish or poultry, reducing cholesterol intake, or becoming vegetarian. A change in economic status, such as loss of a spouse or unemployment, might alter food choices. Political policies that effect subsidies for farmers and economic difficulties that cause a sharp rise in meat prices, diseased cattle, drought conditions, high feed prices, might also reduce beef and dairy consumption.

Each generation creates new, so-called cultural icons: pop stars, athletic heroes, logos, images associated with advertising products. But no other artifact, animal or image, with the possible exception of gold, has endured so long as a desired object as has that of the cow. The end of civilization, like the eye of the hurricane, a calm center of stillness before the swift advent of a catastrophe, may not be much different from its humble origins: the search for a perpetual food supply.

The owl, the Roman goddess Minerva's symbol of wisdom, is able to rotate its head 180 degrees to see in every direction. Like the owl, we need to be able to see into the past, not just because the past is prologue, a useful introduction to life messages, but because it is instructive, a way of understanding the present. Whoever we are, however we perceive ourselves, cattle have been an integral part of the world's art, literature, mythology, religion, food and entertainment.

NOTES

Introduction

1. From the inimitable *NYT* columnist, Maureen Dowd, "Herd on the Street, Mad Cows and Udder Fools." *The New York Times*, (2001, April 11), A27.

2. Roberts, J. *Antiquities of West Cork*. (Skibbereen: West Cork, Ireland, no date).

3. From the popular, long-running New York Times bestseller, Eric Schlosser's *Fast Food Nation, The Dark Side of the All-American Meal*. (Boston: Houghton Mifflin, 2001).

4. P. L. Brown, "In California Bullfights, The Final Deed is Done with Velcro." (*NYT*, June 27, 2001), A1.

5. McCurdy, J. "Banged-up Bull Rider Taking 'em 1 at a Time." *Arizona Republic*, (October 19, 2001), C3. These introductory references are intended to demonstrate the traits of cows and bulls appearing daily in the culture. In the first five chapters I show bovine origins and in the second part of the book contemporary bovine issues. Cf. a newspaper story I wrote after attending a bull riding rodeo event: "Bullriding Speaks to Something Ancient in Us." *Standard Examiner* (September 2, 1999).

6. A. Scioscia, Steers and Queers. *Arizona New Times*, 2001, 32(7), 30-36.

7. Originally Al Andalus in Arabic when the Moors ruled southern Spain from the 8th century until 1492. About the size of South Carolina, Andalusia is generally arid, poorer than Spain's north, and dotted with wetlands at the delta of the Guadalquivir River near Cadiz.

8. There are also female fighting bulls raised in Andalusia, smaller than the males, with shorter and thinner horns, a longer neck, a small dewlap under the neck, and no visible udders. They act like steers and are often exhibited in bullrings in southern France, like the quaint village of St. Marie de la Mare in the swamp region of the Camarck where I watched such an event a few years ago. They are not killed like the bulls but fight until their fighting spirit is exhausted.

9. Particularly *The Sun Also Rises*. (New York: Simon & Schuster, 1926) and *Death in the afternoon*. (New York: Scribner, 1932). "The stretch of ground from the edge of town to the bullring was muddy. There was a crowd all along the fence that led to the ring, and the outside balconies and the top of the bullring were solid with people. I heard the rocket and knew I could not get into the ring in time to see the bulls come in, so I shoved through the crowd to the fence. I was pushed close against the planks of the fence. Between the two fences of the runway the police were clearing the crowd along. They walked or trotted on into the bullring. Then people commenced to come running. A drunk slipped and fell. Two policemen grabbed him and rushed him over to the fence. The crowd was running fast now. There was a great shout from the crowd, and putting my head through between the boards I saw the bulls just coming out of the street into the long running open. They were going fast and gaining on the crowd. Just then another drunk started out from the fence with a blouse in his hands. He wanted to do capework with the bulls." (Hemingway, 1926, p. 196)

10. A. Shubert, *Death and Money in the Afternoon, A History of the Spanish Bullfight*. (New York: Oxford University Press, 1999), p. 14. See also R. Eder, "Defeating Death by Writing About It in the Bullring." (*NYT*, March 30, 2001), B38. "A bull that has been cut will become less fierce and intractable, but he will not lose his strength, he will be as good as ever for work." (Xenophon, c.430–355 BCE, *The Education of Cyrus*, p. 245).

11. *Illustrated London News*. (1849, Jan.–June). London: William Little, v.14, p. 375-76. An interesting source I happened to find in a rented cottage in Ireland on the Bay of Bantry in County Cork.

12. Hemingway. 1926, *op. cit.*

13. R. Trow-Smith, *Life from the Land*. (New York: Longmans, 1967).

14. V. Porter, *Cattle, A Handbook of Breeds of the World*. (New York: Facts on File, 1991).

15. J. Pukite, *A Field Guide to Cows, How to Identify and Appreciate America's 52 breeds.* (New York: Penguin Books, 1966).

16. A. Sanders, *The Cattle of the World, Their Place in the Human Scheme—Wild Types and Modern Breeds in Many Lands.* (Washington, DC: The National Geographic Society, 1926).

17. J. Diamos. "Holy Bull is One of Six Inducted into the Hall." (*NYT*, August 7, 2001), C16.

18. Beef slaughterhouses, apart from Upton Sinclair's *Jungle* expose in 1910, have been the topic of several 1990s books including Jeremy Rifkin's *Beyond Beef* (Dutton, 1992), Gail Eisnitz's *Slaughterhouse*, and Schlosser's *Fast Food Nation*.

Chapter One

1. Besides the reconstructed cave at Lascaux in the Dordogne valley, I entered three other caves in the region and in each saw examples of the painted figures of ancient aurochs.

2. J. Clottes, *The Cave of Niaux.* (Boulogne: Castelet, 1981). After walking about a mile into the interior, a place of utter darkness and, upon reaching the interior where the painting are located the guide turns the artificial lights back on revealing the still luminous colors of the wall paintings is a sudden artistic delight.

3. J. Campbell, *Primitive Mythology, The Masks of God.* (New York: Penguin Books, 1987). These Paleolithic ancestors used natural pigments—yellow, red and brown ochers ground with powder and mixed with animal fats. They outlined the drawings with charcoal to give the painting body and a distinctive outline and used bone lamps with reindeer fat as fuel and juniper twigs as wicks as light when painting bison, woolly mammoth, reindeer, elk, wild bull (aurochs), horses, stags, and bears, their joint inhabitants on the land. The drawings of humans, in contrast to the animal paintings, are usually stick figures and distorted images.

4. Contemporary cow art satirizes art criticism. Mark Tansey's 1981 painting called *The Innocent Eye Test*, is an oil on canvas hanging in the Metropolitan Museum of Art in New York City. It depicts a cow staring at a painting of cows lounging in a pastoral scene underneath a tree. Elderly and distinguished male judges or museum employees are holding the painting and undoing the draping around the wall-sized painting not yet in a frame. The cow's reaction is to be the ultimate criterion of whether or not the painting hangs or get stored.

5. J. Diamond, "Evolution, Consequences and Future of Plant and Animal Domestication." *Nature*, 2002 August 8, 418(6898): 700-7, and H. Pringle, "Reading the Signs of Ancient Animal Domestication." *Science*, 1998, 282: 1448. D. G. Bradley has academic sources of cattle domestication at: D. G. Bradley, *et al.*, "A Microsatellite Survey of Cattle from a Centre of Origin: the Near East." *Molecular Ecology*, 1999 Dec. 8(12), 2015-22, and D.G. Bradley, *et al.*, "Mitochondrial Diversity and the Origins of African and European cattle." *PNAS*, 1996, May 14, 93(10), 5131-5; and "Evidence for Two Independent Domestications of Cattle." *PNAS*, 1994 Mar 29, 91(7), 2757-61.

6. J-M Chauvet, E. B. Deschamps, & C, Hillaire, *Dawn of Art: The Chauvet Cave, The Oldest Known Paintings in the World.* New York: Henry N. Abrams, 1996). It was discovered the same year I was visiting cave paintings in France.

7. Ian Tattersall, *Becoming Human, Evolution and Human uniqueness.* (Harcourt Brace & Co., 1998), pp. 14-28.

8. *Diodorus on Egypt.* (Trans. E. Murphy). (Jefferson, NC: McFarland and Co., 1985), p. 27. Diodorus, a Sicilian who died c. 21 BCE, wrote over 40 volumes in Greek about major regions of the Roman Empire.

9. My principal source of Xenophon is *The Education of Cyrus* translated by H. G. Dakyns. (London: J. M. Dent & Sons, no date).

10. Charles Bergman's *Orion's Legacy, A Cultural History of Man as Hunter.* (New York: Dutton, 1996), pp. 67-83.

11. Obed Borowski, *Every Living Thing, Daily Use of Animals in Ancient Israel.* (Walnut Creek, CA: AltaMira Press, 1998), p. 59.

12. Conversations I had with Arab colleagues while teaching at Zayed University, United Arab Emirates in 2002.

13. Troy, C., *et al.*, "Genetic evidence of near-eastern origins of European cattle." *Nature,* 2001, 410, 1088-90.

14. R. T. Loftus, "Evidence for two independent domestications of cattle." *PNAS*, 1994, 91: 2757-61.

15. Troy *et al., op. cit.*, 1088-90. This type of test, incidentally, has also revealed the common ancestry of humans traceable to about 200,000 years ago from a common African female. See B. M. Fagan, *The Journey from Eden, The Peopling of our World.* (London: Thames & Hudson, 1990). Cf. also K. J. McNamara, *Shapes of time, The Evolution of Growth and Development.* (Baltimore: The Johns Hopkins Press, 1997), pp. 59-60 and Rozen, S., *et al.,* "Abundant Gene, Conversion Between Arms of Palindromes in Human and Ape Y Chromosomes." *Nature* 2003, June 19, 423, 825–837.

16. MacHugh, D. E. *et al.*, "Microsatellite DNA Variation and the Evolution, Domestication and Phylogeography of Taurine and Zebu Cattle (*Bos taurus* and *Bos indicus*)." *Genetics*, 1997, 146, 1071-1086.

17. J. Reade, J. *Mesopotamia.* (London: The British Museum Press, 1991) and Moorey, P. R. S. *The Ancient Near East.* (Oxford: The Ashmoleum Museum, 1994). The British have never recovered from the colonial expressions that defined their former empire. The Far East, the Near East and the Middle East are east of what and where? The answer is Greenwich, or London in general. But the politically correct terms today should define the geography and not the capitol or directions from imperial governance, and they are Asia, Western Asia and Far Western Asia.

18. Amoroso, E. C., Jewell, P. A. "The Exploitation of Milk-Ejection Reflex by Primitive Peoples." In A. E. Mourant, F. E., Zeuner, (Eds.) *Man and Cattle.* (London: Royal Anthropological Institute, 1963), 126-137.

19. Borowski, 1998, *op cit.*, pp. 190-192.

20. Reade, 1991, *op. cit.*

21. Jacquetta Hawkes (wife of J. B. Priestley, amateur archaeologist, Ministry of Education official in Britain during WWII, and specialist for UNESCO in the 1950s allowing her ample time for travel when the world was a less bellicose place). *Dawn of the Gods, Minoan and Mycenaean Origins of Greece.* (New York: Random House, 1968), p. 122. They wrote a book together, *Journey Down a Rainbow,* (London: Heinemann-Cresset, 1955) a travelogue adventure in the American West.

22. S. Lavietes, "2 archaeologists, Robert Braidwood, 95, and his wife, Linda Braidwood, 93, die." (*NYT*, January 17, 2003), p. 25. A Google search of Catal Huyuk also reveals archaeological evidence. And see D. Leeming, *Jealous Gods and Chosen People.* (New York: Oxford University Press, 2004), 33-35.

23. J. Oates. *Babylon.* (London: Thames and Hudson, 1986), 191 & 194-96.

24. W. W. Malandra, *An Introduction to Ancient Iranian Religion.* (Minneapolis: University of Minnesota Press, 1983).

25. Zoroaster. *The Hymns of Zarathustra.* (Tr. J. Duchesne-Guillemin). (Boston: Beacon Press, 1963).
How is he to obtain cattle which brings prosperity, O Wise One,
He who desires it, together with his pastures?
Those who among the many that behold the sun,
Live uprightly according to Righteousness. (Yasna 50, p. 29)

26. Malandra, 1983, *op.cit.*, p. 38.

27. The earliest writings about Sumerians are by C. Leonard Wooley, *The Sumerians.* (W. W. Norton, 1965) who jointly excavated Ur beginning in 1928 for the University of Pennsylvania and British Museum.

28. D. O. Lodrick, D. O. (1981). *Sacred Cows, Sacred Places, Origins and Survivors of Animal Homes in India.* (Berkeley: University of California Press, 1981).
29. I draw upon information about the Canaanites largely from J. N. Tubbs' scholarly study, *Canaanites.* (London: British Museum Press, 1998).
30. D. G. Tulloch in "The Buffaloes of Asia." In Cockrill, W. R. *The Husbandry and Health of the Domestic Buffalo.* (Rome: FAO of the United Nations, 1974).
31. *Ibid.*, p. 516.
32. Philo's (64-141 CE) work, once discredited, has been elevated in stature since the finds of Ebla, the 1974 archaeological site of numerous previously unknown Canaanite writings confirming his descriptions. H. W. Attridge & R. A. Oden, *Philo of Byblos, The Phoenician History.* (Washington, DC: CBAA, 1981). The quote is on p. 55.
33. *Revelations* 17:4-6.
34. Both Reade, 1991, *op. cit.* and Diamond, *op. cit.,* 1999, speak to this point.
35. D. K. Sharpes, *Advanced Educational Foundations for Teachers, The History, Philosophy and Culture of Schools.* (New York: Routledge, 2001). Joan Oates's *Babylon* (Thames & Hudson, 1986) was informative for the writing, art and archaeology of this period.
36. Oates, *op. cit.*, pp. 127-33, & 144-61.
37. *Ibid.*, p. 120.
38. *Ibid.*, p. 37.

Chapter Two
1. 1 *Kings* 7:24-26. I have relied on biblegateway.com website for the majority of quotations in this chapter and this concordance web site translates into many languages and bible versions.
2. Tad Szulc, T. (2001). "Abraham, Journey of faith." *NG,* 2001, 200(6), pp. 90-129.
3. *Genesis* 13:2. The writings of Zoroaster who may have lived in a similar generation as Abraham, reveal simpler references.
The ordinance of sprinkling the water of the cattle,
For the welfare of the ox,
And the milk for the welfare of men desiring food,
This has the Wise Lord, the Holy One
Fashioned by his decree." (Zoroaster, op.cit., p. 59).
No companion is there for the ox
That is free from hatred. Men do not understand
How the great deal with the lowly.
Of all beings he is the strongest
To whose aid I come at his call. (Ibid., p. 57).
4. *Judges* 6:24-26.
5. *Exodus* 11: 5.
6. In *Nehemiah* 10.
7. *Exodus* 29:11-13.
8. *Leviticus* 4:6-8. The sacrifice of cows continued throughout Hebrew history as fragments of the Dead Sea Scrolls indicate. "Also, with regard to the purity of the heifer that purifies from sin (the Red Heifer): he who slaughters it and he who burns it and he who gathers its ashes and he who sprinkles (the water of purification from) sin—all these are to be pure with the setting of the sun..." See Eisenman, R. H. & Wise, M., *The Dead Sea Scrolls Uncovered.* (New York: Barnes & Noble Books, 1994), p. 194. One of the Dead Sea Scroll fragments (4Q276-277) contain specific preparation procedures for the slaughter of the Red Heifer. *Ibid.*, 210-212.
9. *Leviticus* 8:14-16.
10. *Isaiah* 66:2-4.
11. *1 Kings*: 8, and *2 Chronicles*: 7, but exaggerated figures and famine might have resulted.

12. *John* 2.

13. *Acts* 14:12-14.

14. *Hebrews* 9:12-14.

15. *Genesis* 45:10. See also Borowski, 1998, *op. cit.*

16. *Genesis* 49:26

17. Alison Roberts, *Hathor Rising, The Serpent Power of Ancient Egypt.* (Totnes, UK: Northgate Publishers. 1995), p. 146.

18. *I Kings* 12:28.

19. Among other places, in *Exodus 29, Deuteronomy 18 and 21, Leviticus 7, Isaiah 1 and Ezekiel 39.*

20. *Deuteronomy.* 21.

21. *Genesis.* 15:9.

22. I *Kings,* 1:9.

23. C. Auge, & J-M Dentzer, *Petra, The Rose-Red City.* (London: Thames & Hudson, 1999), pp. 47-49.

24. I. Finkelstein, I. & N. A. Silberman, *The Bible Unearthed, Archaeology's New Vision of Ancient Israel and the Origin of Sacred Texts.* (New York: The Free Press, 2001), 229 ff.

25. *Deuteronomy*, 21, 1-9.

26. Tubb, 1998, *op. cit.*, pp. 13-16.

27. Rifkin, 1992, *op. cit.*, 16-19.

28. G. Pettinato, *The Archives of Ebla, An Empire Inscribed in Clay.* (Garden City, NY: Doubleday and Co., 1981), 184 ff.

29. Pettinato, *op. cit.,* p. 198.

30. *Ibid.*, p. 255.

31. J. C. de Moor, *An Anthology of Religious Texts from Ugarit.* (Leiden: E. J. Brill, 1987).

32. A. R. Petersen, *The Royal God, Enthronement Festivals in Ancient Israel and Ugarit?* (Sheffield, UK: Sheffield Academic Press, 1998).

33. De Moor, 1987, op. cit., p. 114-16. Because King Minos of Crete refused to sacrifice a white bull to Poseidon, the god caused Minos' wife Pasiphae to have lust towards the bull. Daedalus built a device whereby she was to be clothed in a Bull's skin and so the bull mounted her and she bore the Minotaur, a creature with the head of a bull and body of a man who periodically devoured seven youths and seven maidens in tribute until the courage of Theseus and the cleverness of Ariadne managed to kill the Minotaur and free the kingdom from the beast held in the labyrinth. There are clear parallels between the Canaanite myth, the Cretan myth and the Europa myth of the Greeks. More about this myth in Chapter Four.

34. *Ibid..* p. 140.

35. *Ibid.*, p. 186.

36. *Ibid.*, p. 172.

37. *Ibid.* p. 173.

38. *Ibid.* p. 194.

39. *Judges* 2:10.

40. I. Cornelius, *The Iconography of the Canaanite Gods Reshef and Ba'al.* (Fribourg, Switzerland: The University Press, 1994).

41. W. Khayyata, *Guide to the Museum of Aleppo.* (Aleppo, Syria: Arab National Printing Press, 1977), 53.

42. *Deuteronomy* 11: 10-11.

43. F. E. Woods, *Water and Storm Polemics Against Baalism in the Deuteronomic History.* (New York: Peter Lang, 1994).

44. *I Kings* 18: 18-40. The episode reads as if it is the recording of an event, but its symbolism is too transparent and the murder of the priests of Baal, although pertinent to the establishment of the cult of Yahweh, certainly glorifies religious intolerance and injustice and goes against the prohibition of not killing. The prophets would continue railing against the

bull and calf cult: "Thy calf, O Samaria, is cast off, my wrath is kindled against them...for itself also is the invention of Israel: a workman made it and it is no god: for the calf of Samaria shall be turned into spiders' webs." (*Hosea* 8:5)

45. U. Oldenburg, *The Conflict Between El and Ba'al in Canaanite Religion.* (Leiden: E. J. Brill, 1969), p. 64.

46. I *Kings,* 18:1.

47. Among them Woods, 1994, *op. cit.*

48. *Jeremiah* 14:22.

49. In *Judges* 2:13; 3:7; 10:6.

50. O. Keel, & C, Uehlinger, *Gods, Goddesses, and Images of God in Ancient Israel.* (Minneapolis: Fortress Press, 1992), p. 191.

51. *Ibid.*, p. 280.

52. Finkelstein & Silberman, *op. cit.*, Borowski *op. cit.* and Tubb *op. cit.* all make this point.

53. Among the first and most notable is Sigmund Freud in "Moses and Monotheism," in *The Complete Psychological Works of Sigmund Freud.* v.23. (London: The Hogarth Press, 1954).

54. Notably J. Kirsch, *Moses, a Life.* (New York: Ballantine Books, 1998).

55. *Deuteronomy* 34:4.

56. *Exodus* 32, 1-10.

57. *Ibid.*, 32:28.

58. The death of heretics from this time forward has a biblical sanction, an idea ignored or forgotten in the haste to condemn Islamic suicide bombers and so-called martyrs as irreligious.

59. *Numbers, 23,1-2. Numbers, 23, 22.*

60. I *Kings* 12:28-30.

61. I *Kings* 12:32.

62. *Psalm* 106:19-20. Even though the Koran is by tradition the work of one man, and hence one of the most influential books in history, and despite its subsequent reverential care, it is even more chronologically and thematically disorganized than other books in the Bible. As a scriptural parallel to the Bible and in keeping with references to the affinity of a community with livestock of the day, the *Koran (or Qur'an)* also contains notations to cattle. One describes how God created cows for the benefit of humankind. The usefulness of cows receives a divine blessing in the Islamic heritage. I used the R. Bell edition. *The Qur'an.* (Edinburgh: T & T Clark, 1937, 1960)

And the cattle--
he created them for you; in them is warmth, and uses
various, and of them you eat, and there is beauty
in them in you, when you bring them home to rest
and when you drive them forth abroad to pasture,
and they bear your loads unto a land that you
never could reach. (16: 5-7, in Bell, p. 249)

And when we appointed with Moses forth nights
then you took to yourselves the Calf after him
and you were evildoers. (2:51 in Bell, p. 8)

They have hearts, but understand
not with them;
they have eyes, but perceive
not with them;
they have ears, but they hear
not with them.
They are like cattle. (7:178 in Bell, p. 155)

One would search in vain for references, for example, to Egyptian, Mesopotamian or Canaanite culture in the *Koran*. Mohammed was illiterate. The cattle references re-emphasize the importance of this animal and its blessings to an agricultural and semi-nomadic society, but, unlike Hebrew writers, without any historical or literary allusions.

63. A newsworthy tidbit noted in A. Fraser, *The Bull*. (New York: Charles Scribner's Sons, 1972), p. 42.

Chapter Three

1. *Wall Paintings of the Tomb of Nefertari, Scientific Studies for Their Conservation*. (Cairo, Egypt: The Egyptian Antiquities Organization, 1987). The Narmer Palette found 60 miles south of Luxor dates from 3,100 BCE. Narmer is shown as a raging bull knocking down city walls and goring inhabitants who resist. (Friedman, R. 2003). "City of the Hawk, Deciphering the Narmer Palette." *Archaeology, 56*(6), 52-53.

2. See the well-written and comprehensively researched book by T. D. Barnes, *Athanasius and Constantius, Theology and Politics in the Constantinian Empire*. (Cambridge: Harvard University Press, 1963).

3. Books on pre-dynastic Egypt, like the history and culture of Egypt, are legion but one I found useful is J. E. M. White, *Ancient Egypt, Its Culture and History*. (New York: Dover Publications, 1970).

4. Contained in Document 1, #22.3.16 in the Metropolitan Museum in New York City.

5. P. R. S. Moorey, *op. cit.*

6. Herodotus, *The Histories*. (V.1 & 2) (London: Henry Colburn & Richard Bentley, 1830), V. 1, pp. 185-88. Herodotus is an excellent source of Persians and Egyptians, and although he often just repeats the myths and fables he has heard, he also brings a healthy skepticism to what he hears.

7. E. A. W. Budge, *Egyptian Religion, Egyptian Ideas of the Future Life*. (London: Routledge and Kegan Paul, 1899, 1980), 33ff. E. A. Wallis Budge, a prodigious scholar author of 27 books, was Keeper of Egyptian and Assyrian antiquities in the British Museum in the late 19th century and wrote extensively on Osiris and Egyptian gods, translated the Egyptian *Book of the Dead* and the *Papyrus of Ani*, complied a hieroglyphic dictionary and a biography of Alexander the Great, among other literary and linguistic works.

8. *Treasurers of Tutankhamun* (no author). (New York: The Metropolitan Museum of Art, 1976).

9. J. E. M. White, *Ancient Egypt, Its Culture and History*. (New York: Dover Publications, 1970).

10. C. W. Towne, & E. N. Wentworth, *Cattle and Men*. (Norman: University of Oklahoma Press, 1955), pp. 43-55.

11. Herodotus, *op. cit.*, p. 206.

12. C. J. Bleeker, *Hathor and Thoth, Two Key Figures of the Ancient Egyptian religion*. (Leiden: E. J. Brill, 1973), p. 39.

13. Herodotus, 1830, *op. cit.*, p. 190.

14. Herodotus, 1830, *op cit.*, V. II, p. 25 ff., and E. A. W. Budge, *Osiris and the Egyptian Resurrection*. (New York: Dover Publications, 1973),397-404. The emperor Julian was notified in 362 that a new Apis bull had been located. The last of the Roman historians, Ammianus Marcellinus describes the Apis bull and its ceremony. (Ammianus Marcellinus, *The Later Roman Empire A.D. 354-378*. (New York: Penguin Books, 1986), 251. The ceremony and its rituals were still active as late as the 4th century CE when cities like Alexandria were primarily Christian.

15. J. H. Taylor, *Death and the Afterlife in Ancient Egypt*. (Chicago: University of Chicago Press, 2001).

16. K. Weeks, *The Lost Tomb*. (New York: William Morrow and Co., 1998), 259-63.

17. *Ibid.*, p. 260-61.
18. Diodorus, 1985, *op. cit.*, p. 110.
19. Herodotus, 1830, *op. cit.*, p. 189.
20. Taylor, 2001, *op. cit.*
21. Virgil, *The Eclogues, Georgics and Aeneid.* (London: Henry Colburn and Richard Bentley, 1830), p. 60. The quote is from the Wrangham translation and the first lines from the First Eclogue. (*The Aeneid* translation is by John Dryden). Virgil is also available at: classics.mit.edu/Virgil/georgics.html.
22. *Dan.* 6:28, *Ezra*, and *2 Chronicles.*
23. Herodotus, *op. cit.*, p. V.2, p. 25-27
24. The clever and satirical Voltaire, in "The White Bull," in Voltaire, *The Best Known Works of Voltaire.* (New York: The Book League, 1940), pp. 96-118.
25. For this section I relied on the work of Alison Roberts, *op. cit.*
26. A. Zivie, "A Pharaoh's Peace Maker," *NG*, 2002, 202(4), 27-31.
27. Bleeker, 1973, *op. cit.*, p. 83.
28. A. W. Shorter, *The Egyptian Gods.* (London: Routledge & Kegan Paul, 1937), 130-31. Shorter was the Assistant Keeper of the Egyptian and Assyrian antiquities in the British Museum in the generation after Budge.
29. J. E. M. White, *Ancient Egypt, Its Culture and History.* (New York: Dover Publications, 1970).
30. R.M. & J. J. Janssen, *Growing up in Ancient Egypt.* (London: The Rubicon Press, 1990).
31. Roberts, 1995, *op. cit.*, p. 9.
32. Bleeker, 1973, *op. cit.*, p. 40.
33. Budge, 1967, *op. cit.*, p. 115.
34. *Ibid.*, p. 216.
35. Bleeker, *op. cit.*, p. 45.
36. Roberts, *op. cit.*, p. 27.
37. Herodotus, v. 1, p. 265.
38. Recall the quote at the beginning of Chapter Two and found in *1 Kings* 7:24-26: "The Sea stood on twelve bulls, three facing north, three facing west, three facing south and three facing east. The Sea rested on top of them, and their hindquarters were toward the center." Awe at the power of bulls beyond human control could easily be substituted for a lack of understanding of the laws of gravity, thermodynamics and relativity.
39. Budge, 1973, *op. cit.*, pp. 397-400

Chapter Four

1. F. Durando, *Ancient Greece, The Dawn of the Western World.* (New York: Barnes and Noble, 1997), p. 112. The most lasting historical influence on me has been Will and Ariel Durant and the monumental and exhaustive scholarship, *The Story of Civilization* in 11 volumes, each one enough for the career of less dedicated historians. For this book the first four volumes, all published by Simon & Schuster, are significant: *Our Oriental Heritage.* (1935), *The Life of Greece* (1939), *Caesar and Christ* (1944), *The Age of Faith* (1950, 1966).
2. The classic study of Greek myths has always been the American Thomas Bulfinch's (1796-1867) *The Age of Fable or Beauties of Mythology.* (New York: New American Library, 1962), and in a newer English edition, *The Golden Age of Myth and Legend.* (Hertfordshire, UK: Wordsworth Reference, 1993). Edith Hamilton's *Mythology* (New York: Mentor Books, 1940, 1962) ranks high on the list. L. Burn, *Greek myths.* (London: British Museum Press, 1990), makes a contribution. The Greeks were literary masters in portraying the rivalry between women, the jealousy and revenge they could exact from their competition and the devilish schemes they could use in getting even with men. They were also keen to show how the passions of the gods and goddesses resembled human emotions and how divine power could be used to deceive divinities and mask their love with animal metamorphosis,

like Zeus assuming the form of a swan in order to seduce Leda, or vegetative transformations, like the purple color of the mulberries stained forever with the blood of the lovers Pyramus and Thisbe, or Zeus assuming the form of a cow to steal Europa.

3. The 832–line poem, and locations of manuscripts about Hesiod is available at http://sunsite.berkeley.edu/OMACL/Hesiod/works.html.

4. From the work of the Hungarian scholar, C. Kerenyi, *The Gods of the Greeks*. (London: Thames and Hudson, 1951, 1992).

5. Horace, *The Odes of Horace*. (Tr. H. R. Henze). (Norman: University of Oklahoma Press, 1961), III, 27.

6. Herodotus, *op. cit.*, p. 193.

7. A colossal statute of Hercules, sculpted by Romans in the 4th century after an earlier Greek model, with its exaggerated muscles, looks as if it could have been the model for Michelangelo's David or Moses or both. It resides in the National Archaeological Museum in Naples. Also cf. Stephano DeCaro, *National Archaeological Museum of Naples*. (Naples: National Archaeological Museum, 1999), p. 29. After Hercules slew the many-headed Hydra monster by firing arrows coated with fiery pitch at her, he cut her open and dipped all his arrows in her venom, an act which puts him first in line, at least in literature, of using biological and chemical weapons warfare.

8. E. Hamilton, *op. cit.*, pp. 224-43. Also L. Burn, *Greek Myths*, *op. cit.*, for stories about Hercules.

9. Xenophon, *The Education of Cyrus*. (London: J. M. Dent & Sons, no date), p.268.

10. Guzzo, P. G. & d'Ambrosio, *Pompeii*. (Napoli: L'erma di Bretschneider, 1998), and Guibelli, G. *Herculaneum* (Napoli: Carcavallo, no date).

11. Guzzo, *op. cit.*, p. 40.

12. The painting may be autobiographical since at the time Picasso had impregnated his girl-mistress, Marie-Therese as his marriage with Olga was ending.

13. John Keats, *Ode On a Grecian Urn*. In J. Keats, *Selected Poems*. (New York: Gramercy Books, 1993), p. 109.

14. I am indebted to L. Cottrell, *The Bull of Minos, The Discoveries of Schliemann and Evans*. (New York: Facts on File Publications, 1953), and to J. Hawkes, *Dawn of the Gods*. New York: Random House, 1968), and Durant, 1966, *op.cit.*

15. Cottrell, *op. cit.*, p. 70.

16. Durando, 1997, *op. cit.*, p. 23.

17. Homer, *The Odyssey*. (New York: Anchor Books, 1963), p. 43.

18. Euripedes best conveys this classic sense and poetic phrases that belie the bovine myths. An example from Euripdes' *Iphigeneia at Aulis* will suffice.

Chorus:
Oh Paris, they took you as a baby
to grow up herding white heifers on Mount Ida,
making on reeds a barbarous
music, a thin echo
of the Phrygian pipes of Olympus.
The milk-laden cattle
never stopped grazing when the goddesses
stood forth for you to judge their beauty.
(Euripedes, *The Tragedies*. (London: A. J. Valpy, 1832), p. 166).

Chapter Five
1. A. T. Walker, *Caesar's Gallic Wars*. (Chicago: Scott, Foresman, 1907), p. 347.

2. Many of the classics are available online through university and course websites, as is Virgil's *Georgics* at http://classics.mit.edu/Virgil/georgics.4.iv.html. Cf. also Durant, 1944, *op. cit.*, 235-44.

3. It isn't just poetry but music too that can resonate with cattle. In no other classical music I can think of does the instrumental sound of cattle drawing an oxcart resemble cows or oxen as does *Pictures at an Exhibition*, Modest Mussorgsky's illustrious orchestral composition, one that came quickly to him in his anguished psychological state, a man of insecurity who died of alcoholism at age 46. The work was inspired by an exhibition of a deceased friend's paintings. Victor Hartmann had died at age 39 and Mussorgsky was anxious to pay tribute to Hartmann's artistry and talent. One of Mussorgsky's scores from 10 scenes, but scored in 14 musical sections, is known as "Bydlo," or cattle. In the scene a Polish peasant drives an oxcart. The wheels rotate rhythmically, with a regular beat, conveyed by the plodding sound by the orchestral brass section.

4. *The Odes and Epodes of Horace*. (London: George Routledge & Sons, 1872), pp. 4 & 196. He wrote about his friends, his supposed lovers, his patron Maecenas, but no more touching poetry than about the charms of rustic life, the exquisite, subtle and delicate references to cattle, a lively art hidden in the multiple meanings, expressed in the same metrics, alternate trimeter and dimeter iambics in the epodes. Translations over the years trying to squeeze into Horace's meters appear stilted.

5. Durant, 1944, p. 245n.

6. *The Odes of Horace*. (Norman: University of Oklahoma Press, 1961), p. 136.

7. Horace, 1872, *op. cit.*, p. 149.

8. *The Odes and Epodes of Horace, op. cit.*, p. 196. Juvenal, the Roman satirist has graphic references to cows:

Would you feel more horror, or think it more appalling a portent,
If a woman dropped a calf, or a cow gave birth to a lamb?
But the frisky animal set aside for Tarpian Jove
Is pulling the long rope tight and fiercely tossing its head
For that's a spirited calf, just ready for temple and altar,
Ripe for a sprinkling of wine. He's ashamed to tug any more
At his mother's tats, and he butts at oaks with his budding horns.

Juvenal, *The Satires*. (Tr. Niall Rudd). (Oxford: The Clarendon press), p. 13, p. 107, and cf. Juvenal, *D. Iunii Iuvenalis, Fourteen Satires of Juvenal*. (Cambridge: Cambridge University Press, 1970, 1898).

9. *Psalm* 93:1.

10. Horace, 1872, *op. cit.*, p. 33.

11. Two primary texts: V. J. Walters, *The Cult of Mithras in the Roman Provinces of Gaul*. (Leiden: E. J. Brill, 1974) and M. J. Vermaseren, *Mithras, the Secret God*. (New York: Barnes & Noble, 1963).

12. M. P. Speidel, *Mithras-Orion, Greek Hero and Roman Army God*. (Leiden: E. J. Brill, 1980).

13. If this legendary story sounds too familiar, associated with Jesus, it is because the heroic legends of ancient literature, many only recently discovered, have often been repeated in biblical scripture in a new context.

14. A conclusion proposed by D. Ulansey, "Solving the Mithraic Mysteries," *BAR*, 1994, 20(5), 40-53.

15. R. Turcan, *Mithras Platonicus, Recherches sur L'Hellenisation Philosophique de Mithra*. (Leiden: E. J. Brill, 1975).

16. F. Cumont, *The Mysteries of Mithra*. (New York: Dover Publications, 1956).

17. See Moses and the story in *Exodus* 17, 5-6.

18. Vermaseren, *op. cit.*, p. 67.

19. *2 Kings*, 2, 11.

20. The Holy Communion of Zoroastrians is the Hom or Haoma, a liquid derived from the fat of the bull, Hadhayans, who has been slain by Soshyans, the hero and Savior of his people, who is not a god. This consecrated liquid is the elixir of eternal life and insures bodily

immortality and is performed during the raising of the dead. "While the resurrection of the dead proceeds, Soshyans and his helpers will perform the sacrifice of the raising of the dead, and in that sacrifice the bull Hadhayans will be slain, and from the fat of the bull they will prepare the white Hom (Haoma) (the drink of) immortality, and give it to all men. And all men will become immortal for ever and ever." (R. C. Zaehner, *The Teachings of the Magi, A Compendium of Zoroastrian beliefs.* (New York: Oxford University Press, 1956), pp. 148-49. All this has a clear parallel to the Eucharist in the Catholic mass. Instead of the Zoroastrian fat of the bull, Catholics are drinking the blood of the Lamb of God. A saying of Zarathustra, a 6th century BCE Persian religious 0figure, speaking to his pupils, notes: "He who will not eat of my body and drink my blood, so that he will be made one with me and I with him, the same shall not know salvation" (Vermaseren, 1963, *op.cit.*, 104). The Eucharistic ceremonies of Christianity, and similar quotes from John's Gospel of sayings of Jesus, have remarkable similarities to the Mithraic and Zoroastrian religious mysteries.

21. Eusebius, *The History of the Church from Christ to Constantine.* (New York: Dorset Press, 1965).

22. J. J. Norwich, *A Short History of Byzantium.* (London: Penguin Books, 1968).

23. Julian, *The Works of Emperor Julian.* v.2. (London: William Heinemann, 1913), p. 57.

24. E. Sauer, E. (1996). *The End of Paganism in the North-western Provinces of the Roman Empire, the Example of the Mithras Cult.* (Oxford, UK: Hadrian Books, 1996).

25. S. Bokonyi, "The development of stockbreeding and herding in medieval Europe." In Sweeney, D., *Agriculture in the Middle Ages.* (Philadelphia: University of Pennsylvania Press, 1995), pp. 41-61.

26. Gerald of Wales. *The History and Topography of Ireland.* New York: Penguin Books, 1982), 73-74.

27. *Ibid.*, p. 74. *The Táin* is the great epic of Ireland, the *Iliad* or *Beowulf* of Irish poetry, the most intriguing and oldest sagas of Western Europe written in Gaelic as early as the seventh century. The Cattle Raid of Cooley is our interest in this episode in a long war between Connacht and Ulster, caused by the desertion by Conor mac Nessa, king of Ulster, of his wife Queen Maeve of Connacht. She then married Eochaid Dala, but she fell in love with her grand-nephew Aillil. Aillil killed Eochaid and replaced him as her consort. Maeve invaded Ulster to steal the Brown Bull of Cooley so she would be equal in wealth with her husband, Aillil, who owned the White-horned Bull. Cúchulain defended Ulster alone, because the Ulster warriors were afflicted by a curse. Maeve brought the Brown back to Connacht. When the White-horned Bull saw the Brown, they fought and killed each other. There are likely elements of truth in the epic account of the struggle of two enormous bulls in Ireland (Erin) and one is the possibility of cattle raids between tribes and clans throughout the island intertwined with human and heroic struggles over power, wealth and land.

28. *Ibid.*, p. 74.

29. C. W. Towne & E. N. Wentworth, *op. cit.*

Chapter Six

1. A. Majors, *Seventy Years on the Frontier.* (Chicago: Rand McNally, 1893), 194 ff.

2. The University of Colorado adopted the buffalo as its athletic mascot in 1934. One named Mr. Chips roamed the sidelines of football games in the 1950s, but a buffalo named Ralph began the tradition of running the buffalo through the stadium. Renamed Ralphie, after an alert fan noticed that the buffalo was female, the buffalo was at every Colorado home game from 1966 to 1978.

3. C. Martin, *The Saga of the Buffalo.* (New York: Hart Publishing Co., 1973), pp. 39-40. A reliable and compact history. The movie "A Man Called Horse" (1970) starring Richard Harris vividly depicts just such a ceremony.

4. *Ibid.*

5. K. Q. Seelye, "Yellowstone Bison Thrive, but Success Breeds Peril." *The New York Times*, January 26, 2003, A16.

6. J. Blunt, "When the Buffalo Roam." *The New York Times*, February 18, 2003, A27.

7. Martin, *op. cit.*, pp. 58-62.

8. J. C. Fremont, *Narratives of Exploration and Adventure*. (New York: Longmans, Green & Co., 1956), p. 234.

9. S. E. Ambrose, *Undaunted Courage, Meriwether Lewis, Thomas Jefferson and the Opening of the American West*. (New York: Simon & Schuster, 1996), 176ff.

10. B. DeVoto, *The Journals of Lewis and Clark*. (Boston: Houghton Mifflin, 1953),76. It's the journal entry for January 5, 1805 while the party was wintering in the Mandan camp. Cf. also R. G. Thwaites, *Original Journals of the Lewis and Clark Expeditions, 1804-1806*. (New York: Antiquarian Press, 1959), p. 244-45.

11. *Ibid.*, p. 121; Ambrose, 1996, *op. cit.*, 227.

12. W. Irving, *A Tour of the Prairies*. (Norman: University of Oklahoma Press, 1956), 173.

13. *Ibid.*, p. 141-42.

14. H. Greeley, *An Overland Journey from New York to San Francisco in the summer of 1859*. (New York: Alfred Knopf, 1964), 69.

15. Fremont, 1956, *op. cit.*, 103.

16. G. F. Spaulding, G. F. (1968). *On the Western Tour with Washington Irving, The Journal and Letters of Count de Pourtales*. (Norman: University of Oklahoma Press, 1968).

17. E. E. Dale, *Cow Country*. (Norman: University of Oklahoma Press, 1943).

18. A. P. Vivian, *Wanderings in the Western Land*. (London: Sampson, Low, Marston, Searle, & Rivington, 1879), 294. The Inter-Tribal Bison Cooperative is an affiliation of 51 Native American tribes located in Rapid City, South Dakota established to restore the American bison to the range and promote its benefits to native peoples. Bison graze on native prairie grasses and ranchers do not add chemicals, growth hormones or stimulants or grains.

19. W. F. Cody, *Life and adventures of Buffalo Bill*. (New York: Willey Book Company, 1927), 115.

20. B. Still, *The West, Contemporary Records of America's Expansion Across the Continent, 1607-1890*. (New York: Capricorn Books, 1961), 197ff.

21. Located at www.ars.usda.gov/is/qtr/q297/adp297.htm.

22. www.huntingaz.net for Arizona hunts. The Center for Bison and Wildlife Health at Montana State University has a repository of documentation on the American bison and sponsors research and conferences. The North American Bison Cooperative is an association of over 400 bison ranchers processing over 12,000 bison annually for domestic consumption. States like Missouri and Michigan have bison associations as do western regions.

23. One of those amusing feature stories carried by newspapers, this one in *The New York Times*, July 27, 2000, p. B4.

24. J. U. Terrell, *Land Grab*. (New York: Dial Press, 1972), pp. 4-10. Terrell wrote over 20 books on the West but *Land Grab* was a polemical notebook about all the opprobrious episodes everyone would rather forget because so much of the economic and social progress in the West was been built on the crimes of predecessors.

25. S. Romero, "Where Two Cultures collide." *The New York Times*, Oct. 27, 2002, TR10.

26. P. I. Wellman has two books relevant to this topic: *The Trampling Herd*. (New York: Carrick & Evans, 1939), and *Glory, God and Gold*. Garden City, NY: Doubleday, 1954).

27. Dale, 1943, *op. cit.*, 13.

28. *Annual Report of the Secretary of the Interior for the Fiscal Year Ending June 30, 1890*. (Washington, DC: Government Printing Office, 1890). Indians were not too pleased either. Cf. also M. A. Link, *Treaty Between the United States of America and the Navajo Tribe of Indians*. (Las Vegas: KC Publications, 1968).

29. C. LeDuff, "Range War in Nevada Pits U.S. Against Two Shoshone sisters." *The New York Times*, Oct. 31, 2002), A16.

30. S. Sturluson, *The Prose Edda*. (New York: The American-Scandinavian Foundation, 1916). I am indebted to Prof. Lotte Schou of Denmark for calling my attention to this remarkable Norse legend. Thomas Carlyle comments on the Icelandic origins of *The Prose Edda* in a lecture entitled *The Hero as Divinity*. "One other thing we must not forget, it will explain a little, the confusion of those Norse Eddas. They are not one coherent system of thought, but properly the summation of several successive systems ...What history it had, how it changed from shape to shape, by one thinker's contribution after another, till it got the final shape...no man will ever know." (Thomas Carlyle, *Sartor Resartus and Lectures on Heroes*. (London: Chapman and Hall, 1858), 201).
31. J. Gleiter, J. & K. Thompson, *Paul Bunyan and Babe the Blue Ox*. (New York: Torstar Books, 1985).
32. In 2002 Ted Turner opened the first of his restaurant chain of grills featuring bison burgers, outlets for his prized bison raised on his ranches throughout the West. By 2003 there were 11 such restaurants in the South and West. G., Fabrikant & S. Strom, "Bison Burgers, for Humanity's Sake." *The New York Times*, October 5, 2003, Section 3, 1.

Chapter Seven
1. R. W. Fox, "The Whole Nine Yards." *The New York Times Book Review*, December 1, 2002, p. 30.
2. Wallace Stegner's books are the most literary, filled with facts and anecdotes of the land steal and the farmers' dilemmas. *Beyond the Hundredth Meridian, John Wesley Powell and The Second Opening of the West*. (New York: Penguin Books, 1954, 1992) and *The Sound of Mountain Water*. (New York: Doubleday, 1969) are two I consulted.
3. D. K. Sharpes, "Princess Pocahontas, Rebecca Rolfe (1595-1617)."*American Indian Culture and Research Journal*, 1995, 19 (4), 231-40.
4. A. de Tocquerville, *Democracy in America* (ed. by R. D. Heffner). (New York: Mentor Book/New American Library. 1956), and Still, *op. cit.*, p. 261.
5. D. Sharpes, *Education and the U.S. Government*. (New York: St. Martin's Press, 1988), 16.
6. *Annual Report of the Commissioners of the General Land Office*. (Washington, DC: Government Printing Office, 1885), 3. Tables of land sales are on pp. 4-5. The General Land Office was the agency overseeing homestead and other land giveaways.
7. John L. O'Sullivan, editor of the *Democratic Leader* in 1845, available online at: http://www.mtholyoke.edu/acad/intrel/osulliva.htm.
8. *Report of the Secretary of the Interior 1875*. (Washington, DC: Government Printing Office, 1875), iii.
9. H. S. Drago, *Great American Cattle Trails, The Story of the Old Cow Paths of the East and the Longhorn Highway of the Plains*. (New York: Dodd, Mead & Co., 1965), 34-35.
10. P. W. Gates, *California Ranchos and Farms, 1846-1862*. (Madison, WI: State Historical Society of Wisconsin, 1967), pp. 85-134 for interesting observations from personal notes of the history of early Los Angeles.
11. J. Hall, *The West, Its Soil, Surface and Productions*. (Cincinnati: Derby, Bradley & Co. Publishers, 1848), 13.
12. Still, 1961, *op. cit.*, 17-25.
13. Horace Greeley, who advised young men to go West actually went West himself. *An Overland Journey from New York to San Francisco in the Summer of 1859*. (New York: Alfred Knopf, 1964), 53. A Plains Song sums up the reality.
"Give me the plains
Where it seldom rains
And the wind forever blows;
It is there I would be
Though there's never a tree
And the farther you look

The less you see
For there's where the beefsteak grows."
14. E. Dick, *The Story of the Frontier.* (New York: Tudor Publishing Co., 1941). 95.
15. M. Lind, "The New Continental Divide." *The Atlantic Monthly,* 2003, 291(1), 86-88.
16. W. P. Webb, *The Great Plains.* (New York: Grosset & Dunlap, 1931). 107.
17. P. I. Wellman, *The Trampling Herd.* (New York: Carrick & Evans, 1935), 57, 58.
18. Webb, 1931, *op. cit.,* 211.
19. Terrell, 1972, *op. cit.,* 193.
20. *Report of the Secretary of Agriculture for the year 1890.* (Washington, DC: Government Printing Office, 1890), 302.
21. Dale, 1943, *op. cit.,* 90.
22. *Ibid.,* 92.
23. J. MacDonald, *Food from the Far West.* (New York: Orange Judd Company, 1878). British financiers wanted a direct observation by one of their own of the cattle supply situation and this book is that report.
24. *Ibid.,* 278.
25. *Ibid.,* 36.
26. J. Rifkin, *Beyond Beef, The Rise and Fall of the Cattle Culture.* (New York: Dutton, 1992). Also see M. Frink, W. T. Jackson, & A. W. Spring, *When Grass was King.* (Boulder: University of Colorado Press, 1956), a classic study of the western landscape and its history. Terrell, 1972, *op. cit.,* 222, & W. von Richthofen, *Cattle-raising on the Plains of North America.* (Norman: University of Oklahoma Press, 1964), 61. The German Baron von Richthofen was the grandfather of the World War I ace known as the Red Baron.
28. Dale, 1943, *op. cit.,* 97. Dee Brown, *The American West.* (New York: Touchstone, 1994), 295.
29. R. Aldridge, *Life on a Ranch.* (New York: Argonaut Press, 1966). First published in 1884 with humorous if droll descriptions only inexperience brings. His accounts of hunting for deer with a "fowling piece" and not a rifle and whipping rattlesnakes dead inspire more humor than admiration.
30. Dale, 1943, *op. cit.,* 101-103.
31. R. R. Dykstra, *The Cattle Towns.* (New York: Atheneum, 1979).
32. F. L. Paxton, *History of the American Frontier 1763-1893.* (New York: Houghton Mifflin, 1924).
33. www.archives.gov.
34. In Still, 1961, *op. cit.,* p. 223. A song sung on the trail entered popular culture.
>Whoopee ti yi yo, get along little dogies,
>It's your misfortune and none of my own.
>Whoopee ti yi yo, get along little dogies,
>For you know Wyoming will be your new home.
35. E. Eakin, "Reopening a Mormon Murder Mystery." (*The New York Times,* October 12, 2002), A19.
36. M. Twain, *Roughing It.* (New York. Harper & Row, 1871, 1913), 310. Twain's account of his journey West and meeting Brigham Young in Salt Lake City.
37. Greeley, 1964, *op. cit.,* 190.
38. D. H. Bain, "The Great Utah Mystery." *The New York Times Book Review,* September 7, 2003), 14–15. John D. Lee was the great, grandfather of Stewart Udall, three-term Arizona congressman and Secretary of the Interior under Presidents Kennedy and Johnson. On September 3, 1877, William "Buffalo Bill" Cody acted in a new border drama in the Bowery Theater in New York called "May Cody, or Lost and Won," based on the incidents of the Mountain Meadows Massacre and life among the Mormons. (W. F. Cody, *Life and Adventures of Buffalo Bill.* New York: Willey Book Company, 1927, 304).
39. Still, 1961, *op. cit.,* 143ff.

40. Dykstra, 1979, *op. cit.*

41. The tough life on the frontier is described graphically in A. Majors, *Seventy Years on the Frontier.* (Chicago: Rand McNally, 1893), and in Teddy Roosevelt, *Theodore Roosevelt, An Autobiography.* (New York: Charles Scribner's Sons, 1925). The storm of 1886 lasted 10 days and at times reached 46 degrees below zero. See Dee Brown, *The American West.* (New York: Touchstone, 1994), 330 ff.

41. A. Majors, 1893, *op. cit.,* 278.

42. Still, 1961, *op. cit.,* 255.

43. www.nps.ars.usda.gov/programs/nrsas.htm.

44. H. Greeley, *op. cit.,* 109.

45. *Ibid.* Greeley, Colorado, a cow town if there ever was one (even though Northern Colorado University is located there), where Horace attempted to start a communal living project, is named after him.

46. Terrell, *op. cit,.* 199.

47. L. Price, L. (2002). "Cowboy Boosterism, Old Cowtown Museum and the Image of Wichita, Kansas." *Kansas History*, 2002, 24(4), 301-17.

48. *Report of the Commissioner of Agriculture for the year 1870.* (Washington, DC: Government Printing Office, 1871), 39ff., table 47-48, markets, 49, dairy data 310-28. The compiled data from early settlement is useful for comparative purposes in later years.

49. T. Lea, *The King Ranch.* (Boston: Little, Brown and Company, 1957). Kingsville, Texas is named after King.

50. Stegner, 1992, *op. cit.,* 322.

51. J. W. Powell, *Report on the Lands of the Arid Region of the United States.* (Cambridge, MA: Belknap Press. 1878, 1962).

52. *Ibid.,* opening sentence, 1.

53. *Ibid.,* 88.

54. Stegner, 1992, *op. cit.,* 315.

55. J. C. Fremont, *Narratives of Exploration and Adventure.* (New York: Longmans, Green & Co., 1956), 232-33. Included is his description of the buffalo: "Unwieldly as he looks, the buffalo bull moves with a suddenness and alertness that makes him at close quarters a dangerous antagonist." The city of Fremont, California, just north of San Jose on the eastern side of San Francisco Bay is named after him. See also Larry McMurtry, "Mountain Man." *The New York Review of Books,* (October 9, 2003), L, 15, 20-22.

56. C. W. Upham, *Life, Explorations and Public Services of John Charles Fremont.* (Boston: Ticknor and Fields, 1856).

57. Cody, 1927, *op. cit.,* 306. He does not mention how many cattle he purchased. Beginning in 1883 he began his Wild West Show with an entourage of buffalo, bears, Indians, 50 cowboys, who engaged in lasso throwing, daring rides and shooting exhibitions by women. The Wild West Show toured Europe triumphantly three times visiting most countries, the 1902 tour lasting four years.

58. Roosevelt, 1925, *op. cit.* See also Dee Brown, *The American West, op. cit.,* 322.

59. *Ibid.,* 119.

60. *Ibid.,* 396.

61. P. F. Starrs, *Let the Cowboy Ride, Cattle Ranching in the America West.* (Baltimore: Johns Hopkins Press, 1998).

62. A. P. Vivian, *Wanderings in the Western Land.* (London: Sampson Low Marston, Searle, & Rivington, 1879), 115.

Chapter Eight

1. P. I. Wellman, *op. cit.,* pp. 16-17. Dee Brown, *op. cit.,* 42.

2. J. E. Rouse, *The Criollo, Spanish Cattle of the Americas.* (Norman: University of Oklahoma Press, 1977), 43ff.

3. A. H. Sanders, *The Cattle of the World, Their Place in the Human Scheme—Wild Types and Modern Breeds in Many Lands.* (Washington, DC: The National Geographic Society, 1926), 119.

4. www.cancernutritioncenter.com.

5. I consulted four texts on Kino: his first biographer, H. E. Bolton, *The Padre on Horseback, A Sketch of Eusebio Francisco Kino, S. J._* (Chicago: Loyola University Press, 1932, 1986), F. J. Smith, J. L. Kessell, & F. J. Fox, *Father Kino in Arizona.* (Phoenix: Arizona Historical Foundation, 1966), F. C. Lockwood, *With Padre Kino on the Trail.* (Tucson: University of Arizona Press, 1934), and E. J. Burrus (1961). *Correspondencia del P. Kino con los generales de la compania de Jesus, 1682-1707.* (Mexico City: Testimonia Historica, 1961), No.5. Kino's birth name was Chini in Italian.

6. Smith, *et al., op. cit.,* 104-05. His first mission in Sonora was Nuestra Señora de los Dolores. The mission at San Xavier del Bac, for example, just south of present day Tucson, in 1698 had 160 houses, more than 800 people and 30 head of cattle.

7. *Ibid.,* 88.

8. Lockwood, *op. cit.,* 119.

9. D. K. Sharpes, *Advanced Educational Foundations for Teachers, The History, Philosophy and Culture of Schooling* (New York: Routledge, 2002), 265.

10. Towne & Wentworth, 1955, *op. cit.,* 140. John Winthrop Governor of the colony, described in his journal a young man caught "in buggery with a cow upon the Lord's Day," and had the inarticulate cow killed in front of the repentant young man who was then also executed without compunction. (C. Crain, "The Puritan Dilemma." *The New York Times Book Review,* September 21, 2003), p. 11.

11. P. W. Bidwell, & J. I. Falconer, *History of Agriculture in the Northern United States.* (New York: Peter Smith, 1941).

12. *Ibid.,* p. 22.

13. Towne & Wentworth, 1955, *op. cit.,* p. 133.

14. H. S. Drago, *Great American Cattle Trails, The Story of the Old Cow Paths of the East and the Longhorn Highway of the Plains.* (New York: Dodd, Mead & Co., 1965), 6-7.

15. Bidwell & Falconer, 1941, *op. cit.,* 388. Over the threshold of the butcher's guild in Nuremberg, Germany, a reclining longhorn stone ox, placed in 1599 at the entrance to the fleischbank, the meat hall, was visible until World War II. Cattle were the preferred meat of choice, supplemented with local deer, ducks, geese and turkeys by colonists only when they were faced with food shortages.

16. Towne & Wentworth, 1955, *op. cit.,* 142.

17. Cody, 1927, *op. cit.,* 47.

Chapter Nine

1.E. Levine, "The Burger Takes Center Stage." *NYT,* January 15, 2003), D1.

2.Several investigators and reporters, like Upton Sinclair at the beginning of the 20th century, have exposed the flaws in the beef slaughtering business. Three of the most popular and readable are: Jeremy Rifkin, *Beyond Beef, The Rise and Fall of the Cattle Culture.* (New York: Dutton, 1992), Gail Eisnitz, *Slaughterhouse, The Shocking Story of Greed, Neglect, and Inhumane Treatment Inside the U.S. Meat Industry.* (Amherst, NY: Prometheus Books, 1997), and Eric Schlosser, *Fast Food Nation, The Dark Side of the All-American Meal.* (Boston: Houghton Mifflin, 2001). When faced with nasty terms like slaughter house that might connote unpleasant images of animal death, guts being ripped out, and places where rodents, insects and stray birds might contaminate the meat, it's better to give the offending place a French name, like croutons for stale bread. Abattoir is such a name for a slaughterhouse. M. Petersen & C. Drew reported in "As Inspectors, Some Meatpackers Fall Short" (*NYT,* October 10, 2003, p. A1) that the USDA inspectors are still finding feces on meat carcasses moving down the processing line at Shapiro Packing which processes 1,200 cattle a day. Although prodded into more vigorous action by the alarm from restaurants, fast food and supermarket chains, government inspection is persistent but enforcement powers still weak.

3. U. Sinclair, *An Upton Sinclair Anthology*. (Los Angeles: Published by the Author, 1934).

4. Rifkin, Eisnitz, and Scholosser, *op. cit.*

5. C. Drew, "Plant's Sanitation may have Link to Deadly Bacteria." (*NYT*, December 11, 2002), A24.

6. www.fsis.gov.

7. *Annual Reports of the Department of Agriculture for the fiscal year ended June 30, 1900.* (Washington, DC: Government Printing Office, 1900), 3.

8. http://www.mdac.state.ms.us/Library/AgencyInfo/Laws/Meat/LawsMeatInspection.html

9. The FAO published 19 volumes in 1992, updating its food and safety standards. Volume Ten contains the codes of practice and guidelines for processing meat. (*Codex Alimentarius, Volume Ten, Meat and Meat Products Including Soups and Broths*. FAO, World Health Organization, Rome, 1993).

10. The main sources for this and other disturbing statistics are from CDC, the U.S. Department of Agriculture and the NAS publications. The CDC website is at: http://www.cdc.gov/mmwr/preview/mmwrhtml/mm5129a1.htm

11. The website for the NAS: http://books.nap.edu/books/0309086272/html/1.html#pagetop

12. Eisnitz, *op. cit.* p. 61-62.

13. D. Barboza, "Sara Lee Corp. Pleads Guilty in Meat Case." *NYT*, June 23, 2001b), A7. The Food and Safety Inspection Service provided a later notice on its website in the summer, 2002. "Traceback microbiological samples were collected by FSIS on June 24, 2002. They were reported positive for *E. coli* O157:H7 on June 29, 2002. On July 18, 2002, the recall was expanded following a scientific and technical review of plant practices and company records by FSIS. As a result of the FSIS investigation it was revealed that the Recall 055-2002 needs to be expanded to include the products produced between selected dates from April 12, 2002 through July 11, 2002, because these products may be contaminated with *Escherichia coli* O157:H7."

14. G. Winter, "Beef Processor's Parent no Stranger to Troubles." (NYT, July 20, 2002), A9.

15. E. Becker, "Government in Showdown in Bid to Shut Beef Processor." (*NYT*, January 23, 2003), A16. Also cf. M. Petersen & C. Drew, "As Inspectors, Some Meatpackers Fall Short" (*NYT*, October 10, 2003), A1.

16. *Ibid.*, A16 & A1.

17. Scholosser, *op. cit.*, 205.

18. S. Day, "Prosecutors in Smuggling Case Against Tyson Contend Trial is about 'Corporate Greed." *NYT*, February, 6, 2003), A22.

19. P. Belluck, "Big Business of Cattle Theft is a Growing Threat to Ranchers." (*The New York Times*, March 27, 2001), p.A1. See also Kilborn, P. T. (2003, November 16). "Cattle Rushed to Market as the Price of Beef Soars." *NYT*, YT18.

20. R. E. Milloy, "The Old West Yields to Urbanization." (*NYT*, March 28, 2001a), A10.

21. *The Politics of Meat*. (Frontline, Public Broadcasting System, April 24, 2002) at: *www.pbs.org/wgbh/pages/frontline/shows/meat/politics*.

22. "Curse of Factory Farms." (*NYT*, August 30, 2002), A18. Iowa Quality Beef plant opened in Tama, Iowa July 21st, 2003 and immediately employed 420 people. The plant had been closed twice previously creating wholesale unemployment. IBP closed the plant in 1999 donating the plant and land to the city of Tama whose own beef-packing plant went bankrupt in 2001. The current plant now processes 1,700 head of cattle per day. (*Iowa Farm Bureau Spokesman*, Vol. 80(8), October 25, 2003, 1-3).

23. "Schools to be Allowed to Serve Irradiated Meat." *NYT*, October 27, 2002), A21. Schools may have been allowed by the Dept. of Agriculture but most states were reluctant to purchase such irradiated beef, most schools not even responding to surveys about whether or not they

would use it. M. Burros, "Schools Seem in No Hurry to Buy Irradiated Beef. (*NYT*, October 8, 2003), D1.

24. Michael Pollan, "When a Crop Becomes King." (*NYT*, July 19, 2002), A27, "The (Agri)Cultural Contradictions of Obesity." (*The New York Times Magazine*, October 12, 2003), 41-48, and his *A Botany of Desire*. (New York: Random House, 2001). Some sustainable cattle ranchers, like George Kahrle of Three Forks Montana near the headwaters of the Missouri river, have reverted to grass-fed beef exclusively, relying on lower labor costs, less hay, fewer antibiotics, but with higher profits. J. Robbins, "Balancing Cattle, Land and Ledgers." (*NYT*, October 8, 2003), D8.

25. *Report of the Commissioner of Agriculture for the year 1870*. (Washington, DC: Government Printing Office, 1871). For Texas cattle trade see pp. 346-52 and p. 506, and for dairy data see pp. 310-28. For every pound of butter there are 15-20 pounds of skim milk and about 3 pounds of buttermilk. For every 10 pounds of cheese there are about 9 pounds of whey, though actual amounts may vary depending on the process. Cf. H. E. Alvord, "Utilizations of by-products of the dairy." *Yearbook of the United States Department of Agriculture, 1897*. (Washington, DC: Government Printing Office, 1898), 509-28.

26. Report, 1870, *op. cit.*, 310-28.

27. *Annual Reports of the Department of Agriculture for the fiscal year ended June 30, 1900*. (Washington, DC: Government Printing Office, 1901), table on p. xv, Texas fever pp. xxi-xxii, and dairy info on xxii-xxiii.

28. D. Barboza, "America's Cheese State Fights to Stay that Way." (*NYT*, June 28, 2001), C1.

29. *Ibid.*

30. R. Leeson, "Why we need N.E. dairy compact." (*The Providence Journal*, August 31, 2001), B4.

31. R. Apple, "A new Normandy, North of the Golden Gate." (*NYT*, November 28, 2001), E1.

32. Jews have dietary restrictions against the mixture of meat and milk. (*Exodus* 23:19, 34:26 and *Deuteronomy* 14:21). Flesh and cheese cannot be served together on the same table. A person can tie up meat and cheese together in the same package provided they do not touch each other. If a drop of milk falls into meat cooking and that milk can be said to flavor the cooking, then none of it can be eaten. A. Hertzberg, *Judaism*. (New York: George Braziller, 1962), 102.

33. B. Stamler, "Got Sticking Power? A Tagline and a Mustache Stop a Slide in Milk Consumption." (*NYT*, July 30, 2001), C11.

34. N. Barnard, N. (2002, summer). "Doctor in the House." (*Peta's Animal Times*, summer, 2002), 17.

35. J. E. Brody, "Drink your Milk: A Refrain for all Ages, Now More Than Ever." (*NYT*, January 7, 2003), D5.

36. S. Killman & A. Merrick, "Could the High Price of Milk be a Byproduct of Supermarket Mergers?" (*WSJ*, August 15, 2000), B1.

37. *Ibid.*

38. http://www.nps.ars.usda.gov

39. B. Harden, "Livestock Testing Chases Cheats from State Fairs." (*NYT*, August 12, 2002), C2.

40. K. S. Huang, "Price and Income Affect Nutrients Consumed from Meats." *Food Review*, 1996, 19(1), 37–40.

41. According to a federal judge in Manhattan, bringing a suit against McDonald's accusing it of deceiving consumers about the high levels of fat, sugar and cholesterol in its products is insufficient reason for individuals to blame a restaurant for obesity. The court ruled that it is not the duty of the law to protect people from their own excesses. B. Weiser, "Big Macs Can Make you Fat? No Kidding, Federal Judge Rules." (*NYT*, January 23, 2003), A23. Michael

Pollan's examination of the contradictions in obesity in the U.S. and federal farm policy that encourages over-production of corn and similar products to maintain profits for farmers is not exaggerated. It is not just cheap beef but cheap processed food that has led to obesity problems as larger portions, the direct consequence of over-production, rather than reduced prices are the staple at fast food restaurants. (Pollan, 2003, op. cit., 41-48).

42. A. Revken, "Forget Nature. Even Eden is Engineered." (*NYT*, August 20, 2002), D1-11.

43. J. Christensen, "Environmentalists Hail the Ranchers: Howdy Pardners? (NYT, September 10, 2002), D3.

44. On the web at: http:/www.ars.usda.gov/is/AR.

45. www.ars.usda.gov/is/AR/archive/nov98/tan1198.htm

46. *Ibid.*

47. This research is part of Rangeland, Pasture, and Forages, an Agricultural Research Service Program on the World Wide Web at http://www.nps.ars.usda.gov.

Chapter Ten

1. D. Grady, D. (2003, December 31). "U.S. Imposes Stricter Safety Rules for Preventing Mad Cow Disease." *NYT*, A1.

2. "Preliminary FoodNet Data on the Incidence of Food borne Illnesses—Selected Sites, United States, 2002," at http://www.cdc.gov/mmwr/preview/mmwrhtml/mm5215a4.htm

3. J. Diamond, *Guns, Germs and Steel, The Fate of Human Societies.* (New York: W. W. Norton, 1999), 196-97.

4. L. Eiseley, *The Invisible Pyramid.* (New York: Charles Scribner's Sons, 1970), 75–77. Loren Eiseley (1907-1977), Professor at the University of Pennsylvania from 1949 until his death, was one of those rare literary and scientific figures who perceptively understood the relationship of humanity with nature, a naturalist who wrote as mystically and poetically as any since Lucretius' *On the Nature of Things*, though Konrad Lorenz would certainly make the list. K. L. Lorenz, *King Solomon's Ring, New Light on Animal Ways.* (London: Metheun & Co., 1952).

5. http://bmj.bmjjournals.com/collections/bse/prions.htm. The Mad Cow controversy was international news for months as new countries found its noxious presence in beef or among consumers. I relied on periodic reports in this developing controversy from 1999 to 2003 on websites and journal articles and on *The New York Times* and chiefly from reporters Elizabeth Becker, Sandra Blakeslee and Warren Hoge who covered these stories extensively. One website location of the Mad Cow disease is: http://www.cyber-dyne.com/~tom/mad_cow_disease.html

6. http://news.bbc.co.uk/1/hi/uk/370863.stm

7. W. Hoge, "5 Britons' 'Mad Cow' Deaths Traced to Butchering Method." (*NYT*, March 22, 2001), A10.

8.   On the website of the NIH at: http://ehpnet1.niehs.nih.gov/qa/106-3focus/

9. J. S. Griffith, "Self-replication and Scrapie." (*Nature,* 215, 1967), 1043–1044.

10. http://www.mad-cow.org/griff.html. According to the *PNAS* (79 (17), 1982, 5220-5224): "The properties of the scrapie agent distinguish it from both viroids and viruses and have prompted the introduction of the term "prion" to denote a small proteinaceous infectious particle that resists inactivation by procedures that modify nucleic acids." Also see the documentation of Dr. Stanley Prusiner who won the Nobel Prize in Medicine for his work on prions at http://www.nobel.se/medicine/laureates/1997/prusiner-autobio.html. By 2004 researchers in Italy had discovered a variant of the prions in made cow disease making testing of all cattle more imperative. The most current study report by Couzin in *Science* in 2004 noted that artificial prions, those mis-shaped proteins, or strings of amino acids, have been created and that they produce the same kind of deadly infection in mice they do in cows. See also S. Blakeslee, "Expert Warned that Mad Cow was Imminent." *NYT*, 2003, December 25, A1.

11. "The Science of BSE." (*The Economist*, March 14, 1998), 21–23.

12. In an episode I watched on British TV while living in Oxford.

13. R. Hernandez, "Citing Mad Cow, F.D.A. Panel Backs Blood
Donor Curbs." (*NYT*, June 29, 2001), A23.

14. http://www.fda.gov/oc/bse/contingency.html and the website for The Food and Drug
Administration www.fda.gov for brochures and reports.

15. C. Southley, "Fresh Charges of BSE Cover-up." *Financial Times* (United Kingdom),
September 3, 1996, and for complete BBC coverage which had 211 stories on the progress of
the problem, see: http://news.bbc.co.uk/1/hi/world/monitoring/media_reports/1190938.stm

16. The British government's full text was at: http://www.bseinquiry.gov.uk/

17. S. Daley, "France Told to End Ban on Imports of British Beef." (*NYT*, December 14,
2001), A16. The FAO website at: www.fao.org then search "mad cow" for various press
releases over the past few years.

19. web site: www.mad-cow.org a clearinghouse of information.

20. "U.S. Cautious on Mad Cow, Seizes Flock of Sheep." *NYT*, March 22, 2001), A12.

21. S. Blakeslee, "Estimates of Future Death Toll from Mad Cow Disease Vary Widely."
(*NYT*, October 30, 2001), D6.

22. E. Becker, "U.S. Mad Cow risk is Low, a Study by Harvard Finds." (*NYT*, December 1,
2001), A10, and the Harvard Study at "Preliminary Foodnet Data on the Incidence of
Foodborne Illnesses—Selected Sites, United States. *Morbidity and Mortality Weekly Report*,
April 19, 2002, 51(15), 325-29. Or go to:
http://www.cdc.gov/mmwr/preview/mmwrhtml/mm5115a3.htm21.21.223.23. S. Blakeslee,
"Experts Consider How to Stop a Variant of Mad Cow Disease." (*NYT*, August 7, 2002),
science.

24. C. Krauss, & S. Blakeslee, "Case of Mad Cow in Canada Prompts U.S. to Ban its Beef."
(*NYT*, May 21, 2003), A1.

25. F. L. Garcia, R. Zahn, R. Riek, K. Wuthrich, "NMR Structure of the Bovine Prion
Protein." PNAS, July 18, 2000. v.97, no. 15, 8334–8339. "Although there are characteristic
local differences relative to the conformations of the...hamster prion proteins, the bPrP
structure is essentially identical to that of the human prion protein. On the other hand, there
are differences between bovine and human PrP in the surface distribution of electrostatic
charges, which then appears to be the principal structural feature of the "healthy" PrP form
that might affect the stringency of the species barrier for transmission of prion diseases
between humans and cattle."

26. CDC website has over 3,800 citations for E.coli at www.cdc.gov.

27. The Food Safety and Inspection service's website at: http://www.fsis.usda.gov/

28. E. Becker, "Agricultural Chief Disavows Plan to Eliminate Test
on School Beef." (*NYT*, April 6, 2001), A1.

29. D. Grady, "On an Altered Planet, New Diseases Emerge as Old Ones Re-Emerge." (*NYT*,
August 20, 2002), D2.

30. "Containing Foot-and-Mouth Disease." (*NYT*, March 23, 2001), A20.

31. S. Lyall, "Wrapping British Farmers in Growing Isolation." (*NYT*, March 10, 2001), A3.

32. A series by A. Cowell, in *The New York Times* highlighted England's problem: "Trying
to Stem Foot-and-Mouth, Britain Buries Carcasses." (March 27, 2001, A23),"Britain
Reluctantly Considers Animal Vaccination." (March 28, 2001, A1), "Foot-and-Mouth
Damages English Tourism Too." (March 16, 2001, A4), and "Cattle Disaster Still Felt in
Britain." (September 10, 2002, W1).

33. W. Hoge, "Livestock Epidemic Widens its Menace for British Farms." (*NYT*, March 24,
2001), A1.

34. Cowell, *op. cit.*, March 16, 2001, A4.

35. B. Lavery, "Despite Safeguards, Foot-and-Mouth Disease Spreads to Ireland." (*NYT*,
March 23, 2001), A6.

36. L. Kaufman, "Feeling the Pinch of Luxury Leather." (*NYT*, March 29, 2001), C1.
37. P. Reaney, "Britain Faces Long Road to Curb Foot-and-Mouth. *The Globe and Mail* (Toronto), September 8, 2001), A16.
38. C. Clover, "Quantitative 'Failure' Led by Farm Epidemic." (*The Daily Telegraph* (London), May 8, 2002), 13.
39. In an interview broadcast on National Public Radio on March 16, 2001.
40. F. Houston, "A Life's Work up in Smoke." (*Arizona Republic*, March 25, 2001), B11.
41. C. Marquis, & D. G. McNeil, "Meat from Europe is Banned by U.S. as Illness Spreads." (*NYT*, March 14, 2001), A1.
42. E. L. Andrews, "Dutch Farmers Facing Mass Foot-and-Mouth Slaughter." (*NYT*, April 6, 2001), A4.
43. Lavery, *op. cit.* A6.
44. M. Landler, "Less Alarm in Hong Kong over Disease of Livestock." (*NYT*, April 8, 2001), A9.
45. J. Brooke, "In Japan, Beef Business Sinks in a Sea of Skepticism." (*NYT*, November 4, 2001), A10.
46. E. Becker, "What if the Buffalo Roam into Foot-and-Mouth?" NYT, May 6, 2001, wk16.
47. N. Wade, "How a Patient Assassin Does its Deadly Work." *NYT*, October 23, 2001, D1.
48. R. E. Milloy, "Anthrax Hides Along Cattle Trails of the Old West." *NYT*, October 29, 2001, A8.
49. M. Wines, "Russians are Unfazed by Anthrax, A Common Rural Problem." NYT, October 10, 2001, A4.
50. www.nature.com/biotech. The article is "Cloned transchromosomic calves producing immunoglobulin." (August 12, 2002).
51. A. Pollack, "Joint Venture Clones Cattle with Human Antibodies." *NYT*, August 12, 2002, C2.
52. J. Robbins, "Unsolved Mystery Resurfaces in Montana: Who's Killing Cows?" NYT, September 17, 2001, B1.

Chapter Eleven
1. P. Robertshaw, *Early Pastoralists of South-Western Kenya*. (Nairobi: British Institute in Eastern Africa, 1990).
2. J. Huxley, *Africa View*. (New York: Harper & Bros., 1931).
3. V. Porter, Cattle, *A Handbook of Breeds of the World*. (New York: Facts on File, 1991).
4. A. B. Smith, *Pastoralism in Africa, Origins and Development Ecology*. (London: Hurst and Co., 1992).
5. Huxley, *op. cit.,* 53.
6. C. Dundas, *Kilimanjaro and Its People*. (London: Frank Cass & Co., 1968).
7. G. J. Klima, *The Barabaig, East African Cattle-Herders*. (New York: Holt, Rinehart and Winston, 1970).
8. R. L. Swarns, "Cattle Prices are Up, So is Buying a Bride's Hand." (*NYT*, October 3, 2001), A4.
9. U. Almagor, *Pastoral Partners, Affinity and Bond Partnership Among the Dassanetch of South-West Ethiopia*. (Manchester, UK: Manchester University Press, 1978).
10. *Ibid.*, 25-26.
11. *Ibid.*, 142.
12. J. Adamson, *Born Free, A Lioness in Two Worlds*. (New York: Pantheon Books, 1961).
13. K. Arhcm, *Pastoral Man in the Garden of Eden, the Maasi of the Ngorongoro Conservation Area, Tanzania*. (Uppsala: Uppsala Research Reports in Cultural Anthropology, 1985). I have relied on this Swedish study for much of what I learned about the Maasi people and their cattle culture.
14. P. Rigby, *Cattle, Capitalism, and Class*. (Philadelphia: Temple University Press, 1992).

15. K. Holland, *On the Horns of a Dilemma, The Future of the Maasi.* (Montreal: McGill University, Department of Anthropology, 1987). The title of this book is a popular phrase but redundant. The Greek word for horn is "lemma" and thus dilemma means "two horns."

16. D. P. Crandall, *The Place of Stunted Ironwood Tress, A Year in the Lives of the Cattle-Herding Himba of Namibia.* (New York: Continuum, 2000). An interesting account of one man's ethnographic study of this tribal group.

Chapter Twelve

1. D. O. Lodrick's book, *Sacred Cows, Sacred Places, Origins and Survivors of Animal Homes in India.* (Berkeley: University of California Press, 1981), a study I found informative and which I relied on extensively. I have written about my own experiences in India (D. K. Sharpes, *An Asian Enquiry, Religion and Culture from Israel to Borneo* (Bognor Regis, UK: Anchor Publications, 1986) a defunct publishing house, and commentary in "Trees, The Providers." (*The India Magazine*, February, 1985). Additional books I drew upon were Indira Gandhi's *India* (London: Hodder & Stoughton1975), historian Francis Watson's, *A Concise History of India* (London: Thames & Hudson, 1974), and keen observers like Ved Mehta's *Portrait of India* (London: Weidenfield and Nicolson, 1967), John Keay's *Into India* (London: John Murray, 1982), and V. S. Naipaul's, *India: A Wounded Civilization* (New York: Penguin Books, 1977). The stories and writings that appealed most to me were those of the gifted artist Rabindranath Tagore, the first Asian to receive the Nobel Prize for Literature, particularly *Sadhana, The Realization of Life.* (London: Macmillan, 1921). Lastly, Muhatma Gandhi's own writings were inspirational: M. K. Gandhi, *How to Serve the Cow.* (Ahmedabad: Navajivan Publishing Co., 1954).

2. Arising at slightly different eras in different parts of the world, the Neolithic age postdates stone tool making and predates urban civilization with its literary refinements, but does highlight farming and animal husbandry. Cf. F. R. Allchin, *Neolithic Cattle-Keepers of South India.* (Cambridge, UK: Cambridge University Press, 1962). A useful study of archaeological remains of entrants into India during the Neolithic period.

3. W. J. Hatch, *The Land Pirates of India.* (Philadelphia: J. B. Lippincott, 1928).

4. Ralph T. H. Griffith's 1896 translation of the *Rig Veda* at http://www.sacred-texts.com/hin/rigveda/rv01004.htm. *The Rig Veda*, a group of hymns, are the oldest Hindu writings but the *Bhagavad Gita* and the *Upanishads* also yield insights into Hindu mystical thought. Cf Juan Mascaro's translations of *The Bhagavad Gita.* (New York: Penguin Books, 1962), and *The Upanishads.* (New York: Penguin Books, 1965).

5. L. Renou,(ed.)., *Hinduism.* (New York: George Braziller, 1962) and the late British scholar R. C. Zaehner's classic study *Hinduism.* (New York: Oxford University Press, 1962).

6. S. Prabhavananda, & F. Manchester, *The Upanishads.* (Hollywood, CA: Vedanta Press, 1975), p. 11.

7. Collected from notes I took while at the Victoria and Albert Museum in London in 2000. More is located at: http://www.jainism.org/

8. Lodrick, *op. cit.,* p. 2.

9. M. Edwards, "Marco Polo, Journey Home (Part III)." (*NG*, 2001, 200(1), 28–47.

10. Lodrick, *op. cit.,* 1, 68, 75.

11. *Ibid.,* 13-28, 43-70, 71-104, 105-169.

12. *Ibid.,* 199.

Chapter Thirteen

1. Columbus's journal is historically significant if filled with too much fantasy, medieval cosmology, royal obsequiousness and wishful thinking about the western cultural icon of gold. The journal's translation I used is J. Cummins, *The Voyage of Christopher Columbus, Columbus' Own Journal of Discovery* (New York, St. Martin's Press, 1992), 93ff. The best

modern biography of Columbus I think is by F. Fernandez-Armesto, *Columbus*. (New York: Oxford University press, 1991).

2. J. E. Rouse, *The Criollo, Spanish Cattle of the Americas*. (Norman: University of Oklahoma Press, 1977).

3. P. H. Smith, *Politics and Beef in Argentina*. (New York: Columbia University Press, 1969).

4. L. S. Jarvis, *Livestock Development in Latin America*. (Washington, DC: The World Bank, 1986). 3-4, 10-13.

5. www.tradewatch.org.

6. The reporting of *New York Times* reporter L. Rohter during this breaking financial crisis has been instrumental. His three stories: "Once Secure, Argentines Now Lack Food and Hope." (2001, March 2), yt6. "Disease is Big Setback for Brazil Cattle Region." (June 20, 2001), W1. "Brazil's Wild and Woolly Side: Meet the Caubois." (August 23, 2001), A4.

7. Smith, 1969, *op. cit.*

8. The FAO's Volume Ten contains the international codes of practice and guidelines for processing meat. (*Codex Alimentarius, Volume Ten, Meat and Meat Products Including Soups and Broths*. FAO. (World Health Organization, Rome, 1993).

9. *Animal and Plant Health Inspection Service: Importation of Beef from Argentina*. (Washington, DC: Government Accounting office: U.S. Dept. of Agriculture, 1996). (Microfiche: GA 1.13 OGC 97-52).

10. C. Torres, "U.S. Takes Cautious Stance on Argentine Beef Imports. (*WSJ*, August 14, 2000), A12.

11. "Argentina Kept Mum on Disease." (*Arizona Republic*, March 17, 2001), A24. But for the background on Argentina and its beef industry cf. two studies by A. Carreras, *La Aftosa en la Argentina, un Desafio Competitivo*. (Buenos Aires: Amara Argentina de Consignatarios de Ganado, 1993) and *El Comercio de Ganados y Carnes en La Argentina*. (Buenos Aires: Editorial Hemisferio sur S.A, 1986).

12. J. Rich, "Brazil Postpones its Beef Dreams." (*NYT*, March 14, 2001),W1.

13. M. D. Faminov, *Cattle, Deforestation and Development in the Amazon*. (Wallingford, UK: CAB International, 1998). A distressing study of the ecological relationship between animal production, forest preservation and the impact on the human environment.

14. *Ibid.*, p. 48.

15. J. Rifkin, *Beyond Beef, The Rise and Fall of the Cattle Culture*. (New York: Dutton, 1992) among others.

16. T. Weiner, "Growing Poverty is Shrinking Mexico's Rain Forest. (*NYT*, December 8, 2002), yne12.

17. Rohter, August 23, 2001, *op. cit.*, A4.

Chapter Fourteen

1. C. Smith, "Milk Flows from Desert at a Unique Saudi Farm." (*NYT*, December 31, 2002), yne A4.

2. W. R. Cockrill, "The Working Buffalo." In W. R. Cockrill, *The Husbandry and Health of the Domestic Buffalo*. (Rome: FAO, 1974), 313-28.

3. J. E. Frisch & J. E. Vercoe, "Improvement of the Productivity of the Swamp Buffalo in S. E. Asia." In (NA) *The Use of Nuclear Techniques to Improve Domestic Buffalo Production in Asia*. Proceedings of the Final Research Co-ordination Meeting Organized by the Joint FAO/IAEA Division of Isotope and Radiation Applications of Atomic Energy for Food and Agricultural Development. (Vienna: IAEA, 1984).

4. Cockrill, 1974, *op. cit.*, 834

5. P. Mahadevan, "Improving Domestic Buffalo Production in Asia." In *The Use of Nuclear Techniques to Improve Domestic Buffalo Production in Asia. Proceedings of the Final Research Co-ordination Meeting Organized by the Joint FAO/IAEA Division of Isotope and*

Radiation Applications of Atomic Energy for Food and Agricultural Development. (Vienna: IAEA, 1984).

6. "The Use of Nuclear Techniques to Improve Domestic Buffalo Production in Asia." *Proceedings of the Final Research Co-ordination Meeting on the Use of Nuclear Techniques to Improve Domestic Buffalo Production.* Vienna: IAEA, 1990).

7. Cockrill, *op.cit.*, 39n.

8. W. R. Cockrill, *The Buffaloes of China.* (Rome: FAO, 1976). Cockrill conducted this research project on water buffalo in China in 1974 when China was still in the throes of the so-called Cultural Revolution, a movement that paralyzed the country and traumatized citizens when they weren't killed. Cockrill's research activities were conducted entirely on state-run communes, the only agricultural operations allowed. There were no private farmers or private landholders. As a result, water buffalo were the state property of the socialist government. Cockrill fell into the public relations trap of extolling these communist ventures by parroting communist party-line slogans, not realizing that within two years of his visit the Cultural Revolution would cease and China would open up economically to the global village. Here is some of what Cockrill wrote:

"Today there are attacks on the "four olds" in China: old thought, old culture, old customs, old habits. Many ancient beliefs and philosophies are now being discredited and discarded. Before long much of the mythology and folklore concerning buffaloes will be forgotten and will exist only in museums and libraries." (p. 3). Cockrill is a good example of a water buffalo researcher who doesn't see political indoctrination for what it is and who, absent a sense of history and culture and the context for conducting research, swallows political pills and slogans unthinkingly like buffalo chew grass. Even China within a few years, and especially after Mao's death, would disavow its past social and political mistakes and return the country from communist collectivization to economic recovery through privatization of agriculture. I witnessed Chinese farmers selling their produce for their profit in open stalls with government approval in 1988. Indeed, the theme of this book has been to propose the opposite of what the Cockrill quote presumes, that a country's culture cannot be discarded nor eradicated, that a part of it indeed does exist in "museums and libraries," but that the story of the cow and buffalo, and its relationship with humans, is a global story which no political regime can extinguish.

9. According to Jimmie Weng, Managing Director of JIKE Imports and Exports in Australia, Global Ostrich Investments attracted more than $6 million (Australian dollars) from Chinese investors for ostrich farms in 1997. Australian ostrich farms have more than 80,000 ostrich, and 10,000 were sold to China in 1997 for breeding purposes. In 1992 no one in China had heard of an ostrich. By 1997, there were over 300 ostrich farms in China and over 30,000 ostrich bred for consumption. Moreover, there were 300,000 ostrich killed in South Africa in 1996, most of the meat intended for China.

10. J. W. Longworth, C. G. Brown, & S. A. Waldron, *Beef in China, Agribusiness Opportunities and Challenges.* (St. Lucia: University of Queensland Press, 2001).

11. Cockrill, 1974, *op. cit.*, 325.

12. "Bull Clone surprises scientists," (*Gulf News*, May 1, 2002), 20.

13. S. Strom, "In Japan, A Steak Secret that Rivals Kobe. (*NYT,* July 18, 2001), B11.

Epilogue

1. Jared Diamond's companionable and scholarly book, a veritable *vade mecum*, is *Guns, Germs and Steel, The Fate of Human Societies.* (New York: W. W. Norton, 1999).

2. Cf. C. W. Towne & E. N. Wentworth, *Cattle and Men.* (Norman: University of Oklahoma Press, 1955).

3. R. H. Waterston, "The Mouse Genome." *Nature*, 2002, 420, 520-62.

4. In late 2003 it was announced that the poodle had its genome deciphered.

5. *Agricultural Statistics 2001*. (Washington DC: U.S. Government Printing Office, 2001), VII-1-17.

6. *Agricultural Decisions (July-December, 2000)*. V59. (Washington DC: U.S. Department of Agriculture, 2000).

7. See the EPA website for policy statements, research documents and linkages to other sites regarding agricultural waste. http://www.epa.gov/sectors/agribusiness/index.html See also R. Manning, "Against the Grain, A Portrait of Industrial Agriculture as a Malign Force." *The American Scholar*, 73(1), 2004, 13-35.

8. Research conducted by Debra L. Donahue, Professor of Law at the University of Wyoming. According to her studies, echoing the Powell Report of over a hundred years earlier, livestock, mostly cattle: introduce and spread non-native plant species and disease; compete with native species for habitat; destroy sensitive native plants; contribute to soil compaction, drying, and excessive erosion; disrupt aquatic systems; and alter hydrological patterns. Donahue argues that removing livestock would be the single most effective strategy for conserving native biodiversity on arid public lands. The bulk of rangelands managed by the Bureau of Land Management (BLM) qualify as arid or semiarid. According to Department of Interior reports (1994), "watershed and water quality would improve to their maximum potential" if livestock were removed from public lands. The U.S. Forest Service concluded that livestock grazing is the number one cause of species endangerment in arid regions of the West, such as the Colorado Plateau and Arizona Basin. Grazing was identified as the number one cause of pollution of surface waters in the western states and the principal cause of desertification in North America.

9. Eddie Alford, a biologist for the U.S. Forest Service, knows the effects of overgrazing only too well as it is still in evidence today in the Tonto National Forest in Arizona. The tableau is lifeless of flora, gaunt, devoid of shrubs and grasses. Officials and politicians are quick to blame drought for the desiccated conditions that have forced removal of cattle herds from the region. But Forest Service personnel have different views and assign blame to ranchers, who have over-stocked the rangelands with grazing cattle, exceeding their permits, and public officials for turning a blind eye to mismanagement of the land and watershed. Alford claims the land hasn't fully recovered from the overgrazing a century earlier. Ranchers say the land has always looked as it has and that's why so few cattle graze on it now. Cf. J. Cart, "Overgrazing Cited for Tonto Forest Woes." (*The Scottsdale Tribune*, October 27, 2002), A24.

10. R. Manning, "Against the Grain, A Portrait of Industrial Agriculture as a Malign Force." *The American Scholar*, 73(1), 2004, 20.

11. W. Stegner, *The Sound of Mountain Water*. (New York: Doubleday, 1969).

12. G. Kolata, "24 Cow Clones, All Normal, Are Reported by Scientists." (NYT, November 23, 2001), A17.

13. Principally McDonald's, Wendy's, Hardees, Burger King, Carl's Jr., In-N-Out Burger, Jack in the Box. The table at http://www.sfsu.edu/~shs/dpm/fast.htm describes the ounces, calories, fat and sodium content of all the beef sold at fast food outlets. Also cf. Schlosser, *Fast Food Nation, The Dark Side of the All-American meal*. (Boston: Houghton Mifflin, 2001).

15. A. Cohen, "The McNugget of Truth in the Lawsuits Again in Fast-Food Restaurants." (*NYT*, February 3, 2003), A28.

16. You can check out dietary guidelines issued by the U.S. Department of Agriculture at: http://www.ars.usda.gov/dgac/2kdiet.pdf.

282

Figure 10: A Mesopotamian Proto Cow From the Archaeology Museum in Aleppo, Syria

REFERENCES

Adamson, J. (1961). *Born Free, A Lioness in Two Worlds.* New York: Pantheon Books.

Adler, J. (2004, January 12). Mad Cow: What's Safe Now? *Newsweek*, 43ff.

Agricultural Decisions (July-December, 2000). V. 59. Washington DC: U.S. Department of Agriculture.

Agricultural Statistics 2001. Washington DC: U.S. Government Printing Office.

Aldred, C. (1988). *Akhenaten, King of Egypt.* London: Thames & Hudson.

Aldridge, R. (1966). *Life on a ranch.* New York: Argonaut Press.

Allchin, F. R. (1962). *Neolithic cattle-keepers of south India.* Cambridge, UK: Cambridge University Press.

Allen, L. (2005, February). Back Home on the Range. *Smithsonian*, 35(11), 32–34.

Almagor, U. (1978). *Pastoral partners, affinity and bond partnership among the Dassanetch of south-west Ethiopia.* Manchester, UK: Manchester University Press.

Ammianus Marcellinus. (1986). *The Later Roman Empire (A.D. 354–378).* New York: Penguin Books.

Amoroso, E. C., Jewell, P. A. (1963). The exploitation of milk-ejection reflex by primitive peoples. In Mourant, A. E., Zeuner, F. E., Eds. *Man and cattle.* London: Royal Anthropological Institute, 126-137.

Alvord, H. E. (1897). Utilizations of by-products of the dairy. *Yearbook of the United States Department of Agriculture, 1897.* Washington, DC: Government Printing Office.

Ambrose, S. E. (1996). *Undaunted courage, Meriwether Lewis, Thomas Jefferson and The opening of the American West.* New York: Simon & Schuster.

Andrews, E. L. (2001, April 6). Dutch farmers facing mass foot-and-mouth Slaughter. *The New York Times*, A4.

Animal and plant health inspection service: Importation of beef from Argentina. (1996). Washington, DC: Government Accounting Office: U.S. Dept. of Agriculture. (Microfiche: GA 1.13 OGC 97-52).

Annual Report of the Commissioners of the General Land Office. (1885). Washington, DC: Government Printing Office.

Annual Report of the Secretary of the Interior for the fiscal year ending June 30, 1890. Washington, DC: Government Printing Office.

Annual Reports of the Department of Agriculture for the fiscal year ended June 30, 1900. Washington, DC: Government Printing Office.

Apple, R. W. (2001, November 28). A new Normandy, North of the golden gate. *The New York Times*, E1.

Argentina kept mum on disease (2001, March 17). *Arizona Republic*, A24.

Arhem, K. (1985). *Pastoral man in the garden of eden, the maasi of the ngorongoro conservation area*, Tanzania. Uppsala: Uppsala Research Reports in Cultural Anthropology.

Auge, C. & J-M Dentzer (1999). *Petra, The rose-red city.* London: Thames & Hudson.

Attridge, H. W. & Oden, R. A. (1981). *Philo of Byblos, The Phoenician History.* Washington, DC: The Catholic Biblical Association of America.

Bain, D. H. (2003, September 7). The Great Utah mystery. *The New York Times Book Review*, 14-15.

Barboza, D. (2001a, June 28). America's cheese state fights to stay that way. *The New York Times*, C1.

Barboza, D. (2001b, June 23). Sara Lee corp. pleads guilty in meat case. *The New York Times*, A7.

Barnard, N. (2002, summer). Doctor in the house. *Peta's Animal Times*, 17.

Barnes, T. D. (1993). *Athanasius and Constantius, Theology and politics in the Constantinian empire.* Cambridge, MA: Harvard University Press.

Becker, E. (2004, February 18). Jury Awards Ranchers $1.28 Billion from Tyson. *The New York Times,* C1.

Becker, E. (2003, January 23). Government in showdown in bid to shut beef processor. *The New York Times,* A16.

Becker, E. (2002a, October 17). Parents of sickened children ask for tighter food rules. *The New York Times,* A18.

Becker, E. (2001d, December 1). U.S. mad cow risk is low, a study by Harvard finds. *The New York Times,* A10.

Becker, E. (2001c, May 6). What if the buffalo roam into foot-and-mouth? *The New York Times,* wk16.

Becker, E. (2001b, April 6). Agricultural chief disavows plan to eliminate test on school beef. *The New York Times,* A1.

Becker, E. & Marquis, C. (2001a, March 26). As livestock disease spread, questions arise on U.S. defenses. *The New York Times,* A1.

Bell, R. (1960, 1937). *The Qur'an.* Edinburgh: T & T Clark.

Belluck, P. (2001, March 27). Big business of cattle theft is a growing threat to ranchers. *The New York Times,* A1.

Bergman, C. (1996). Orion's legacy, a cultural history of man as hunter. New York: Dutton.

Biblical Archaeology Review, (1993). 30,000-Year-Old Sanctuary Found at Har Karkom, 19(1), 38.

Bidwell, P. W., Falconer, J. I. (1941). *History of agriculture in the northern United States.* New York: Peter Smith.

Blakeslee, S. (2004, 7/30). Study lends Support to Mad Cow Theory. *The New York Times,* A9.

Blakeslee, S. (2001a, March 13). A 24-hour lab meeting on mad cow disease. *The New York Times,* D3.

Blakeslee, S. (2001b, October 30). Estimates of future death toll from mad cow disease Vary widely. *The New York Times,* D6.

Blakeslee, S. (2002, August 7). Experts consider how to stop a variant of mad cow disease. *The New York Times,* science.

Blakeslee, S. (2003, December 25). Expert Warned that Mad Cow was Imminent. *The New York Times,* A1.

Blakeslee, S. (2004, January 27). Human Brains Examined for Clues About Mad Cow. *The New York Times,* D5.

Bleeker, C. J. (1973). *Hathor and Thoth, Two key figures of the ancient Egyptian religion.* Leiden: E. J. Brill.

Blunt, J. (2003, February 18). When the buffalo roam. *The New York Times,* A27.

Boisselier, J. (1993). *The Wisdom of the Buddha.* London: Thames and Hudson.

Bolton, H. E. (1986, 1932). *The Padre on Horseback, A Sketch of Eusebio Francisco Kino,* Chicago: Loyola University Press.

Bokonyi, S. (1995). The development of stockbreeding and herding in medieval Europe. In Sweeney, D., *Agriculture in the Middle Ages.* Philadelphia: University of Pennsylvania Press, 41-61.

Borowski, O. (1998). *Every living thing, Daily use of animals in ancient Israel.* Walnut Creek, CA: AltaMira Press.

Braudel, F. (1993). *A History of civilizations.* New York: Penguin Books.

Brody, J. E. (2003, January 7). Drink your milk: A refrain for all ages, now more than ever. *The New York Times,* D5.

Bronowski, J. (1973). *The Ascent of Man.* Boston: Little, Brown.

Brooke, J. (2001, November 4). In Japan, beef business sinks in a sea of skepticism. *The New York Times*, A10.

Brown, D. (1994). *The American West*. New York: Touchstone.

Brown, P. L. (2001, June 27). In California bullfights, the final deed is done with velcro. *The New York Times*, A1.

Bulfinch, T. (1962, 1855). *The Age of Fable*. New York: New American Library.

Bulfinch, T. (1993). *The golden age of myth and legend*. Hertfordshire, UK: Wordsworth Reference.

Budge, E. A. W. (1973). *Osiris and the Egyptian Resurrection*. New York: Dover Publications.

Budge, E. A. W. (1967, 1895). *The Book of the Dead, The papyrus of Ani*. New York: Dover Publications.

Budge, E. A. W. (1899, 1980). *Egyptian religion, Egyptian ideas of the future life*. London: Routledge & Kegan Paul.

Budge, E. A. W. (1983). *Egyptian language, Easy lessons in Egyptian hieroglyphics*. New York: Dover Publications.

Bulfinch, T. (1962). *The age of fable or beauties of mythology*. New York: New American Library.

Bulfinch, T. (1993). *The golden age of myth and legend*. Hertfordshire, UK: Wordsworth Reference.

Bull Clone surprises scientists. *Gulf News*, May 1, 2002, p. 20.

Burn, L. (1990). *Greek myths*. London: British Museum Press.

Burns, J. F. (2003, January 26). The killing of Iraq's ancient marsh culture. *The New York Times*, wk14.

Burros, M. (2002, May 29). The greening of the herd. *The New York Times*, 29 WELL.

Burros, M. (2001, May 8). Experts see flaws in U.S. mad cow safeguards. *The New York Times*, D7.

Burrus, E. J. (1961). *Correspondencia del P. Kino con los generales de la compania de Jesus, 1682-1707*. Mexico City: Testimonia Historica, No. 5.

Cahill, T. (1998). *The gifts of the Jews, How a tribe of desert nomads changed the way everyone thinks and feels*. New York: Nan A. Talese/Anchor Books.

Campbell, J. (1988). *The power of myth*. (with Bill Moyers). New York: Doubleday.

Campbell, J. (1987). *Primitive mythology, the masks of god*. New York: Penguin Books.

Carlyle, T. (1857). *Sartor Resartus and Lectures on Heroes*. London: Chapman and Hall.

Carreras, A. (1993). *La aftosa en la argentina, un desafio competitivo*. Buenos Aires: Amara Argentina de Consignatarios de Ganado.

Carreras, A. (1986). *El comercio de ganados y carnes en lar argentina*. Buenos Aires: Editorial Hemisferio sur S.A.

Cart, J. (2002, October 27). Overgrazing cited for Tonto forest woes. *The Scottsdale Tribune*, A24

Cather, W. (2003, 1913). *O Pioneers!* New York: Barnes and Noble Classics.

Chauvet, J-M, Deschamps, E. B. & Hillaire, C. (1996). *Dawn of art: the Chauvet cave, the oldest known paintings in the world*. New York: Henry N. Abrams.

Christian, S. (2001, August 12). Revival is born where cattle concerned. *The New York Times*, 32.

Christensen, J. (2002, September 10). Environmentalists hail the ranchers: Howdy pardners? *The New York Times*, D3.

Chilton, L., Chilton, K., Arango, P., Dudley, J., Neary, N., Stelzner, P. (1984). *New Mexico*. Albuquerque: University of New Mexico Press.

Clottes, J. (1981). *The cave of niaux*. Boulogne: Castelet.

Clover, C. (2002, May 8). Quantitative 'failure' led by farm epidemic. *The Daily Telegraph* (London), 13.

Clottes, J. (2001, August). France's magical ice age art, Chauvet cave. *National Geographic,* 200(2), 104-121.

Cockrill, W. R. (1974a). The working buffalo. In Cockrill, W. R. (1974). *The husbandry and health of the domestic buffalo.* Rome: Food and Agricultural Organization of the United Nations.

Cockrill, W. R. (1974b). *The husbandry and health of the domestic buffalo.* Rome: Food and Agricultural Organization of the United Nations.

Cockrill, W. R. (1976). *The buffaloes of China.* Rome: Food and Agricultural Organization of the United Nations.

Cody, W. F. (1927). *Life and adventures of buffalo bill.* New York: Willey Book Company.

Cohen, A. (2003, February 3). The McNugget of truth in the lawsuits again in fast-food restaurants. *The New York Times,* A28.

Containing foot-and-mouth disease (2001, March 23). *The New York Times,* A20.

Cornelius, I. (1994). *The iconography of the Canaanite gods reshef and ba'al.* Fribourg, Switzerland: The University Press.

Cott, J. (1994). *Isis and Osiris, Exploring the goddesss myth.* New York: Doubleday.

Cottrell, L. (1953). *The bull of Minos, The discoveries of Schliemann and Evans.* New York: Facts on File Publications.

Couzin, (2004). Biomedicine: An End to the Prion Debate? Don't Count on It. *Science,* 305: 589.

Cowell, A. (2002, Sept. 10). Cattle disaster still felt in Britain. *The New York Times,* W1.

Cowell, A. (2001a, March 16). Foot-and-mouth damages English tourism too. *The New York Times,* A4.

Cowell, A. (2001b, March 27). Trying to stem foot-and-mouth, Britain buries carcasses. *The New York Times,* A23.

Cowell, A. (2001c, March 28). Britain reluctantly considers animal vaccination. *The New York Times,* A1.

Crain, C. (2003, Sept. 21). The Puritan Dilemma. *The New York Times Book Review,* 11.

Cumont, F. (1956). *The Mysteries of Mithra.* New York: Dover Publications.

Curse of factory farms. (2002, August 30). *The New York Times,* p. A18.

Crandall, D. P. (2000). *The place of stunted ironwood tress, A year in the lives of the cattle-herding Himba of Namibia.* New York: Continuum.

Dale, E. E. (1943). *Cow country.* Norman: University of Oklahoma Press.

Davey, M. (2003, December 25). Public Splits into 2 Camps on the Eating of Beef. *The New York Times,* A16.

Daley, S. (2001, December 14). France told to end ban on imports of British beef. *The New York Times,* A16.

Datta, D. K. (1980). *Social, moral and religious philosophy of Mohandas Gandhi.* New Delhi: Intellectual Publishing House.

Day, S. (2003, February 6). Prosecutors in smuggling case against Tyson contend trial is about corporate greed. *The New York Times,* A22.

DeCaro, S. (1999). *National Archaeological Museum of Naples.* Naples: National Archaeological Museum.

Deluc, B & G (1990). *Discovering Lascaux.* Bordeaux: Sud Ouest.

De Moor, J. C. (1987). *An anthology of religious texts from Ugarit.* Leiden: E. J. Brill.

De Tocquerville, A. (1956). *Democracy in America* (ed. By R. D. Heffner). New York: Mentor Book/New American Library.

DeVoto, B. (1953). *The journals of Lewis and Clark.* Boston: Houghton Mifflin.

Diamond, J. (1999). *Guns, germs and steel, The fate of human societies.* New York: W. W. Norton.

Diamos, J. (2001, August 7). Holy bull is one of six inducted into the hall. *The New York Times,* C16.

Dick, E. (1941). *The story of the frontier.* New York: Tudor Publishing Co.

Dilke, O. A. W. (1987). *Reading the past, mathematics and measurement.* London: The British Museum.

Diodorus (1985). *Diodorus on Egypt.* (Trans. E. Murphy). Jefferson, NC: McFarland & Co.

Domestic buffalo production in Asia. (1990). *Proceedings of the Final Research Co-ordination Meeting on the Use of Nuclear Techniques to Improve Domestic Buffalo Production.* Vienna: International Atomic Energy Agency.

Dowd, M. (2001, April 11). Herd on the street, Mad cows and udder fools. *The New York Times,* A27.

Dundas, C. (1968). *Kilimanjaro and its people.* London: Frank Cass & Co.

Durando, F. (1997). *Ancient Greece, The dawn of the western world.* New York: Barnes and Noble.

Durant, W. (1935). *Our Oriental Heritage.* New York: Simon & Schuster.

—— (1950). *The Age of faith.* New York: Simon & Schuster.

—— (1966, 1939). *The Life of Greece.* New York: Simon & Schuster.

—— (1944). *Caesar and Christ.* New York: Simon & Schuster.

Drago, H. S. (1965). *Great American cattle trails, The story of the old cow paths of the east and the longhorn highway of the plains.* New York: Dodd, Mead & Co.

Drew, C. (2002, December 11). Plant's sanitation may have link to deadly bacteria. *The New York Times,* A24.

Dykstra, R. R. (1979). *The cattle towns.* New York: Atheneum.

Eakin, E. (2002, October 12). Reopening a Mormon murder mystery. *The New York Times,* A19.

Economist. (2001, April 18). Domestication's family tree. 81-82.

Eder, R. (2001, March 30). Defeating Death by Writing About It in the Bullring. *The New York Times,* B38.

Editors of Funk & Wagnalls (1959). *Builders of America. V.1.* New York: Funk & Wagnalls.

Edwards, M. (2001). Marco Polo, journey home (Part III). *National Geographic,* 200(1), 28–47.

Eisnitz, G. (1997). *Slaughterhouse, The shocking story of greed, neglect, and inhumane treatment inside the U.S. meat industry.* Amherst, NY: Prometheus Books.

Euripedes (1832). *The Tragedies.* (Tr. R. Potter). London: A. J. Valpy.

Eiseley, L. (1970). *The invisible pyramid.* New York: Charles Scribner's Sons.

Eisenman, R. H. & Wise, M. (1994). *The Dead Sea Scrolls Uncovered.* New York: Barnes & Noble Books.

Eusebius (1965). *The history of the church from Christ to Constantine.* (Tr. G. A. Williamson). New York: Dorset Press.

Ellsworth, H. L. (1937). *Washington Irving on the prairie, Or a narrative of a tour of the southwest in the year 1832.* New York: American Book Company.

Fagan, B.M. (1990). *The journey from Eden, the peopling of our world.* London: Thames & Hudson.

Faminov, M. D. (1998). *Cattle, deforestation and development in the Amazon.* Wallingford, UK: CAB International.

Fernandez-Armesto, F. (2001). *Civilizations, Culture, ambition and the transformation of nature.* New York: The Free Press.

Finkelstein, I. & Silberman, N. A. (2001). *The Bible unearthed, Archaeology's new vision of ancient Israel and the origin of sacred texts.* New York: The Free Press.

Fox, R. W. (2002, December 1). The whole nine yards. *The New York Times Book Review,* 30.

Fraser, A. (1972). *The Bull.* New York: Charles Scribner's Sons.

Fremont, J. C. (1956). *Narratives of exploration and adventure.* New York: Longmans.

Freud, S. (1964). Moses and Monotheism. In *The Complete psychological works of Sigmund Freud*. (v. 23) London: The Hogarth Press.

Friedman, R. (2003). City of the Hawk, Deciphering the Narmer Palette. *Archaeology, 56*(6), 52-53.

Frink, M., Jackson, W. T., & Spring, A. W. (1956). *When Grass was King*. Boulder: University of Colorado Press.

Frisch, J. E. & Vercoe, J. E. (1984). Improvement of the productivity of the swamp buffalo in S. E. Asia. In (NA) The use of nuclear techniques to improve domestic buffalo production in Asia. (1984). *Proceedings of the Final Research Co-ordination Meeting Organized by the Joint FAO/IAEA Division of Isotope and Radiation Applications of Atomic Energy for Food and Agricultural Development*. Vienna: International Atomic Energy Agency.

Gandhi, M. K. (1954). *How to serve the cow*. Ahmedabad: Navajivan Publishing Co.

Gandhi, I. (1975). *India*. London: Hodder & Stoughton.

Gascoigne, B. (1971). *The Great Moghuls*. London: Jonathan Cape.

Gates, P. W. (1967). *California ranchos and farms, 1846-1862*. Madison, WI: State Historical Society of Wisconsin.

Gerald of Wales (1982, 1951). *The history and topography of Ireland*. New York: Penguin.

Gilbert, K. S., Holt, J. K., & Hudson, S. (1976). *The treasures of Tutankhamen*. New York: Metropolitan Museum of Art.

Gilberti, H. C. E. (1970). *Historia economica de la ganaderia argentina*. Buenos Aires: Ediciones Solar.

Gilgamesh (1960). *The epic of Gilgamesh* (Trans. N. K. Sandars), Hammondsworth,UK: Penguin Books.

Gleiter, J. & Thompson, K. (1985). *Paul Bunyan and Babe the Blue Ox*. New York: Torstar.

Grady, D. (2003, December 31). U.S. Imposes Stricter Safety Rules for Preventing Mad Cow Disease. *The New York Times*, A1.

Grady, D. (2002, August 20). On an altered planet, New diseases emerge and old ones re-emerge. *The New York Times*, D2.

Gray, J. (1982). *Near eastern mythology*. New York: Peter Bedrick Books.

Greeley, H. (1964). *An Overland Journey from New York to San Francisco in the Summer of 1859*. New York: Alfred Knopf.

Green, B. K. (1969). *Wild cow tales*. Lincoln: University of Nebraska Press.

Gressley, G. (1966). *Bankers and cattlemen*. New York: Knopf.

Guibelli, G. (no date). *Herculaneum*. Napoli: Carcavallo.

Guzzo, P. G. & d'Ambrosio, A. (1998). *Pompeii*. Napoli: L'erma di Bretschneider.

Hall, J. (1848). *The West, Its soil, surface and productions*. Cincinnati: Derby, Bradley & Co.

Hamilton, E. (1993, 1930). *The Greek way*. New York: W. W. Norton.

Hamilton, E. (1962, 1940). *Mythology*. New York: Mentor Books.

Harden, B. (2002, August 12). Livestock testing chases cheats from state fairs. *The New York Times*, C2.

Hart, G. (1990). *Egyptian myths*. London: British Museum Publications.

Hatch, W. J. (1928). *The land pirates of India*. Philadelphia: J. B. Lippincott.

Hawkes, J. (1968). *Dawn of the gods*. New York: Random House.

Healey, J.F. (1990). *Reading the past, the early alphabet*. London: The British Museum.

Hemingway, E. (1926). *The sun also rises*. New York: Simon & Schuster.

Hemingway, E. (1932). *Death in the afternoon*. New York: Scribner.

Hernandez, R. (2001, June 29). Citing mad cow, F.D.A. panel backs blood donor curbs. *The New York Times*, A23.

Herodotus (1830). *The Histories*. (vls.1 & 2) London: Henry Colburn & Richard Bentley.

Hertzberg, A. (1962). *Judaism*. New York: George Braziller.

Hoge, W. (2001a, March 15). As the disease marches on Britain dooms more animals. *The New York Times*, A12.

Hoge, W. (2001b, March 22). 5 Britons' 'mad cow' deaths traced to butchering method. *The New York Times*, A10.

Hoge, W. (2001c, March 24). Livestock epidemic widens its menace for British farms. *The New York Times*, A1.

Holland, K. (1987). *On the horns of a dilemma, The future of the Maasi*. Montreal: McGill University, Department of Anthropology.

Homer. *The Odyssey*. (trans. R. Fitzgerald). (New York: Anchor Books, 1963).

Horace (1961). *The Odes of Horace*. (Tr. H. R. Henze). Norman: University of Oklahoma.

Horace (1872). *The Odes and Epodes of Horace*. (tr. E. Bulwer). London: George Routledge & Sons.

Houston, F. (2001, March 25). A life's work up in smoke. *Arizona Republic*, B11.

http://vassun.vassar.edu/~sttaylor/Cooley/BattleOfBulls.html

Huang, K. S. (1996). Price and income affect nutrients consumed from meats. *Food Review* 19(1), 37–40.

Huxley, J. (1931). *Africa view*. New York: Harper & Bros.

Igler, D. (2001). *Industrial cowboys*. Berkeley: University of California Press.

Illustrated London News. (1849, Jan. to June). London: William Little, 14, 375-76.

Irving, W. (1956). *A tour of the prairies*. Norman, OK: University of Oklahoma Press.

Jarvis, L. S. (1986). *Livestock development in Latin America*. Washington, DC: The World Bank.

Julian. (1913). *The Works of Emperor Julian*. (v. 1-3) (trans. W. C.Wright) London: William Heinemann.

Juvenal. (1991). *The Satires*. (Tr. Niall Rudd). Oxford: The Clarendon press.

Juvenal. (1970, 1898). *D. Iunii Iuvenalis, Fourteen Satires of Juvenal*. (Ed. J. D. Duff, M. Coffey). Cambridge: Cambridge University Press.

Keay, J. (1982, 1973). *Into India*. London: John Murray.

Kaufman, L. (2001, March 29). Feeling the pinch of luxury leather. *The New York Times*, C1.

Keel, O. & Uehlinger, C. (1992). *Gods, Goddesses,and images of god in ancient Israel*. Minneapolis: Fortress Press.

Kerenyi, C. (1992, 1951). *The gods of the Greeks*. London: Thames and Hudson.

Khayyata, W. (1977). *Guide to the museum of Aleppo*. (Tr. S. Shaath). Aleppo, Syria: Arab National Printing Press.

Kilborn, P. T. (2003, November 16). Cattle Rushed to Market as the Price of Beef Soars. *The New York Times*, YT18.

Killman, S. & Merrick, A. (2000, August 15). Could the high price of milk be a byproduct of supermarket mergers? *The Wall Street Journal*, B1.

Kirsch, J. (1998). *Moses, a life*. New York: Ballantine Books.

Klima, G. J. (1970). *The Barabaig, East african cattle-herders*. New York: Holt, Rinehart and Winston.

Kolata, G. (2001, November 23). 24 cow clones, all normal, are reported by scientists. *The New York Times*, A17.

Kraenzel, C. F. (1955). *The great plains in transition*. Norman: University of Oklahoma.

Krauss, C. & Blakeslee, S. (2003, May 21). Case of mad cow in Canada prompts U.S. to ban its beef. *The New York Times*, A1.

Kummer, C. (2003). Back to grass. *The Atlantic Monthly*, 291(4), 138-142.

Landler, M. (2001, April 8). Less alarm in Hong Kong over disease of livestock. *The New York Times*, A9.

Lau, D. C. (1963). *Lao Tzu, tao te ching*. London: Penguin Books.

Lavery, B. (2001, March 23). Despite safeguards, foot-and-mouth disease spreads to Ireland. *The New York Times*, A6.

290

Lavietes, S. (2003, Jan. 17). 2 archaeologists, Robert Braidwood, 95, and his wife, Linda Braidwood, 93, die. *The New York Times*, A23.

Lea, T. (1957). *The King ranch*. Boston: Little, Brown and Company.

LeDuff, C. (2002, October 31). Range war in Nevada pits U.S. against two Shoshone sisters. *The New York Times*, A16.

LeDuff, C. (2003, February 7). U.S. agents seize horses of 2 defiant Indian sisters. *The New York Times*, A16.

Legislation controlling the international beef and veal trade. (1985). Rome: Food and Agricultural Organization of the United Nations.

Leeming, D. (2004). *Jealous Gods and Chosen People*. New York: Oxford University Press.

Lerner, R. E., Meacham, S., & Burns, E. M. (1988). *Western civilizations.* (11th ed.). New York: W. W. Norton.

Levine, E. (2003, January 15). The burger takes center stage. *The New York Times*, D1.

Leeson, R. (2001, August 31). Why we need N.E. dairy compact. *The Providence Journal*, B4.

Lind, M. (2003). The new continental divide. *The Atlantic Monthly*, 291(1), 86-88.

Link, M. A. (1968). *Treaty between the United State of America and the Navajo Tribe of Indians*. Las Vegas: KC Publications.

Lockwood, F. C. (1934). *With Padre Kino on the trail*. Tucson: University of Arizona Press.

Lodrick, D. O. (1981). *Sacred cows, sacred places, origins and survivors of animal homes in India*. Berkeley: University of California Press.

Loftus, R. T. (1994). Evidence for two independent domestications of cattle. *Proceedings of the National Academy of Sciences*, 91: 2757-61.

Longworth, J. W., Brown, C. G., & Waldron, S. A. (2001). *Beef in China, Agribusiness opportunities and challenges*. St. Lucia: University of Queensland Press.

Longworth, J. W. (1983). *Beef in Japan, Politics, production, marketing and trade*. St. Lucia: University of Queensland Press.

Lyall, S. (2001, March 10). Wrapping British farmers in growing isolation. *The New York Times*, A3.

Lytton, E. B. (Trans.) (1872). *The odes and epodes of Horace*. London: George Routledge.

McLaren, G. (2000). *Bog mobs, The story of Australian cattlemen*. Feemantle: Freemantle Arts Centre Press.

McNeil, D. G. (2004, February 17). Research in Italy Turns Up a New Form of Mad Cow Disease. *The New York Times*, A7.

MacDonald, J. (1878). *Food from the far west*. New York: Orange Judd Company.

Mahadevan, P. (1984). Improving domestic buffalo production in Asia. In *The use of Nuclear techniques to improve domestic buffalo production in Asia. (NA) (1984). Proceedings of the Final Research Co-ordination Meeting Organized by the Joint FAO/IAEA Division of Isotope and Radiation Applications of Atomic Energy for Food and Agricultural Development*. Vienna: International Atomic Energy Agency.

Majors, A. (1893). *Seventy years on the frontier*. Chicago: Rand McNally.

Malandra, W. W. (1983). *An introduction to ancient Iranian religion*. Minneapolis: University of Minnesota Press.

Marquis, C. & Mcneil, D. G. (2001, March 14). Meat from Europe is banned by U.S. as Illness spreads. *The New York Times*, A1.

Martin, C. (1973). *The saga of the buffalo*. New York: Hart Publishing Co.

Mascaro, J. (1962). *The Bhagavad Gita*. New York: Penguin Books.

Mascaro, J. (1965). *The Upanishads*. New York: Penguin Books.

McCurdy, J. (2000, Oct. 19). Banged-up bull rider taking 'em 1 at a time. *Arizona Republic*, C3.

McMurtry, L. (2003, October 9). Mountain Man. *The New York Review of Books*, L, 15, 20–22.

McGinn, C. (2002, May 5). Our posthuman future: Biotechnology as a threat to human nature. *The New York Times*, *www.nytimes/2002/05/05/books/review/05MACGINNT@html*.

McDowell, A. (1996). Daily life in ancient Egypt. *Scientific American*, 275(6), 100–105.

McNeil, D. G. (2004, February 3). Man Who Killed the Mad Cow has Questions of His Own. *The New York Times*, D2.

McNeil, D. G. (2004, April 18). Barred from Testing for Mad Cow, Niche Meatpacker Loses Clients. *The New York Times*, A12.

MacHugh, D. E. & M. D. Shriver, R. T. Loftus, P. Cunningham and D. G. Bradley (1997). Microsatellite DNA Variation and the Evolution, Domestication and Phylogeography of Taurine and Zebu Cattle (*Bos taurus and Bos indicus*). *Genetics*, 146, 1071-1086.

Manning, R. (2004). Against the Grain, A Portrait of Industrial Agriculture as a Malign Force. *The American Scholar*, 73(1), 13-35.

Mehta, V. (1967). *Portrait of India*. London: Weidenfield & Nicolson.

Millard, C. S. (2001). Keepers of the faith. *National Geographic*, 200(1), 10-125.

Milloy, R. E. (2001a, March 28). The old west yields to urbanization. *The New York Times*, A10.

Milloy, R. E. (2001b, October 29). Anthrax hides along cattle trails of the old west. *The New York Times*, A8.

Moore, M. (1984, 1935). *The collected poems of Marianne Moore*. Franklin Center, PA: The Franklin Library.

Moorey, P. R. S. (1994). *The ancient near east*. Oxford, UK: The Ashmolean Museum.

Mourant, A. E., Zeuner, F. E., Eds. (1963). *Man and cattle*. London: Royal Anthropological Institute.

Max, D. T. (2004, March 28). The Case of the Cherry Hill Cluster. *The New York Times Magazine*, 50-55.

Naipaul, V. S. (1977). *India, a wounded civilization*. London: Penguin Books.

Norwich, J. J. (1998). *A short history of Byzantium*. London: Penguin Books.

Oates, J. (1986). *Babylon*. London: Thames and Hudson.

Oldenburg, U. (1969). *The conflict between el and ba'al in Canaanite religion*. Leiden: E. J. Brill.

Pattie, T. S. (1979). *Manuscripts of the bible, Greek bibles in the British library*. London: The British Library.

Paxton, F. L. (1924). *History of the American frontier 1763-1893*. New York: Houghton Mifflin.

Petersen, A. R. (1998). *The royal god, Enthronement festivals in ancient Israel and Ugarit?* Sheffield, UK: Sheffield Academic Press.

Petersen, M., & C. Drew (2003, October 10). As Inspectors, Some Meatpackers Fall Short. *The New York Times*, p. A1.

Pettinato, G. (1981). *The archives of Ebla, An empire inscribed in clay*. Garden City NY: Doubleday and Co.

Piper, D. (1995). *Treasures of the Ashmolean Museum*. Oxford: Ashmolean Museum.

Plato. (1961). *Laws*. (v.2). New York: Loeb Classical Library.

Politics of Meat, The. (2002. April 24) Frontline, Public Broadcasting System. *www.pbs.org/wgbh/pages/frontline/shows/meat/politics*.

Pollack, A. (2002, August 12). Joint venture clones cattle with human antibodies. *The New York Times*, C2.

Pollan, M. (2004, January 11). Cattle Futures? *The New York Times Magazine*, 11-12.

Pollan, M. (2002, July 19). When a crop becomes king. *The New York Times*, OP-ED.

Pollan, M. (2003, October 12). The (Agri)Cultural Contradictions of Obesity. *The New York Times Magazine*, Oct. 12, 2003, 41–48.

Pollan, M. (2001). *A Botany of Desire*. New York: Random House.

Porter, V. (1991). *Cattle, A handbook of breeds of the world*. New York: Facts on File.

Powell, J. W. (1962, 1878). *Report on the lands of the arid region of the United States*. Cambridge, MA: Belknap Press.

Prabhavananda, S. & Manchester, F. (1975). *The Upanishads*. Hollywood: Vedanta Press.

Price, L. (2002). Cowboy boosterism, Old cowtown museum and the image of Wichita, Kansas. *Kansas History* 24(4), 301-17.

Pukite, J. (1996). *A field guide to cows, How to identify and appreciate America's 52 Breeds*. New York: Penguin Books.

Race, R. E. (2003). Crossing the species barrier. *Nature*, 423, 118-19.

Rath, S. (2000). *About cows*. Stillwater, MN: Voyageur Press.

Reade, J. (1991). *Mesopotamia*. London: The British Museum Press.

Reaney, P. (2001, September 8). Britain faces long road to curb foot-and-mouth. *The Globe and Mail* (Toronto), A16.

Renou, L. (ed.). (1962). *Hinduism*. New York: George Braziller.

Report of the Commissioner of Agriculture for the year 1862. (1862). Washington, DC: Government Printing Office.

Report of the Commissioner of Agriculture for the year 1867. (1867). Washington, DC: Government Printing Office.

Report of the Commissioner of Agriculture for the year 1869. (1870). Washington, DC: Government Printing Office.

Report of the Commissioner of Agriculture for the year 1870. (1871). Washington, DC: Government Printing Office.

Report of the Secretary of the Interior 1875. Washington, DC: Government Printing Office.

Report of the Commissioner of Agriculture for the years 1881 and 1882. Washington, DC: Government Printing Office.

Report of the Secretary of Agriculture for the year 1890. Washington, DC: Government Printing Office.

Revken, A. (2002, August 20). Forget nature. Even Eden is engineered. *The New York Times*, D1-11.

Rich, J. L. (2001, Mar. 14). Brazil postpones its beef dreams. *The New York Times*, W1.

Rifkin, J. (1992). *Beyond Beef, The rise and fall of the cattle culture*. New York: Dutton.

Rigby, P. (1992). *Cattle, capitalism, and class*. Philadelphia: Temple University Press.

Robbins, J. (2001, Sept. 17). Unsolved mystery resurfaces in Montana: Who's killing cows? *The New York Times*, B1.

Roberts, A. (1995). *Hathor rising, the serpent power of ancient Egypt*. Totnes, UK: Northgate.

Roberts, J. (no date), *Antiquities of West Cork*, West Cork, Ireland: Skibbereen.

Robertshaw, P. (1990). *Early pastoralists of south-western Kenya*. Nairobi: British Institute in Eastern Africa.

Rohter, L. (2003, March 2). Once secure, Argentines now lack food and hope. *The New York Times*, yt6.

Rohter, L. (2001a, June 20). Disease is big setback for Brazil cattle region. *The New York Times*, W1.

Rohter, L. (2001b, August 23). Brazil's wild and woolly side: meet the caubois. *The New York Times*, A4.

Romer, J. (1981). *Valley of the kings, exploring the tombs of the pharaohs*. New York: Henry Holt.

Romero, S. (2002, October 27). Where two cultures collide. *The New York Times*, TR10.

Roosevelt, T. (1925). *Theodore Roosevelt, An autobiography*. New York: Scribners.

Rouse, J. E. (1977). *The criollo, Spanish cattle of the Americas.* Norman: University of Oklahoma Press.

Sandars, N. K. (1972). *The epic of Gilgamesh.* New York: Penguin Books.

Sandars, N. K. (1971). *Poems of heaven and hell from ancient Mesopotamia.* London: Penguin.

Sanders, A. H. (1926). *The cattle of the world, Their place in the human scheme — wild types and modern breeds in many lands.* Washington, DC: The National Geographic.

Sauer, E. (1996). *The end of paganism in the north-western provinces of the roman empire, the example of the Mithras cult.* Oxford, UK: Hadrian Books.

Scandizzo, S. (2002). *Latin American Merchandise Trade and U.S. Trade Barriers.* (*www.latinamericantrade.com*).

Schlosser, E. (2001). *Fast food nation, The dark side of the All-American meal.* Boston: Houghton Mifflin.

Schools to be allowed to serve irradiated meat (2002, October 27). *The New York Times,* A21.

Science of BSE, The. (1998, March 14). *The Economist,* 21-23.

Scioscia, A. (2001). Steers and queers. *Arizona New Times,* 32(7), 30-36.

Seelye, k. Q. (2003, January 26).Yellowstone bison thrive, but success breeds peril. *The New York Times,* A16.

Sharpes, D. K. (2001). *Advanced Educational Foundations for Teachers, The History, Philosophy and Culture of Schools.* New York: Routledge.

Sharpes, D.K. (1999, Aug.2). Bullriding speaks to something ancient in us. *Standard Examiner,* A10.

Shorter, A. W. (1937). *The Egyptian gods.* London: Routledge and Kegan Paul.

Shubert, A. (1999). *Death and money in the afternoon, A history of the Spanish bullfight.* New York: Oxford University Press.

Sinclair, U. (1934). *An Upton Sinclair anthology.* Los Angeles, CA: Published by Author.

Smith, A. B. (1992). Pastoralism in Africa, Origins and development ecology. London: Hurst and Co.

Smith, C. (2002, Dec. 31). Milk flows from desert at a unique Saudi farm. *The New York Times,* yne A4.

Smith, F. J. Kessell, J. L., & Fox, F. J. (1966). *Father Kino in Arizona.* Phoenix: Arizona Historical Foundation.

Smith, P. H. (1969). *Politics and beef in Argentina.* New York: Columbia University Press.

Spaulding, G. F. (1968). *On the western tour with Washington Irving, The journal and letters of Count de Pourtales.* Norman: University of Oklahoma Press.

Speidel, M. P. (1980). *Mithras-Orion, Greek hero and Roman army god.* Leiden: E. J. Brill.

Stamler, B. (2001, July 30). Got sticking power? A tagline and a mustache stop a slide in milk consumption. *The New York Times,* C11.

Starrs, P. F. (1998). *Let the cowboy ride, Cattle ranching in the American West.* Baltimore: Johns Hopkins Press.

Stegner, W. (1969). *The sound of mountain water.* New York: Doubleday.

Stegner, W. (1992, 1954). *Beyond the hundredth meridian, John Wesley Powell and the Second opening of the West.* New York: Penguin Books.

Still, B. (1961). *The West, Contemporary records of America's expansion across the continent, 1607-1890.* New York: Capricorn Books.

Stone, I. (1956). *Men to match my mountains.* Garden City, NY: Doubleday.

Sturluson, S. (1916). *The prose Edda.* (Tr. & Ed., A G. Brodeur). New York: The American-Scandinavian Foundation.

Strom, S. (2001, July 18). In Japan, A steak secret that rivals Kobe. *The New York Times,* B11.

Swarns, R. L. (2001, October 3). Cattle prices are up, so is buying a bride's hand. *The New York Times*, A4.

Szulc, T. (2001). Abraham, Journey of faith. *National Geographic*, 200(6), 90-129.

Tattersall, I. (1998). *Becoming human, Evolution and human uniqueness.* New York: Harcourt Brace & Co.

Taylor, J. H. (2001). *Death and the afterlife in ancient Egypt.* Chicago: University of Chicago Press.

Terrell, J. U. (1972). *Land grab.* New York: Dial Press.

The use of nuclear techniques to improve domestic buffalo production in Asia. (1984). Proceedings of the Final Research Co-ordination Meeting Organized by the Joint FAO/IAEA Division of Isotope and Radiation Applications of Atomic Energy for Food and Agricultural Development. Vienna: International Atomic Energy Agency.

Torres, C. (2000, August 14). U.S. Takes cautious stance on Argentine beef imports. *The Wall Street Journal*, A12.

Towne, C. W., Wentworth, E. N. (1955). *Cattle and men.* Norman: University of Oklahoma Press.

Treasurers of Tutankhamun (no author). (1976). New York: The Metropolitan Museum of Art.

Troy, C., MacHugh, D. E., Bailey, J. F., Magee, D. A., Loftus, R. T., Cunningham, P., Chamberlain, A.T., Sykes, B. C., Bradley, D. G. (2001). Genetic evidence of near-eastern origins of European cattle. *Nature* 410, 1088-90.

Trow-Smith, R. (1967). *Life from the land.* New York: Longmans.

Tubb, J. N. (1998). *Canaanites.* London: British Museum Press.

Tubb, J. N. & Chapman, R. L. (1990). *Archaeology and the Bible.* London: British Museum.

Tulloch, D. G. (1974). The buffaloes of Asia. In Cockrill, W. R. (1974b). *The husbandry and health of the domestic buffalo.* Rome: Food and Agricultural Organization of the United Nations.

Turcan, R. (1975). *Mithras platonicus, recherches sur l'hellenisation philosophique de Mithra.* Leiden: E.J. Brill.

Twain, M. (1871, 1913). *Roughing It.* New York. Harper & Row.

Thwaites, R. G., (1959). *Original Journals of the Lewis and Clark Expeditions, 1804-1806.* New York: Antiquarian Press.

Ulansey, D. (1994). Solving the Mithraic mysteries. *Biblical Archaeology Review*, 20(5), 40–53.

Upham, C. W. (1856). *Life, explorations and public services of John Charles Fremont.* Boston: Ticknor and Fields.

U.S. cautious on mad cow, seizes flock of sheep (2001, Mar. 22). *The New York Times*, A12.

U.S. Department of Interior, Bureau of Land Management. 1994. Rangeland Reform '94: Draft Environmental Impact Statement. Washington, D.C.

Vermaseren, M. J. (1963). *Mithras, the secret god.* New York: Barnes & Noble.

Veyne, P. (1988). *Did the Greeks believe in their own myths?* (Trans. P. Wissing). Chicago: University of Chicago Press.

Virgil, available at: classics.mit.edu/Virgil/georgics.html.

Virgil. (1830). *The Eclogues, Georgics and Aeneid.* London: Henry Colburn & Richard Bentley.

Vivian, A. P. (1879). *Wanderings in the western land.* London: Sampson Low Marston, Searle, & Rivington.

Voltaire (1940). The white bull. In Voltaire, *The best known works of Voltaire.* New York: The Book League, 96-118.

Von Richthofen, W. (1964). *Cattle-raising on the plains of North America.* Norman: University of Oklahoma Press.

Wade, N. (2001, Oct. 23). How a patient assassin does its deadly work. *The New York Times*, D1.

Wald, m. L. & E. Lightblau. December 24, 2003). U.S. Examining mad Cow Case: First for the Nation. *The New York Times*, A1.

Walker, A. T. (1907). *Caesar's Gallic Wars*. Chicago: Scott, Foresman.

Wall paintings of the tomb of Nefertari, scientific studies for their conservation. (1987) Cairo, Egypt: The Egyptian Antiquities Organization.

Walters, V. J. (1974). *The cult of Mithras in the roman provinces of Gaul*. Leiden: E. J. Brill.

Waterston, R. H. (2002). The mouse genome. *Nature*, 420, 520-62.

Watson, F. (1974). *A concise history of India*. London: Thames & Hudson.

Webb, W. P. (1931). *The great plains*. New York: Grosset & Dunlap.

Weeks, K. (1998). *The lost tomb*. New York: William Morrow and Co.

Weiner, T. (2002, Dec. 8). Growing poverty is shrinking Mexico's rain forest. *The New York Times*, yne12.

Weise, E. (2003a, August 13). Food sellers push animal welfare. *USA Today*, D1.

Weise, E. (2003b, August 13). Prodding slaughterhouses. *USA Today*, D5.

Weiser, B. (2003, January 23). Big macs can make you fat? No kidding, federal judge rules. *The New York Times*, A23.

Wellman, P. I. (1939). *The trampling herd*. New York: Carrick & Evans.

Wellman, P. I. (1954). *Glory, God and Gold*. Garden City, NY: Doubleday.

White, J. E. M. (1970). *Ancient Egypt, Its culture and history*. New York: Dover.

Williams, J. A. (1962). *Islam*. New York: George Braziller.

Wines, M. (2001, October 10). Russians are unfazed by anthrax, A common rural problem. *The New York Times*, A4.

Winter, G. (2002, July 20). Beef processor's parent no stranger to troubles. *The New York Times*, A9.

Woods, F. E. (1994). *Water and storm polemics against Baalism in the Deuteronomic history*. New York: Peter Lang.

Wooley, C. L. (1965). *The Sumerians*. New York: W. W. Norton.

Xenophon (no date). *The education of Cyrus* (H. G. Dakyns, trans.). London: J. M. Dent & Sons.

Zaehner, R. C. (1962). *Hinduism*. New York: Oxford University Press.

Zaehner, R. C. (1956). *The teachings of the magi, A compendium of Zoroastrian beliefs*. New York: Oxford University Press.

Zivie, A. (2002). A pharaoh's peace maker. *National Geographic*, 202(4), 27-31.

Zoroaster. (1963). *The Hymns of Zarathustra*. (Tr. J. Duchesne-Guillemin). Boston: Beacon Press.

INDEX